Nanostructured Thin Films and Nanodispersion Strengthened Coatings

NATO Science Series

A Series presenting the results of scientific meetings supported under the NATO Science Programme.

The Series is published by IOS Press, Amsterdam, and Kluwer Academic Publishers in conjunction with the NATO Scientific Affairs Division

Sub-Series

I. Life and Behavioural Sciences	IOS Press
II. Mathematics, Physics and Chemistry	Kluwer Academic Publishers
III. Computer and Systems Science	IOS Press
IV. Earth and Environmental Sciences	Kluwer Academic Publishers
V. Science and Technology Policy	IOS Press

The NATO Science Series continues the series of books published formerly as the NATO ASI Series.

The NATO Science Programme offers support for collaboration in civil science between scientists of countries of the Euro-Atlantic Partnership Council. The types of scientific meeting generally supported are "Advanced Study Institutes" and "Advanced Research Workshops", although other types of meeting are supported from time to time. The NATO Science Series collects together the results of these meetings. The meetings are co-organized bij scientists from NATO countries and scientists from NATO's Partner countries – countries of the CIS and Central and Eastern Europe.

Advanced Study Institutes are high-level tutorial courses offering in-depth study of latest advances in a field.
Advanced Research Workshops are expert meetings aimed at critical assessment of a field, and identification of directions for future action.

As a consequence of the restructuring of the NATO Science Programme in 1999, the NATO Science Series has been re-organised and there are currently Five Sub-series as noted above. Please consult the following web sites for information on previous volumes published in the Series, as well as details of earlier Sub-series.

http://www.nato.int/science
http://www.wkap.nl
http://www.iospress.nl
http://www.wtv-books.de/nato-pco.htm

Series II: Mathematics, Physics and Chemistry – Vol. 155

Nanostructured Thin Films and Nanodispersion Strengthened Coatings

edited by

Andrey A. Voevodin

United States Air Force Research Laboratory,
Wright-Patterson Air Force Base, Ohio, U.S.A.

Dmitry V. Shtansky

Moscow State Institute of Steel and Alloys,
Technological University, Moscow, Russia

Evgeny A. Levashov

Moscow State Institute of Steel and Alloys,
Technological University, Moscow, Russia

and

John J. Moore

Colorado School of Mines, Golden,
Colorado, U.S.A.

Kluwer Academic Publishers

Dordrecht / Boston / London

Published in cooperation with NATO Scientific Affairs Division

Proceedings of the NATO Advanced Research Workshop on
Nanostructured Thin Films and Nanodispersion Strengthened Coatings
Moscow, Russia
8–10 December 2003

A C.I.P. Catalogue record for this book is available from the Library of Cong

ISBN 1-4020-2221-2 (PB)
ISBN 1-4020-2220-4 (HB)
ISBN 1-4020-2222-0 (e-book)

Published by Kluwer Academic Publishers,
P.O. Box 17, 3300 AA Dordrecht, The Netherlands.

Sold and distributed in North, Central and South America
by Kluwer Academic Publishers,
101 Philip Drive, Norwell, MA 02061, U.S.A.

In all other countries, sold and distributed
by Kluwer Academic Publishers,
P.O. Box 322, 3300 AH Dordrecht, The Netherlands.

Printed on acid-free paper

Printed in the Netherlands.

TABLE OF CONTENTS

Part II: RECENT PROGRESS IN THE DEVELOPMENT OF PLASMA DEPOSITION EQUIPMENT

Part III: FILM CHARACTERIZATION AND CONTROL OF STRUCTURE AND PROPERTIES

Part V: BIOCOMPATIBLE THIN FILMS

FOREWORD

This volume contains proceedings of the NATO-Russia Advanced Research Workshop on Nanostructured Thin Films and Nanodispersion Strengthened Coatings (December, 2003, Moscow). During this Workshop leading researchers from twelve countries had presented and discussed most recent developments in the fields of plasma physics and surface engineering related to the preparation and applications of nanostructured thin films and nanodispersion strengthened coatings. These presentations are encompassed in 31 individual chapters. The chapters are assembled in five parts in according to the workshop sessions.

Part I is a compilation of chapters on hard and tribological coatings. The recent advances in this area are significant in that it is now possible to engineer strong, hard, and tough coatings that can operate at temperatures higher than 1200 °C and exhibit 'smart', adaptive characteristics. These coatings are based on an amorphous matrix, e.g. nitrides, carbides, borides, or carbon, in which there is a controlled nucleation and growth of ultra hard nanoparticles of crystalline carbides, nitrides, borides and oxides. The critical feature is the control of both the particle size, i.e., less than 10 nm, and interpartical spacing of a few nanometers. The 'smart' or adaptive characteristic is engineered into the nanostructures using similar sized (less than 10 nm) particles of metallic chalcogenidese, ductile metals, or glass forming elements to provide high lubricity and chemical adaptation at the environment change, e.g., high and low humidities and temperatures.

Part II combines six chapters on the recent progress in the development of plasma deposition processes. Considerable improvement in the increase and control of ion bombardment energy has been achieved using pulsed plasma techniques that involve pulsing the power supply to the magnetron and/or the substrate bias, ion assisted deposition, and the use of hybrid deposition systems that are capable of creating fluxes of ion and neutral species with a range of energies and concentrations. The main concept is to improve the control of the nanostructure and mechanical/tribological properties of the film using the increased ion bombardment energies. The new techniques are applied to increase adatom mobility, stress, hardness, and wear resistance. Using these plasma techniques to modify the nanostructure provides another powerful process parameter for engineering the required structure-property correlations in these coatings that are often multicomponent and multifunctional.

Coupled to the papers presented in Parts I and II is the need to determine the fine nanostructure and properties of these complex thin film systems. The characterization of these complex structures demands more sophisticated characterization techniques to be developed. Part III provides insight into the use of high-resolution transmission electron microscopy (HRTEM) as an essential characterization tool that has to be coupled to nanoindentation and nano-wear and stress measurement techniques. The case for the correlation of these tribological properties with deposition processing parameters and nanostructural features is made clear in this collection of chapters. This philosophy is extended to functional thin films that include the self-assembly of Si-Ge quantum dots for microelectronic applications, soft magnetic thin films for magnetic storage, and ion implantation in metallic ceramic materials for improving high temperature and corrosion resistance.

Part IV is a compilation of chapters that discuss the generation of nanostructued powders and materials using techniques such as self-propagating high temperature synthesis (SHS), electroplating, and CVD. A range of surface engineering technologies is discussed: wear resistant coatings; surface modifications; heat insulating surfaces; the dispersion strengthening influence of diamond used in different media; and the nanophotonic structure of a SBN:Yb/opal material system.

The final section, Part V, examines the role of nanostructured surfaces and thin films in biomaterial applications. The concepts of biocompatibility and bioactivity require a more complete understanding of the interactions that take place on the nanoscale at the interface between an implant and body fluids. Thin films produced by plasma techniques such as unbalanced magnetron sputtering and surface modification using ion beam implantation are discussed in this highly complex environment.

Editors are grateful to all authors for their hard work in developing their presentations into the chapters of this volume, which undoubtedly will serve as a base for new developments and exciting breakthroughs in the nanostrcutured films, coatings and plasma based deposition technologies for myriad of applications.

Andrey A. Voevodin
Dmitry V. Shtansky
Evgeny A. Levashov
John J. Moore

PREFACE

Chapters of this book are compiled of the original works and reviews presented at the NATO-Russia Advanced Research workshop on "Nanostructured Thin Films and Nanodispersion Strengthened Coatings", December 8-10, 2003, Moscow Russia. All chapters were peer-reviewed prior their acceptance for the publication. They arranged in the order of the corresponding presentations during the workshop. The workshop was divided into five sessions:

- Hard and Tribological Coatings;
- Recent Progress in the Development of Plasma Deposition Equipment;
- Film Characterization and Control of Structure and Properties;
- Nanopowders and Nanoparticles in Surface Engineering Technologies;
- Biocompatible Thin Films.

Correspondingly all chapters are arranged within five parts as identified by the Table of Contents.

The workshop was closed by accepting a Memorandum, which summarized main workshop conclusions, reflected on progress in the field of nanostructured coatings and plasma deposition methods, and provided recommendations for future research programs and international cooperation. The text of this memorandum is included into this book.

Workshop directors

Dr. Andrey A. Voevodin, NATO

Prof. Evgeny A. Levashov, Russia

ACKNOWLEDGMENTS

The financial support from the Public Diplomacy Division, Collaborative Programmes Section, NATO, Brussels toward preparation and publication of this book is gratefully acknowledged. This book is a result of an Advanced NATO-Russia research workshop, which was organized under a NATO-ARW grant (JSTC.R-ARW.979948) and co-sponsored by the Ministry of Industry, Science and Technology of Russian Federation and the International Science and Technology Center (grant 07-03). Moscow State Institute of Steel and Alloys (Technological University) is acknowledged for providing resource support and being a host of the workshop. Workshop Directors and Editors would like to express their gratitude to Mrs. Lynne Campbell/Nolan and Dr. Alain H. Jubier of the NATO Public Diplomacy Division for the workshop organizational help and Mrs. Wil Bruins of the Kluwer Academic Publishers for the help with preparation of this NATO series volume.

WORKSHOP MEMORANDUM

"Nanostructured Thin Films and Nanodispersion Strengthening Coatings"

NATO-RUSSIA ADVANCED RESEARCH WORKSHOP

December 8-10, 2003
Moscow State Institute of Steel and Alloys (Technological University)
Russia

NATO Co-Director: Dr. Andrey Voevodin
Russia Co-Director: Prof. Evgeny Levashov
Workshop participants:
NATO

Prof.	J. Bretagne	France
Prof.	J. Colligon	UK
Prof.	M.-P. Delplancke-Ogletree	Belgium
Dr.	F. Gammel	Germany
Prof.	L. Hultman	Sweden*
Dr.	P. Mayrhofer	Austria*
Prof.	J.J. Moore	USA
Prof.	J. Musil	Czech Republic
Dr.	J. Patscheider	Switzerland*
Dr.	F. Rossi	Italy
Dr.	R. Suchentrunk	Germany
Prof.	S. Veprek	Germany
Dr.	A. Voevodin	USA
Prof.	Z. Werner	Poland
Dr.	J. Zabinski	USA

* *NATO partner country*

Russia

Prof.	R.A. Andrievskii
Prof.	A.F. Belyanin
Prof.	V.I. Bogdanovich
Prof.	V.M. Elinson
Prof.	S.N. Grigor'ev
Prof.	V.F. Komarov
Prof.	E.A. Levashov
Prof.	Yu.V. Panfilov
Prof.	N.N. Gerasimenko
Dr.	V.G. Ral'chenko
Prof.	M.I. Samoilovich
Prof.	E.N. Sheftel
Dr.	D.V. Shtansky
Prof.	V.V. Sleptsov
Prof.	B.V. Spitsyn

The RARW reviewed the state-of-the-art of many topics in plasma physics and many related fields by world leaders in nanoscience. Many excellent lectures were presented within five sessions: Hard and Tribological Coatings; Recent Progress in the Development of Plasma Deposition Equipment; Film Characterization, Control of Structure and Properties; Nanopowders and Nanoparticles in Surface Engineering Technologies; Biocompatable Thin Films. Various techniques, scientific equipment and products were presented at the exhibition: "Advanced coating and surface modification technologies and film characterization facilities from Russia". The RARW identified directions for future research, and stimulated the possibility for new close working relationships to be developed between scientists from NATO countries and Russia.

The participants of the workshop reviewed the remarkable worldwide progress and many breakthroughs in this area of fundamental science including successful new industrial applications in the fields of plasma physics, surface engineering, nanostructured thin films and nanodispersion-strengthened coatings. The participants consider that new fundamental knowledge, new materials, and industrial producing methods generated as a result of national and international projects will be of interest to the industrial sector in the USA, European Union and Russia. Further progress in these fields will be more effective if the World Scientific Community can work together to resolve remaining problems.

The RARW has been useful to encourage and facilitate collaboration between research groups in Russia and NATO using all available joint collaboration schemes. These include:

- NATO-Russia Advanced Research Workshops;
- International Science and Technology Center (ISTC);
- Sixth framework program of the European Community for research, technological development and demonstration activities for the period 2002 to 2006 (FP6);
- INTAS;
- US Civilian Research and Development Foundation (CRDF);
- NATO Programme for Security Through Science;
- European Program of Scientific-Technical Integration (Eureka)

The RARW agreed that Russian scientists should be more actively involved in the integrated process with the World Scientific Community including cooperative research, industrial networks, information exchange, knowledge dissemination (scientific and engineering education, technology promotion, integrated post-graduate schools, conferences, seminars etc.). The RARW also agreed that it is very important to concentrate international activities on topics of special interest. The most following interesting and high priority technology fields were identified:

- Nanotechnologies and Nano-Science;
- Science and Engineering of Surfaces and Interfaces;
- Fundamentals of Plasma Processing-Structure-Properties Relationships;
- Biocompatible Thin Films and Interfaces;
- Multifunctional and Functionally Graded Materials;
- Nanopowders and Powder-Based Nanostructured Materials;
- Nanostructured Thin Films and Coating Systems in Practical Applications;
- Ion-Beam and Plasma-Based Deposition Techniques with High Plasma Density and Ionization;
- Techniques for Characterization of Nanostructured Materials, Coatings and Thin Films.

The RARW recommends that these high priority fields should receive special attention from the Ministry of Industry, Science and Technology of the Russian Federation, Russian Foundation of Basis Research, Ministry of Education of Russian Federation, as well as from NATO collaborative programs, various national and international foundations and industry and that research and development in these fields should be supported. Only such immediate support will bring the benefits of these materials to all members of the NATO community.

INTERNATIONAL ORGANIZING COMMITTEE

NATO Co-Director:	Andrey Voevodin
Russia Co-Director:	Evgeny Levashov
Italy:	Francois Rossi
Russia:	Dmitry Shtansky
USA:	John Moore

Chapter 1
SMART NANOCOMPOSITE COATINGS WITH CHAMELEON SURFACE ADAPTATION IN TRIBOLOGICAL APPLICATIONS

A. A. Voevodin and J. S. Zabinski
Air Force Research Laboratory, Materials and Manufacturing Directorate Wright Patterson Air Force Base, OH 45433-7750, U.S.A.

Smart nanocomposite tribological coatings were designed to respond to changing environmental conditions by self-adjustment of their surface properties to maintain good tribological performance in any environment. These coatings have been dubbed "chameleon" because of their ability to change their surface chemistry and structure to avoid wear. The first "chameleon" coatings were made of WC, WS_2, and DLC; these coatings provided superior mechanical toughness and performance in dry/humid environmental cycling. In order to address temperature variation, the second generation of "chameleon" coatings were made of yttria stabilized zirconia (YSZ) in a gold matrix with encapsulated nano-sized reservoirs of MoS_2 and DLC. High temperature lubrication with low melting point glassy ceramic phases was also explored. All coatings were produced using a combination of laser ablation and magnetron sputtering. They were thoroughly characterized by various analytical, mechanical, and tribological methods. Coating toughness was remarkably enhanced by activation of a grain boundary sliding mechanism. Friction and wear endurance measurements were performed in controlled humidity air, dry nitrogen, and vacuum environments, as well as at 500-600 ºC in air. Unique friction and wear performance in environmental cycling was demonstrated.

Keywords: nanostructured materials, composite, coating, hard, tough, low friction, wear reduction, environment cycling.

1. INTRODUCTION

Nanocomposite coatings offer a unique opportunity to design and produce adaptive or smart tribological coatings, which were termed "chameleon" for their ability to resist friction and wear by changing surface chemistry and microstructure in response to environment and loading changes [1,2], much like a chameleon changing its skin color to avoid predators.

Although quite challenging, practical realization of smart coatings is extremely rewarding for tribological pairs subjected to multiple environmental changes, as for example in aerospace applications. The first

A. A. Voevodin et al. (eds.),
Nanostructured Thin Films and Nanodispersion Strengthened Coatings, 1–8.

realization of such coating growth was made using a mixture of oxides and dichalcogenides (PbO/MoS_2, CF_x/WS_2, ZnO/WS_2), which could operate in a broad range of temperatures [3-5]. Advanced multilayer structures were then designed to combine these composites with buried diffusion barrier layers and achieve surface self-adaptation during repeated temperature cycling. Recently, novel wear resistant materials were developed, which combine nanocrystalline carbides (TiC, WC), oxide based ceramics (YSZ and AlON), dichalcogenides (MoS_2, WS_2), and amorphous diamond-like carbon (DLC) into nanocomposite structures [1,2,6].

For example, "chameleon" coatings made of amorphous diamond-like carbon (DLC) matrix with incorporation of nanocrystalline TiC [7,8], WC [9,10], WS_2 [1,11] and laser processed MoS_2 reservoirs [12] had demonstrated an order of magnitude improvement in toughness above that of single phase carbides while maintaining the same level of hardness, a low friction coefficient in cycling from dry to humid environments, and an extremely long life in both ambient and space environments. The surface chemistry, structure, and mechanical behavior of these nanocomposite materials were shown to *reversibly* change in the tribological contact, depending on applied loads and operational environment to maintain low friction and prevent wear.

While keeping the low friction in any environment is important, the coating wear resistance requires an additional blend of both hardness and fracture toughness. Nanocomposite coating designs provide a unique opportunity to achieve the best blend of the toughness, wear resistance, and a low friction performance across multiple environments.

This paper reviews the most recent developments in smart nanocomposite tribological coatings, starting with design criteria and examples of tough tribological nanocomposites and progressing to the design and examples of "chameleon" coatings.

2. TOUGH NANOCOMPOSITE TRIBOLOGICAL COATINGS

An approach with embedding grains of a hard, high yield strength phase into a ductile matrix has been widely explored in macro-composites made of ceramics and metals which are known as cermets [34]. However, if grain sizes and matrix separations are reduced to a nanometer level, dislocation activity as a source of ductility is eliminated and new mechanisms are required to provide toughness enhancement. Uniquely, these nanocomposites contain a high volume of grain boundaries with a crystalline/amorphous transition across grain-matrix interfaces, limiting initial crack sizes and helping to deflect and terminate growing cracks. These mechanisms helped to explain the fracture resistance of novel super-hard composites, where both amorphous ceramic [13-15] and metal [16-18] matrices were used to encapsulate nanocrystalline ceramic grains. More on the super-hard nanocomposite coating design and discussion of the involved mechanisms can be found in available review articles [14,15,17,19,20].

Grain boundary diffusion [21] and grain boundary sliding [21-25] were suggested as mechanisms for improving ductility and providing super-plasticity of single phase ceramic nanocrystalline systems. The most recent research indicates also that high ductility can be more easily achieved in multiphase structures [26] and that grain boundary sliding is a primary mechanism of super-plasticity [27-30]. It was also found that equiaxial grain shapes, high angle grain boundaries, low surface energy, and the presence of an amorphous boundary phase facilitate grain boundary sliding [21,22]. These findings can be expanded into the field of hard wear resistant coatings to introduce ductility and prevent fracture under a high contact load.

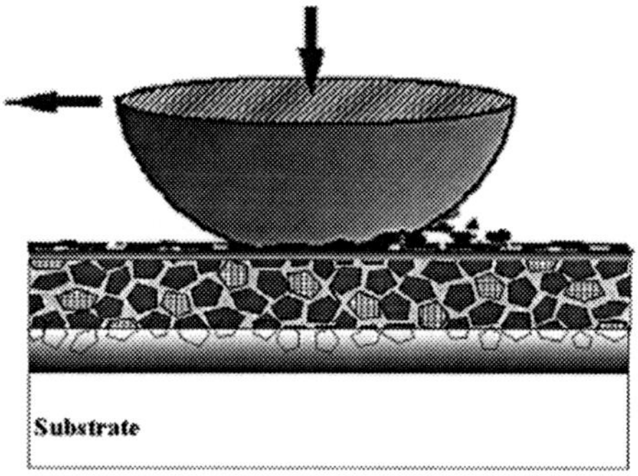

Figure 1. Schematic of a tough nanocomposite coating design, combining a nanocrystalline / amorphous structure with a functionally gradient interface.

In the course of the development of tough tribological nanocomposite coatings the following design concepts were used:

(1) a graded interface layer is applied between the substrate and crystalline/amorphous composite coating to enhance adhesion strength and relieve interface stresses (combination of functional gradient and nanocomposite designs) [31-33];

(2) encapsulation of 3-10 nm sized hard crystalline grains in an amorphous matrix restricts dislocation activity, diverts and arrests macro-crack development, and maintains a high level of hardness similar to super-hard coating designs [14,17];

(3) a large volume fraction of grain boundaries provides ductility through grain boundary sliding and nano-cracking along grain/matrix interfaces [8,34,35].

These design concepts are targeting toughness enhancement through stress minimization, crack deflection, and ductility introduction. They are close but still differ from the design concepts of superhard nanocomposite coatings. The primary differences are viewed in the selection of a matrix phase with a lower elastic modulus, relaxation of the requirement for strong binding between matrix and grains, and selection from a greater range of acceptable grain sizes of nanocrystals phase embedded in an amorphous matrix phase.

Combination of the nanocrystalline/amorphous designs with a functionally graded interface is shown in Figure 1. This design provides both high cohesive toughness and high interface (adhesive) toughness in a single coating. Several examples of tough wear resistant composite coatings have been produced. Two of them combined nano-crystalline carbides with an amorphous DLC matrix designated as TiC/DLC and WC/DLC composites. In another example, nanocrystalline YSZ grains were encapsulated in a mixed YSZ-Au amorphous matrix as shown in Figure 2. In all cases, the large fraction of grain boundary phase provided ductility by activating grain boundary slip and crack termination by nanocrack splitting. This provided a combination of high hardness and toughness in these coatings. Figure 3, compares Vickers indentations made at the highest load of the machine (1 kg). There are no observable cracks in these coatings, even after significant substrate compliance (indentation marks are 9 μm deep into 1 μm thick coatings). The coating hardness was quite high ranging from 18 to 30 GPa, and for most other typical hard coatings at these loads, cracks in the corners of the indentations are expected.

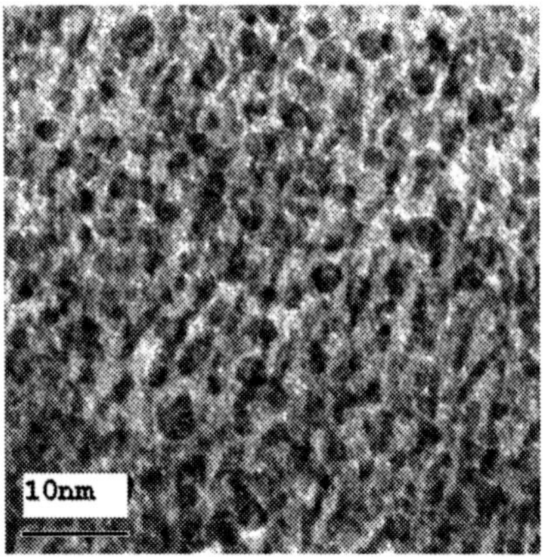

Figure 2. TEM image of an YSZ/Au nanocomposite coating with improved toughness characteristics

Thus, novel nanocomposite designs of tough tribological coatings are very promising. They explore fundamentally different concepts of the toughness improvement, when compared to macrocomposite materials such

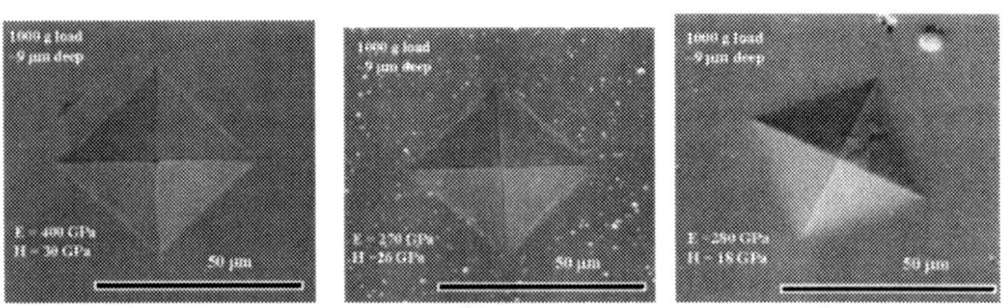

a) b) c)

Figure 3. Knoop and Vickers indenation marks into the surface of 1 μm thick (a) TiC/DLC, (b) WC/DLC, and (c) YSZ/Au tough nanocomposite coatings. Indents were performed with the maximum available load of 1000 gm, providing about 9 μm indentation depth due to the steel substrate deformation. In all cases, there were no cracks at the indention corners, which serve as stress risers.

as cermets. Furthermore, their design can be taken to the next level by realizing that both matrix and nanograins can serve as reservoirs of solid lubricants. These lubricants will be then released in a friction contact in the course of sliding. The active tribological role of matrix and encapsulated grains is a basis for the tough and low friction “chameleon” coatings, which are discussed in the next section.

3. “CHAMELEON” TRIBOLOGICAL COATINGS

The objective of “chameleon” tribological coating designs is to provide a *reversible* self-adjustment of surface chemistry, structure, and mechanical behavior in the friction contact, depending on applied loads and operational environment to maintain low friction and prevent wear.

In order to achieve reversible adaptation, two additional design concepts were developed and combined with the tough nanocomposite coating concepts (design concept numbering is continued here from Section 2):

(4) solid lubricant reservoirs are introduced as amorphous or poorly crystalline inclusions to minimize reduction in composite hardness and elastic modulus, since crystalline solid lubricants are typically very soft [1,36];

(5) friction forces and surface reactions with the environment are used to generate a lubricious transfer film or “skin” at the tribological contact, which can self-adjust with each environmental change [1,36]; *i.e,.* coating components serve as reservoirs to supply material for the “tribo-skin”, where formation of a lubricating film with the required chemistry and structure reduces friction.

Figure 4 presents a schematic of a nanocomposite coating design that exhibits “chameleon” behavior. This design was implemented in the fabrication of the YSZ/Au/DLC/MoS_2 and WC/DLC/WS_2 “chameleon” coatings where an amorphous matrix and a hard nanocrystalline phase (e.g., YSZ or WC) were used to produce optimum mechanical performance and load support. Nanocrystalline and amorphous Au, MoS_2, and DLC were added to achieve chemical and structural

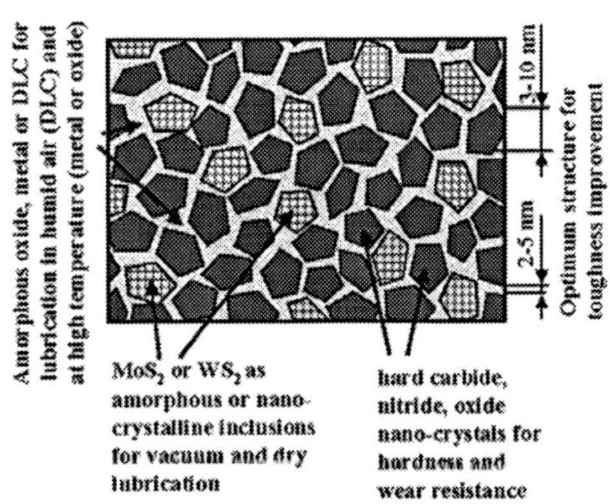

Figure 4. Schematic of a conceptual design of the YSZ/Au/MoS_2/DLC tribological coating with chameleon-like surface adaptive behavior.

adjustment of transfer films formed in friction contacts during dry/humid environment and low/high temperature cycling. "Chameleon" behavior is observed during the following sequence. As deposited, all lubricant phases (WS_2 or MoS_2, DLC, soft metals, and oxides) are either amorphous or poorly crystalline and are buried in the coating where they are sealed from the environment. When sliding commences, wear processes expose the surface components to the environment, and stress and frictional heat cause changes in chemistry and structure.

In low temperature humid environments, a graphitic-like transfer layer is formed by an $sp^3 \rightarrow sp^2$ phase transition of the DLC component, which eventually leads to the formation of some crystalline graphite. This layer provides a low friction coefficient (i.e., 0.1-0.15) and significantly reduces further wear. In humid environments, formation of MoS_2 or WS_2 transfer layers is not favorable, due to their relatively high friction coefficient. Those phases also have a high wear rate and become concentrated in the wear debris as the graphite-like transfer film forms.

When the environment changes to dry (N_2 or air) or vacuum and the temperature remains low, a WS_2 or MoS_2 transfer layer is formed by an amorphous $\rightarrow$ crystalline transformation of dichalcogenide inclusions. Rubbing orients the crystalline material such that the low friction basal plane is parallel to the surface. This provides friction coefficients as low as 0.01, reducing further wear of the composite coating. Graphite from a previous cycle in humid air is removed or covered over in the first several hundred cycles in dry conditions due to its high friction/wear in the absence of intercalation by water molecules. As an example of performance in cycling humidity, Figure 5 shows that the process repeats itself for YSZ/Au/DLC/MoS_2, providing a corresponding switch in tribofilm chemistry and friction response between DLC in humid and MoS_2 in dry conditions.

In high temperature environments (above 500 °C), diffusion of soft metal from the coating to the surface and subsequent crystallization occurs. This results in the formation of an easy-to-shear transfer film for high temperature lubrication with a friction coefficient of about 0.2 [2]. When temperature is cycled back to 25 °C, the friction coefficient stays at about the same level.

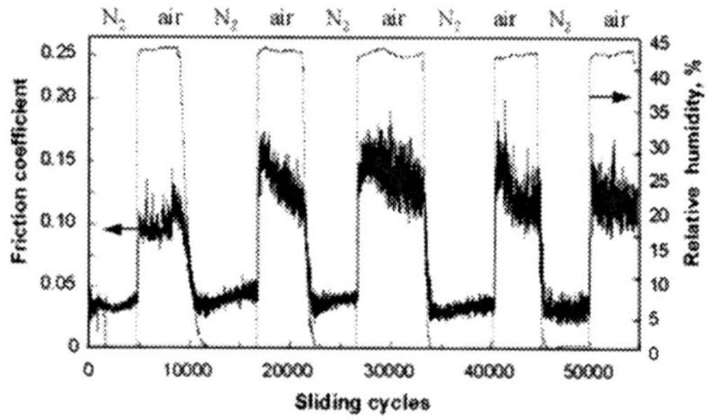

Figure 5. Friction coefficient variation of a "chameleon" YSZ/Au/MoS_2/DLC coating in a test with a cycled relative humidity. Friction response is repeatedly switching from a lubrication provided by MoS_2 in dry environments to lubrication provided by DLC in moist conditions.

Using these mechanisms to provide adaptive lubrication, advanced tribological coatings can be produced that survive multiple dry/humid cycling, temperature cycling, and provide long duration operation in many test environments. Various carbides, nitrides, borides, and oxides may be used as the hard nanocrystalline phase in combination with carbon for lubrication in ambient conditions, MoS_2 and WS_2 for lubrication in dry and vacuum environments, and soft metals (Au, Ag, Ni) or low viscosity glasses (network modified SiO_2) for high temperature lubrication. For example, similar success was achieved with $Al_2O_3/MoS_2/CF_x$ composites and $SiC/Cs_2MoOS_2/MoS_2$/DLC. In the last of these two examples, a Cs-SiO_2 glass transfer film was formed in the friction contact. This transfer film material had a low melting point and provided lubrication above 600 °C [37].

The coating design shown in Figure 4 and the examples provided previously were primarily tailored to aerospace applications, but similar designs can be developed to produce "chameleon" tribological coatings for other environments. The concepts for producing tough, hard, low friction nanocomposite coatings discussed here are universal such that they can be realized using a variety of material systems and are easily tailored to specific applications.

4. CONCLUSIONS

Nanocomposite tribological coatings were designed to respond to changing environmental conditions by self-adjustment of their surface properties to maintain good tribological performance in any environment. These smart coatings have been dubbed "chameleon" because, analogous to a chameleon changing its skin color to avoid predators, the coating changes its "skin" chemistry and structure to avoid wear. Their design concepts combine toughness improvement by nanocrystalline/amorphous structures, functionally gradient interfaces, and incorporation of solid lubricant nano reservoirs to provide tribological self-adaptation in response to environment variations. "Chameleon" tribological behavior is realized by activation of a number of surface-adaptation mechanisms: (i) formation of hexagonal MoS_2 or WS_2 transfer films in dry nitrogen and vacuum tests; (ii) formation of graphite-like carbon transfer films from a DLC matrix in humid air; (iii) formation of transfer films made of soft metal or low melting ceramic glassy phases in 500-600 °C sliding in air. The mechanisms controlling friction could be repeatedly and *reversibly* switched, permitting good performance during environmental cycling, e.g. ambient $\leftrightarrow$ space. The concept of a "chameleon" composite coatings can be realized in different material systems, where nano-reservoirs of solid lubricants are encapsulated inside hard, tough and thermally stable nanocrystalline/amorphous matrices.

REFERENCES

1. A. A. Voevodin, J. S. Zabinski, Thin Solid Films 370 (2000) 223.
2. A. A. Voevodin, J. J. Hu, T. A. Fitz, J. S. Zabinski, J.Vac.Sci.Technol.A 20 (2002) 1434.
3. J. S. Zabinski, M. S. Donley, V. J. Dyhouse, N. T. McDevit, Thin Solid Films 214 (1992) 156.
4. J. S. Zabinski, M. S. Donley, N. T. McDevit, Wear 165 (1993) 103.
5. Zabinski, J. S., Prasad, S. V., McDevit, N. T. Advanced Solid Lubricant Coatings for Aerospace Systems, Tribology for Aerospace Systems, Proceedings of NATO Advisory Group of Aerospace Research and Development (AGARD) Conference on Tribology for Aerospace Systems, Sesimbra, Portugal, May 6-7, 1996, AGARD NATO publication CP 589, printed by Canada Communication Group. Hull, Quebec, 1996.
6. A. A. Voevodin, J. J. Hu, T. A. Fitz, J. S. Zabinski, Surf.Coat.Technol. 146-147 (2001) 351.
7. A. A. Voevodin, S. V. Prasad, J. S. Zabinski, J.Appl.Phys. 82 (1997) 855.
8. A. A. Voevodin, J. S. Zabinski, J.Mater.Sci. 33 (1998) 319.
9. A. A. Voevodin, J. P. O'Neill, S. V. Prasad, J. S. Zabinski, J.Vac.Sci.Technol.A 17 (1999) 986.
10. A. A. Voevodin, J. P. O'Neill, J. S. Zabinski, Thin Solid Films 342 (1999) 194.
11. A. A. Voevodin, J. P. O'Neill, J. S. Zabinski, Surf.Coat.Technol. 116-119 (1999) 36.
12. A. A. Voevodin, J. E. Bultman, J. S. Zabinski, Surf.Coat.Technol. 107 (1998) 12.
13. S. Veprek, S. Reiprich, Thin Solid Films 268 (1995) 64.
14. S. Veprek, J.Vac.Sci.Technol.A 17 (1999) 2401.
15. S. Veprek, A. S. Argon, Journal of Vacuum Science & Technology B 20 (2002) 650.
16. J. Musil, P. Zeman, H. Hruby, P. H. Mayrhofer, Surf.Coat.Technol. 120-121 (1999) 179.
17. J. Musil, Surf.Coat.Technol. 125 (2000) 322.
18. J. Musil, H. Hruby, P. Zeman, R. Cerstvy, P. H. Mayrhofer, C. Mitterer, Surf.Coat.Technol. 142-144 (2001) 603.
19. J. Musil, J. Vlcek, Surf.Coat.Technol. 142-144 (2001) 557.
20. J. Patscheider, T. Zehnder, M. Diserens, Surf.Coat.Technol. 146-147 (2001) 201.
21. J. Karch, R. Birringer, H. Gleiter, Nature 330 (1987) 556.
22. O. D. Sherby, J. Wadsworth, Progr.Mater.Sci. 33 (1989) 169.
23. F. Wakai, Ceram.Int. 17 (1991) 153.
24. T. G. Nieh, J. Wadsworth, F. Wakai, Int.Mater.Rev. 36 (1991) 146.
25. T. G. Langdon, Mater.Sci.Eng.A 166 (1993) 67.
26. C. C. Koch, D. G. Morris, K. Lu, A. Inoue, MRS Bul. 24 (1999) 54.
27. R. W. Siegel, G. E. Fougere, Mater.Res.Soc.Symp.Proc. 362 (2002) 219.
28. T. G. Langdon, Mater.Sci.Forum 189-190 (1995) 31.
29. O. D. Sherby, T. G. Nieh, J. Wadsworth, Mater.Sci.Forum 243-245 (1997) 11.
30. J. Schiotz, F. D. Di Tolla, K. W. Jacobsen, Nature 391 (1998) 561.
31. A. A. Voevodin, J. M. Schneider, C. Rebholz, A. Matthews, Tribol.Int. 29 (1996) 559.
32. A. A. Voevodin, M. A. Capano, S. J. P. Laube, M. S. Donley, J. S. Zabinski, Thin Solid Films 298 (1997) 107.
33. A. A. Voevodin, S. D. Walck, J. S. Zabinski, Wear 203-204 (1997) 516.
34. C. Mitterer, P. H. Mayrhofer, M. Beschliesser, P. Losbichler, P. Warbichler, F. Hofer, P. N. Gibson, W. Gissler, H. Hruby, J. Musil, J. Vlcek, Surf.Coat.Technol. 120-121 (1999) 405.
35. A. A. Voevodin, J. G. Jones, J. J. Hu, T. A. Fitz, J. S. Zabinski, Thin Solid Films 401 (2001) 187.
36. A. A. Voevodin, J. P. O'Neill, J. S. Zabinski, Tribol.Lett. 6 (1999) 75.
37. K. L. Strong, J. S. Zabinski, Thin Solid Films 406 (2002) 174.

Chapter 2
MATERIALS SCIENCE OF WEAR-PROTECTIVE NANOSTRUCTURED THIN FILMS

L. Hultman
Thin Film Division, IFM, Linköping University, 581 83 Linköping, Sweden

Nano-design routes for ceramic thin films by PVD are reviewed. For TiN/NbN superlattices, glide within the layers is the dominant deformation mechanism in support of models for superhardening. Theory presumes plasticity with dislocation hindering at interfaces between phases with different shear modulus. It is shown that the superlattices exhibit crystal rotation near an indent, scratch or wear zone. A concept for age hardening of metastable coating materials is presented for the Ti-Al-N system. Spinodal decomposition with coherent cubic-phase nm-size domains is thus demonstrated, causing an increase in hardness during annealing. Finally, inherently nanolaminated hexagonal $M_{n+1}AX_n$-phases are considered (n=1,2,3; M=transition metal; A group element; X=C/N). These phases combine ceramic and metallic properties. MAX phases in thin film form are reported, including Ti_3SiC_2 and Ti_2AlN. Nanoindentation combined with electron microscopy reveals kink formation and cohesive delamination on the basal planes of these ductile materials.

Keywords: nitrides, carbides, superlattice, nanocomposite, MAX-phases, aging, self-assembly, spinodal decomposition, electron microscopy, nanoindentation.

1. INTRODUCTION

Nanostructured ceramic coatings are attractive in that materials can be synthesized for tailored mechanical properties. Superhard materials are made by hindering plastic deformation and crack propagation by means of defect hardening (via lattice point defects and compressive intrinsic stress) [1], superlattices [2-4] or nanocomposites [5,6]. Model systems for the nanolaminates reviewed here are artificial nitride superlattices (Sect. 2) and inherently nanolaminated MAX-phase ternary ceramics [7,8] (Sect. 3). For nanocomposite films, Sect. 4 discusses the application of self-assembly processes leading to nanostructuring of phases in amorphous Si_3N_4 with nanocrystalline TiN [5] or nc-TiC/C [9] obtained by continuous nucleation and growth with segregation for the partitioning of elements. A thermodynamically different route of formation is by spinodal decomposition during growth or during annealing of as-deposited films. The latter phenomenon has recently been observed to take place in TiAlN coatings and results in age hardening [10,11].

A. A. Voevodin et al. (eds.),
Nanostructured Thin Films and Nanodispersion Strengthened Coatings, 9–21.

2. ARTIFICIAL SUPERLATTICE NITRIDE FILMS

It is reassuring that nitride superlattices (SL) as a type of nanostructure-engineered coating are offered commercially. Discovered in the 80-ies – and thus predating the s.c. nanoscience era – such nanolaminates of transition metal nitrides made by PVD processes are an emerging class of wear-protective coatings [2-4]. Increased hardness (to >50 GPa) as compared to single-layers is reported both for as-deposited single-crystal SL [12-15] and polycrystalline nanolayered thin films [16,17] for, e.g., the material systems TiN/NbN, TiN/VN, TiN/AlN, and their alloyed derivates. Hardness enhancement is explained by dislocation hindering at the interfaces due to differences in shear moduli, and by coherency strain in the lattice-mismatch materials [3,4,18.19]. Figure 1 shows HREM images of a TiN/NbN SL. Layers are well defined from a reactive magnetron sputter deposition process with dual targets. It is also seen that the layers are coherently strained for the 3.5% lattice mismatched system with epitaxial growth of TiN and NbN [12].

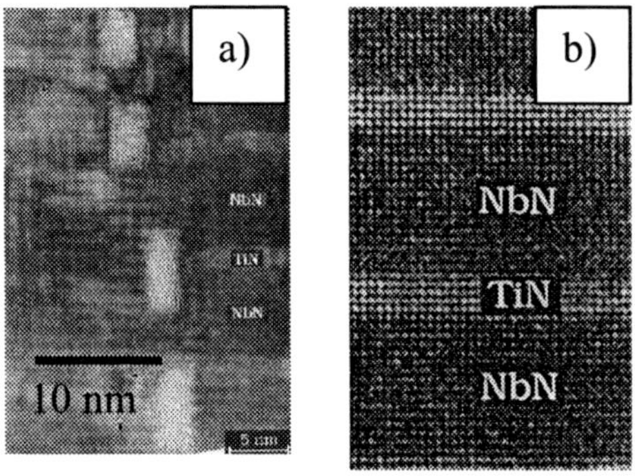

Figure 1. XTEM images of single-crystal TiN/NbN superlattices with a) self-organized nanovoids at layer cusps and b) dense coherently strained layers.

Studies of slip in mechanically deformed zones of single-crystal TiN/NbN and TiN/VN SL was made using TEM and AFM [20,21]. The mechanical deformation in these SL during indentation experiments is elasto-plastic. Figure 2 shows an XTEM image of a TiN/NbN SL after nano-scratching [22]. The induced deformation zone exhibits pile-up at the scratch rim and contains the features: (1) retained SL domains rotated with respect to each other; (2) deformation along column boundaries; and (3) deformation-induced differences in SL period between regions. Deformation is localized to apparent column boundaries that were present in the as-deposited case, but aligned perpendicular to the film surface sample. Such boundaries can be the site for nanosized voids (see Fig.1) [23] formed by a self-organized process under kinetically limited growth conditions (low temperature in combination with shadowing of the deposition flux by surface protrusions). The change in SL period induced by the mechanical deformation suggests transportation of material. HREM imaging revealed bending of the lattice by ordering of glide dislocations. The deformed lattice stays straight within layers, but form knees at layer interfaces indicating that low-angle grain boundaries have formed by these dislocations. The deformation zone in TiN/NbN SL is smaller and more complex compared to that in monolithic TiN. Layers stay relatively intact with no large slip bands, whereas for TiN extensive slip bands were observed on {110} planes, in agreement with the {110}<1-10> slip system for TiN [19,24]. Glide within layers of the SL is the dominating deformation mechanism in support of the prevailing models for superlattice hardening [3,4]. Previously, dislocation confinement to individual layers was reported for metallic multilayers, *cf.*, Cu/Ni [25]. Results from the deformed zone from nanoindents are similar to the nanoscratches [26].

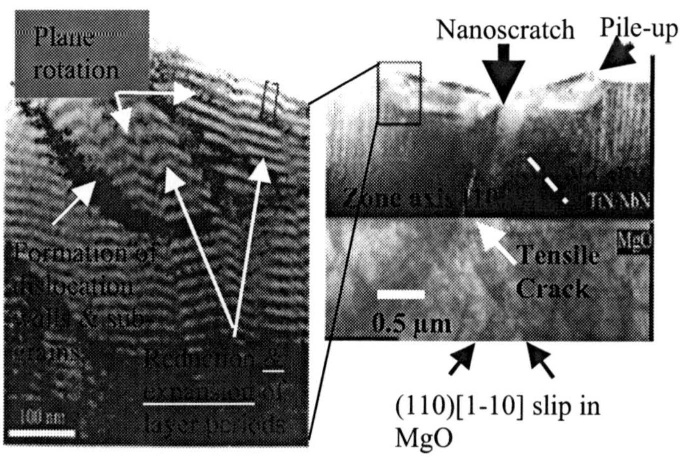

Figure 2. Cross-sectional TEM image of a 12-nm-period (100)-oriented single-crystal TiN/NbN superlattice (deposited on MgO(100) substrate) after nanoscratching with 35 mN Berkovitch diamond tip. From Ref. [22].

Cutting tests of nitride SL were reviewed by Münz [17]. For wear tests of TiAlN/CrN SL, Luo [27,28] found plastic deformation with the dissolution of the layered structure in the ~50 nm near surface region. A distorted SL structure by possible grain rotation was obtained 50-100 nm deeper. Micro and nanocracking occurs at column boundaries as the weakest link in coatings containing residual compressive stress [27,28] or grain boundary porosity [26,29]. It should not be expected that SL hardening in nitrides can be added to defect hardening from a PVD process operated with ion bombardment leading to compressive stress. This is because SL hardening requires dislocation activity and the lattice point defects hinder that process.

While nitride superlattices can exhibit better wear-resistance compared to pure nitride films [17,30], the fracture behavior in the latter is anisotropic with fracture primarily on (110) planes in radial directions from the indent. When indenting nitride SL, concentric cracks occur around the indentation, implying that no predefined fracture planes exists. For TiN/Cu multilayers, hardness was limited by fracture in the relatively brittle TiN layers because of stress concentrations resulting from the compliant Cu layers [31]. We infer that the dislocation pile up together with the observed formation of domains in TiN/NbN SL (Fig. 2) will cause stress concentration, which dictates the fracture behavior and eventually suppresses the material toughness.

The use of SL as protective coatings on cutting tools exposes the structure to elevated temperatures. From industrial interest in increased cutting speed and feeding rate, as well as for dry cutting, there is consequently a demand for coatings that can withstand high temperatures. The SL structure with its high density of interfaces is in a non-equilibrium state and subject to diffusion processes like coarsening. In fact, the TiN-NbN system is miscible [32] and thus exhibits interdiffusion [33]. The lifetime for 8-nm-period TiN/NbN SL at 850 °C was calculated to be 1 h. During annealing the SL hardness was reduced from >30 GPa to 27 GPa. Further annealing to 950 °C resulted in a drop to 23 GPa. Such values are useful input for selecting cutting data in application. Interdiffusion during operation of SL coatings can be avoided by using *immiscible components as* reviewed by Barnett [34]. Enhanced thermal stability was among other systems demonstrated for TiN/AlN [34], NbN/Mo [35] and NbN/W [36] superlattice films.

3. INHERENTLY NANOLAMINATED MAX-PHASE FILMS

$M_{n+1}AX_n$-phases (n = 1, 2, 3) are ductile inherently nanolaminated ternary nitrides and carbides with ~60 known bulk phases [7,8,37-39]. Interest for these is motivated by an unusual combination of metallic and ceramic attributes for the archetypical phase Ti_3SiC_2. From a materials science aspect, research is accelerated by the progress in *Ab initio* calculations of phase stability and properties [40-42]. Properties are typically shared by the other MAX-phases and stem from the highly anisotropic hexagonal crystal structure (space group $P6_3/mmc$) in which the early transition metal atoms and the carbon and/or nitrogen atoms form octahedral edge sharing building (MX) blocks interleaved by pure A-element layers. By different stacking sequences of the MX block between each neighboring A-element layers, the MAX-phases are known in the subgroups M_2AX, M_3AX_2 and M_4AX_3.

Recently, we showed that DC magnetron sputtering can be used to synthesize MAX-phase films from the Ti-Si-C system at temperatures far below those for bulk synthesis [43] and presented three deposition processes for the growth of Ti_3SiC_2: (i) sputtering from a Ti_3SiC_2 compound target, (ii) sputtering from individual Ti and Si targets with co-evaporated C_{60} and (iii) sputtering from all individual elemental targets [44]. It is also possible to grow MAX-phase films of the Ti-Ge-C [45,46] and Ti-Al-C [47] systems as well as the N-based Ti-Al-N [48,49]. Since most MAX phases are equilibrium structures, a preferred synthesis route is to supply the elements in the right proportions and to ensure enough heating for diffusion such that the large unit cells can form. Thin film growth can be complicated by the competing growth of MX phases, e.g., TiC or TiN for which the MAX phases have tie lines [50]. Also, metastable Ti_4SiC_3 films have been demonstrated [44,51] and supported by calculations [52]. The HAADF image in Fig. 3 shows the characteristic nanolaminated stacking of Ti_3SiC_2.

The MAX-phase thin films from the studied materials systems can be synthesized by magnetron sputtering, using high purity elemental targets for the carbides or a Ti_2Al alloy source for the Ti-Al-N system in either high Ar discharges for the carbidic MAX-phases or in a reactive Ar/N_2 mixture for the nitride synthesis. The thin film growth conditions were investigated on $Al_2O_3(0001)$ and/or MgO(111) substrates. An emphasis was placed at substrate temperatures of 800-1000 °C, using seed layering technique.

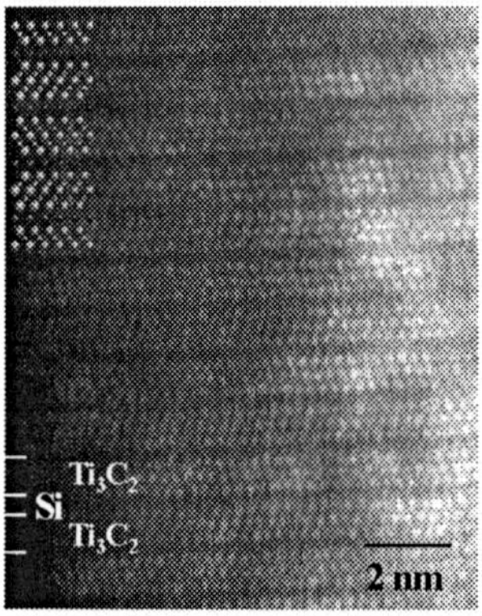

Figure 3. High-angle annular dark field image by scanning TEM from a Ti_3SiC_2 (0001) film, showing a nanolaminated structure in [11-20] projection. Ti_3C_2 and Si layers (dark due to atomic number contras are marked also with a schematic MAX phase structure.

Nanoindentation analysis of ~0.6 µm Ti_3SiC_2 films shows decreasing hardness and elastic modulus values from 23 – 29 GPa and 350 – 375 GPa leveling out to about ~15 GPa and 280 GPa, respectively, as the deformation proceeds with increased maximum indentation loads. Pop-in events occur frequently indicating intermittent plastic deformation [50,53]. The Young's modulus obtained by *Ab initio* calculations of Ti_3SiC_2 in the c-direction is 375 GPa [41] in good agreement with experimental values [50,51]. The AFM image in Fig. 4 shows an indent in a $Ti_{n+1}SiC_n$ film, revealing its interesting deformation behavior with a large plastic deformation and extensive pile up around the indent. The corresponding XTEM image from an indented region shows that cohesive delamination of the basal planes and kink band formation has occurred, thus explaining the formation of pile up seen on the surface of the film [53] as well as supporting the observations made for bulk material [39]. Figure 5 shows a model for the deformation modes observed with shearing, kinking, buckling and delamination; all explained by the generation and glide of dislocations along the basal planes.

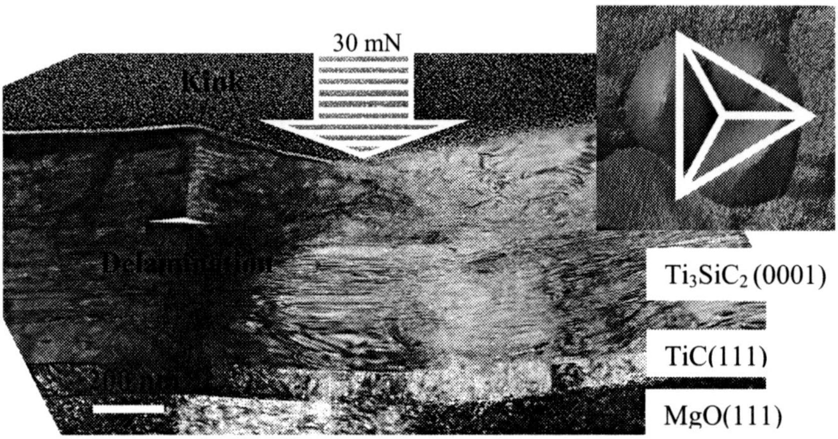

Figure 4. XTEM image obtained from a 30 mN Berkovitch indent to a Ti_3SiC_2 film revealing cohesive delamination of the basal planes and kink band formation. The AFM inset shows the area around the indent with extensive pile up.

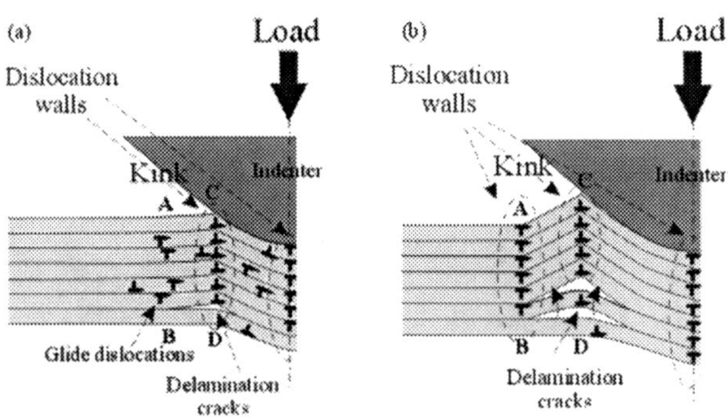

Figure 5. Model for the formation of kink bands in Ti_3SiC_2 during indentation based on observations [53, 54]. (a) Surface radial compression forces the layers to buckle away at the surface to a tilted position. This is possible due to easy delamination and shear between the layers. (b) As the indenter penetrates further, more layers buckle and the delamination crack moves deeper from the film surface. From Ref. 53.

4. NANOCOMPOSITE CERAMIC THIN FILMS

A design concept for nanocomposites was developed by Veprek [5,55,56] and Patscheider [6]. It states that the constituting phases should be selected for strong segregation (immiscibility), strong interface bonding, and have high shear moduli. It is a condition that nm-grains exhibit no dislocation activity [57-60]. The alternative means for plasticity are through block slip (not shown to date) or grain boundary sliding/rotation [61].

Recently, a new type of advanced surface engineering was presented by us [10,11] based on age hardening of as-deposited metastable solid solutions of ceramic coatings. Surprisingly many alloys exhibit miscibility gaps from a limited solid solubility. Spinodal decomposition [62] (requiring negative 2nd derivative of Gibbs free energy and thermal activation for diffusion) is known for bulk. A relevant example is the decomposition in TiMoCN [63]. However, *little is understood for thin films*. Here, both surface and in-depth decomposition can take place in quenched alloys. The shape and properties of the so-formed components might also be possible to control due to anisotropic conditions. E.g., *surface-initiated* spinodal decomposition of the Ti-Al-N takes place *during growth*, to form a rod-like nanostructure of fcc-TiN- and fcc-AlN-rich domains with a period of 2-3 nm (see Fig. 6) [64,65].

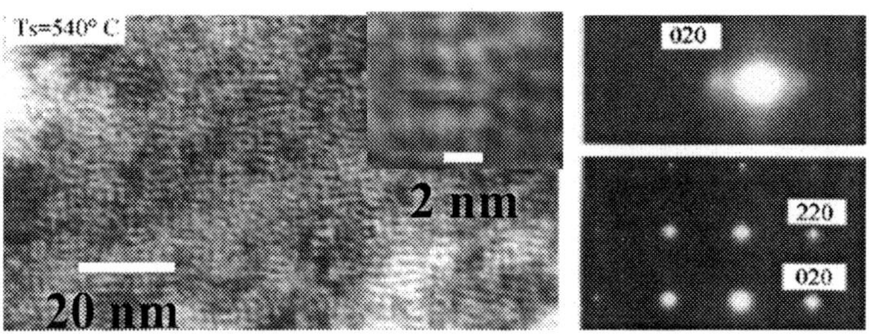

Figure 6. Plan-view TEM image with selected area electron diffraction patterns revealing surface-initiated spinodal decomposition during growth of a $Ti_{0.5}Al_{0.5}N$ (001) film at T_S = 540 °C by dual target reactive magnetron sputtering. High-magnification of the 020 reflection shows satellite peaks along [010] direction in a cross-sectional view and along both [010] and [100] in plan-view. From Ref. [64].

Low-temperature ion-assisted (20-50 eV) PVD growth processes are needed to stabilize thin film alloys with compositions within the miscibility gap in a metastable state [64,66,67]. This function relies on the fact that the processes operate at kinetically limited conditions and with ion-bombardment-induced recoil mixing of surface atoms and thus work against thermodynamically-driven segregation. Phase field arguments for a range of pseudo-binary nitrides systems including TiN-AlN were presented early by Holleck [68].

Age hardening in metallurgy is studied since 100 years. It has been speculated that the hardness increase for some ternary ceramic coating materials at elevated temperatures is caused by phase separation and spinodal decomposition [69-74]. However, a critical review reveals that there is a lack in explaining the responsible microstructural changes and providing supporting observation. Our group confirmed spinodal decomposition and increased hardness in metastable $Ti_{0.34}Al_{0.66}N$ *thin films* at 700-1000 °C. In the as-deposited state, films show a homogeneous $Ti_{1-x}Al_xN$ solid solution of NaCl-structure. This material is metastable and tends to decompose into the stable phases fcc-TiN and hcp-AlN. This was motivated by recent *ab initio* calculations [75]. It can be inferred from [75] that this nitride system is apt to decompose spinodally due to a negative 2nd derivative of the Gibbs energy and that there is an energy gain going from an alloy to TiN- and AlN-rich cubic components.

A hardness increase in arc-deposited c-$Ti_{0.34}Al_{0.66}N$ films is observed at 600-1000 °C (see Fig. 7), indicating *age hardening*. This effect does not occur for TiN [76], which instead softens at $T_a \geq 400$°C, due to stress recovery and recrystallization [77]. During annealing, exothermic reactions are seen in differential scanning calorimetry (DSC) [11] including the sequential thermal activation of the processes 1) recovery including lattice defect annihilation with residual stress relaxation, 2) Al-N segregation from the matrix, 3) Ti-N segregation, and 4) h-AlN precipitation and grain growth. The activation energies for recovery and the formation of TiN domains had values of 3.4 eV. Recovery and spinodal decomposition are diffusion-controlled within one phase, thus explaining the similar values. For the fcc-hcp AlN transformation and subsequent recrystallization, additional nucleation is needed resulting in E_a = 3.6 eV. Considering the similarity of these values with that reported for surface diffusion of Ti on TiN of 3.5 eV [78], we infer that spinodal decomposition can be a defect-assisted process.
This was the *first evidence for spinodal decomposition* *in thin films* and the nitrides. Evidence was presented earlier by us for *surface-initiated* spinodal decomposition *during growth* of $Ti_{0.5}Al_{0.5}N$ films at $T_s \geq 540$°C. The much lower T_a for that process is explained by the difference in activation energy for surface and bulk diffusion.

To assess the phase transformations, we performed *Ab initio* studies [79] to compare the phase stability of AlN in the candidate phases; wurtzite, NaCl, and zinc blende (see Table 1). It is clear that during transformation, the nucleation of zb-AlN is unlikely considering a positive energy difference and a large volume mismatch. The driving force for transformation into w-AlN is strong, however, with a large expansion in the cells. This corroborates the high-pressure experiments by Xia et al. [80] that gave 20% larger molar volume for w-AlN compared to NaCl-AlN. The lattice strain gives a large nucleation energy barrier for w-AlN. It is concluded that the metastable system initially gains energy by decomposing into TiN- and AlN-rich structure for which there is little volume mismatch. The present realization of the molar volume mismatch points to eventual cohesive failure of $Ti_{1-x}Al_xN$ material during precipitation and growth of w-AlN.

Effectively the (Ti,Al)N coatings harden as a nanocomposite adapting in strength to the annealing. The reasons for this lies in coherence strain from the 2.8% lattice mismatch between such volumes ($a_{c\text{-}TiN}$=4.24 Å, $a_{c\text{-}AlN}$=4.12 Å), but also hardening from a difference in shear modulus, which hinder dislocation movements. We propose the term *'auto-organization'* for the thin film phenomena as a special case of *self-organization.* The revelation that *thin films* can be age-hardened through spinodal decomposition has a huge

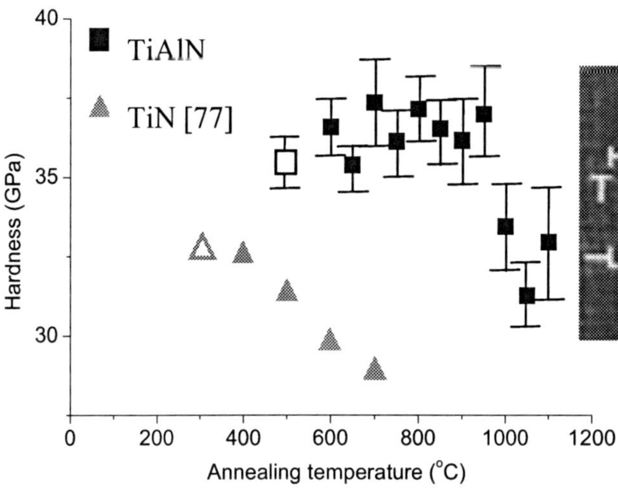

a)

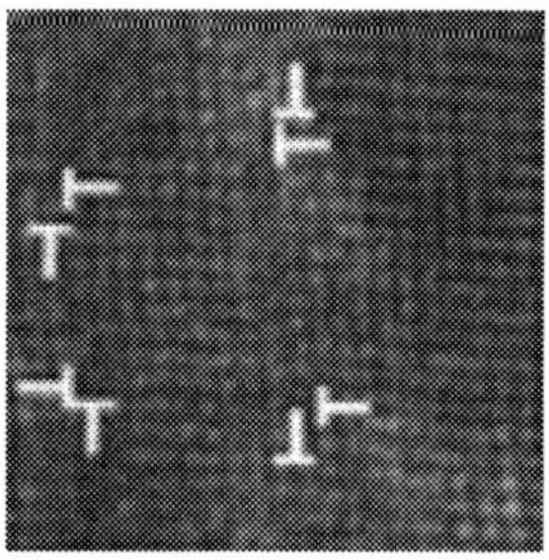

b)

Figure 7. a) Influence of isothermal (2h) annealing on hardness of arc-deposited $Ti_{0.34}Al_{0.66}N$ coatings. Results for a TiN coating is shown for comparison. Open symbols refer to as-deposited samples. b) HREM image of a $Ti_{0.34}Al_{0.66}N$ coating annealed at 1100°C showing the [001] projection of $Ti_{1-x}Al_xN$ lattice with dissociated {110}<110> misfit dislocations due to relaxation of coherency strains between c-AlN and TiN.

Table 1. *Ab initio* calculation results for AlN polytypes. From [11,79].

Polytype	Energy difference [meV]	Volume per formula unit [Å3]
NaCl-AlN (rock salt; cubic)	0	16.30
Zb-AlN (sphalerite; cubic)	8	20.70
w-AlN (wurtzite; hcp)	-71	20.87

technological importance. We submit that the commercial success of $Ti_{1-x}Al_xN$ coatings is not only based on its superior oxidation resistance [81-89], but also on the ability for self-adaptation to the thermal load in cutting operations as revealed here. In fact, such coated milling inserts exhibit an improved performance in cutting operations compared to TiN in machining of cast iron where the rake face heats to ≥900°C. The empiric findings by industry may now obtain a scientific explanation.

5. OUTLOOK

By focusing on the nm-scale, formidable questions can be posed for materials structure-property relationships. E.g., -what is the composition of a phase in a nanocomposite? -Through Gibbs-Thomson effect, the concentration of solute in a matrix increases as $X_r/X_\infty \approx 1+1/r$[nm]. Thus, we expect non-stoichiometric phases with different elastic constants than bulk (sic!). Wide research opportunities are thus offered for a materials research laboratory featuring recently available nm-probe FEG-TEM/EDX/EELS, *thin-film* DSC, *in-situ* furnace XRD, and *in situ* TEM-nanoindentation. Also, the use of *ab initio* calculation and MD methods should be applied to determine elastic modulus values of constituent phases including those for compositional gradients and interfaces. Other challenges for the advancement of nanostructured wear-resistant coatings include making real time observations of dislocation glide in superlattice films – including non-isostructural systems - as well as plasticity and fracture in the nanocomposites at the relevant 1-10 nm feature sizes. The thin film structures of interest as prepared by PVD methods like arc deposition and unbalanced magnetron sputtering are typically containing high levels of compressive stress. This and other conditions may complicate interpretation by nanoindentation experiments or simulations unless artifacts from pile-up at the indenter are controlled and bonding models developed for defects. For the case of thermal stability in nanolaminates and composites, the diffusion responsible for coarsening or interdiffusion are another *terra incognita* and should be addressed for the ceramics. The outlined approach for age hardening in thin films coatings can be expected to be of vast importance for the design of next generation wear resistant coatings by advanced surface engineering. *This is nanoscience & nanotechnology at work.*

ACKNOWLEDGMENTS

The work was supported by the Swedish Research Council (VR), and the Swedish Foundation for Strategic Research (SSF).

REFERENCES

1. L. Karlsson, L. Hultman, and J.-E. Sundgren, Thin Solid Films 371 (2000) 167.
2. U. Helmersson, S. Todorova, S.A. Barnett, J.-E. Sundgren, L.C. Markert, and J.E. Greene, J. Appl. Phys. 62 (1987) 481.
3. S .A. Barnett in "Physics of Thin Films", ed. M.H. Francombe and J.L.Vossen, (Academic Press, San Diego, 1993) Vol. 17, p 1.
4. J.S. Koehler, Phys. Rev. B2 (1979) 547.
5. S. Veprek, J.Vac.Sci.Technol. A17 (1999) 2401.
6. J. Patscheider, MRS Bulletin March 2003, 180.
7. M. W. Barsoum, Progress in Solid State Chemistry 28 (2000) 201.
8. H. Nowotny, Progress in Solide State Chemistry, 2 (1970) 27.
9. A.A. Voevodin and J.S Zabinski, J. Mater. Sci. 33 (1998) 319.
10. A. Hörling, L. Hultman, M. Odén, J. Sjölén, and L. Karlsson, J. Vac. Sci. Technol. A20 (2002) 1815.
11. P. H. Mayrhofer, A. Hörling, L. Karlsson, J. Sjölén, C. Mitterer, L. Hultman, Appl. Phys. Lett. 83 (2003) 2049.
12. P.B. Mirkarimi, S.A. Barnett, K.M. Hubbard, T.R. Jervis, and L. Hultman, J. Mater. Res. 9 (1994) 1456.
13. M. Shinn, L. Hultman, and S.A. Barnett, J. Mater. Res. 7, 901 (1992).
14. P.B. Mirkarimi, L. Hultman, S.A. Barnett, Appl. Phys. Lett. 57 (1990) 2654.
15. Q. Li, I.W. Kim, S.A. Barnett, and L.D. Marks, J. Mater. Res. 17 (2002) 1224.
16. H. Ljungcrantz, C. Engström, L. Hultman, M. Olsson, X. Chu, M.S. Wong and W.D. Sproul, J. Vac Sci. Technol. A16 (1998) 3104.
17. W.-D. Münz, MRS Bulletin March 2003, 173.
18. X. Chu and S. A. Barnett, J. Appl. Phys. 77 (1995) 4403.
19. L. Hultman, M. Shinn, P.B. Mirkarimi, S.A. Barnett, J. Cryst. Growth 135 (1994) 309.
20. M. Odén, H. Ljungcrantz and L. Hultman, J. Mater. Res 12 (1997) 2134.
21. L. Hultman, C. Engström, J. Birch, M.P. Johansson, M. Odén, L. Karlsson, and H. Ljungcrantz, Z. Metallkunde 90 (1999) 803.
22. L. Hultman, C. Engström, and M. Odén, Surf. Coat. Technol., 133-134 (2000) 227.
23. L. Hultman, L.R. Wallenberg, M. Shinn, S.A.Barnett, J. Vac. Sci. Technol. A10 (1992) 1618.
24. Z. Wokulski, Phys. Stat. Sol. (a) 120 (1990) 175.
25. P.M. Anderson, T.Foecke, and P.M.Hazzeldine, Materials Res. Soc. Bull. 24 (1999) 27.
26. J.M. Molina-Aldareguia, S.J. Lloyd, W.J. Clegg, M. Odén, and L. Hultman, Phil. Mag. A 82 (2002) 1983.
27. Q. Luo, W.M. Rainforth, and W.D. Münz, Wear 225 (1999) 74.
28. Q.Luo et al , Scripta Mater. 45 (2002) 399.
29. N.J.M. Carvalho, J. Th. M. De Hosson, MRS Proc. 2001.
30. M. Nordin: PhD Thesis (Faculty of Sci. and Techn. No.509; Uppsala University 2000).
31. Henrik Ljungcrantz, PhD Thesis (Linköping University, 1995).
32. H. Holleck, Binare und Ternare Carbid und Nitridsysteme der Ubergangsmetalle (Gebrüder Borntraeger, Berlin-Stuttgart, 1984).
33. C. Engström, L. Hultman, J. Birch, J.R.A. Carlsson, C. Lavoie, and C. Cabral, J. Vac Sci Technol A17 (1999) 2920.
34. S.A. Barnett, A. Madan, I. Kim, and K. Martin, MRS Bulletin, March 2003, 169.
35. C. Engström, Madan,Birch, Nastasi, Hultman, Barnett, J.Mater.Res. 15 (2000) 554.
36. A. Madan, W. Yun-Yu, S.A. Barnett, C. Engstrom, H. Ljungcrantz, L. Hultman, M. Grimsditch. J. Appl. Phys. 84 (1998) 776.
37. W. Jeitschko, H. Nowotny, Monatshefte für Chemie, 98 (1967) 329.
38. M. W. Barsoum and T. El-Raghy, J. Am. Ceram. Soc, 79 (1996) 1953.
39. M. W. Barsoum, American Scientist 89 (2001) 334.

40. R. Ahuja, O. Eriksson, J. M. Wills B. Johansson, Appl. Phys. Lett, 76 (2000) 2226.
41. B. Holm, R. Ahuja and B. Johansson, Appl. Phys. Lett, 79 (2001) 1450.
42. Y. Zhou, Z. Sun, X. Wang and S. Chen, J. Phys. Condens. Matter, 13 (2001) 10001.
43. J.-P. Palmquist, U. Jansson, T. Seppänen, P. O. Å. Persson, J. Birch, L. Hultman and P. Isberg, Appl. Phys. Lett., 81 (2002) 835.
44. T.Seppänen, J.P.Palmquist, P.O.Å. Persson,J.Emmerlich, J.Molina,J.Birch,U. Jansson, P.Isberg, L.Hultman, SCANDEM Proc. ISSN 1455-4518, Tampere, Finland (2002).
45. H. Högberg, P. Eklund, J. Emmerlich, J. Birch and L. Hultman, unpublished results.
46. H. Högberg, Proc. 4th Asian-Europ. Int. Conf. Plasma Surf. Eng. , Sept. 28 – Oct 3, 2003, Jeju City (Korea).
47. J.-P. Palmquist, O. Wilhelmsson and U. Jansson, unpublished.
48. T. Joelsson, A. Hörling and L. Hultman, submitted.
49. T. Joelsson, Licentiate Thesis No. 1029, (Linköping 2003).
50. J. Emmerlich, J. P. Palmquist, H. Högberg, J. M. Molina-Aldareguia, Zs. Czigány, Sz. Sasvàri, P. Persson, U. Jansson, L. Hultman, unpublished.
51. J.-P. Palmquist, S. Li, P. O. A. Persson, J. Emmerlich, O. Wilhelmsson, H. Högberg, M. Katsnelsson, B.Johansson, R.Ahuja, O.Eriksson, L. Hultman U. Jansson, Submitted
52. B. Holm, R. Ahuja, S. Li and B. Johansson, J. Appl. Phys., 91 (2002) 9874.
53. J. M. Molina-Aldareguia, J. Emmerlich, J.-P. Palmquist, U. Jansson and L. Hultman, Scripta Materialia, 49 (2003) 155.
54. M.W Barsoum, Nature Materials 2 (2003) 1077.
55. Veprek et al. Thin Solid Films 268 (1995) 64.
56. M. Jilek and S.Veprek, Vacuum 67 (2002) 443.
57. K.S. Kumar, H. Van Swygenhoven, S. Suresh, Acta Materialia, 51 (2003) 5743.
58. J. Schiøtz, F.D. Di Tolla, K.W. Jacobsen, Nature, 391 (1998) 561.
59. A. Voevodin, J. Zabinski, Thin Solid Films 370 (2000) 223.
60. P. H. Mayrhofer, G. Tischler, C. Mitterer, Surf. Coat Techn. 142-144 (2001) 78.
61. K.L. Merkle, L.J. Thompson, Mater. Lett. 48 (2001) 188.
62. J. W. Cahn, Acta Met. 9, 795 (1961).
63. E.Rudy, J. Less-Common Metals 33 (1973) 43.
64. F. Adidi, I.Petrov, L. Hultman, U. Wahlström,T. Shimizu, D. McIntyre, J.E. Greene, J.E. Sundgren, J. Appl. Phys. 69 (1991) 6437.
65. L. Hultman, Key Engin. Mat. 103, 181 (1995).
66. R.Kaspi, L.Hultman, S.A.Barnett, J. Vac. Sci. Technol.. B13, (1995) 978.
67. T. Seppänen, G.Z. Radnoczi, S. Tungasmita, L. Hultman, J. Birch, submitted.
68. H. Holleck, Surf. Coat. Technol. 36 (1988) 151.
69. O.Knotek and A. Bariami, Thin Solid Films 174 (1989) 51.
70. R.A. Andrievski, I.A. Anisimova, V.P. Anisimov, Thin Solid Films 205 (1991) 171
71. R.Prange, R.Cremer, D. Neuschütz, Surf. Coat. Technol. 133 (2000) 208.
72. J. Patscheider. T. Zhender and M. Diserens, Surf. Coat. Technol. 146 (2001) 201.
73. P.Holubar, M. Jilek and M. Sima, Surf.Coat.Technol. 133-134 (2000) 145.
74. H.D.Männling, D.S. Patil, K. Moto, M. Jilek and S.Veprek, Surf. Coat. Technol. 146 (2001) 263.
75. H.W. Hugosson, H.Högberg, M.Algren, M. Rodmar, T. Selinder, J.Appl.Phys. 93 (2003) 4505.
76. L. Hultman, Vacuum 57 (2000) 1.
77. L. Karlsson, G. Ramanath, M.P. Johansson, A. Hörling, L. Hultman, Acta Mater. 50 (2003) 5103.
78. S. Kodambaka, V.Petrova, A. Vailionis, P. Desjardins, D.G. Cahill, I. Petrov and J.E. Greene, Thin Solid Films 392 (2001) 164.
79. T.Larsson, L.Karlsson, L.Hultman, unpubl. J.M.Wills supplied FP-LMTO code [H. Skriver, LMTO Method, Solid State Sci., Springer (1984)] in LDA-DFT. Exchange correlation function by Hedin-Lundqvist [J.Phys. C4 (1971) 2064]. Cell dimensions varied uniformly with theoretical size obtained from min. total energy. c/a ratio and z

coordinates of N atoms allowed to relax for wurtzite. Energy convergence was reached for all structures with respect to the number of k points used.

80. Q. Xia, H. Xia, A. Ruoff, *J. Appl. Phys.* 73 8198 (1993).
81. G. Beenoh-Marchwicka, L. Kröl, P. Posadowski, Thin Solid Films 82 (1981) 313.
82. W.-D. Münz , J. Göbel, Proc. 7th Int. Conf. Vacuum Metallurgy, Linz, Austria 1985.
83. O. Knotek, W. Bosch, T. Leyendecker, Proc. 7th Int. Conf. Vac. Metall., Linz, 1985.
84. O. Knotek, M. Böhmer T. Leyendecker, J. Vac. Sci. Technol. A4 (1986) 2695.
85. H. Jehn, S.Hoffmann, V.-E. Rückborn W.-D.Münz, J. Vac. Sci. Technol.A4(1986) 2701.
86. W.-D. Münz, J. Vac. Sci. Technol. A4 (1986) 2717.
87. S. Hoffman and H. Jehn, Surf. Interface Anal. 12 (1988) 329.
88. D. McIntyre, J. E. Greene, G. Håkansson, J.-E. Sundgren, and W.-D. Münz, J. Appl. Phys. 67 (1990) 1542.
89. W.-D. Münz, T.Hurkmans, G.Keiren, T.Trinh, J. Vac. Sci. Technol. A11 (1993) 2583.

Chapter 3
RECENT PROGRESS IN SUPERDHARD NANO-COMPOSITES: PREPARATION, PROPERTIES AND INDUSTRIAL APPLICATIONS

Stan Veprek
Institute for Chemistry of Inorganic Materials, Technical University Munich, Lichtenberstr. 4, D-85747 Garching, Germany

Advantages and drawbacks of different techniques for the preparation of superhard coatings are briefly summarized with emphasis on high thermal stability and oxidation resistance. It is shown that the hardening by energetic ion bombardment yields coatings with a too low thermal stability whereas the hardness and nanostructure of the superhard nanocomposites prepared according to our generic design principle remain stable up to ≥ 1100 °C. The lack of success of some researchers to reproduce our results is due to either an incorrect choice of the deposition conditions (too low nitrogen pressure and/or deposition temperature), or impurities, or a combination of those. The recent progress in the development of the coating technology and industrialization of the nc-$(Al_{1-x}Ti_x)N$/a-Si_3N_4 coatings is briefly summarized. The new **La**teral **R**otating arc **C**athode, **LARC®** that uses **Virtual Shutter®** for pre-cleaning of the cathodes is presented together with an example of excellent cutting performance of these new coatings.

Keywords: superhard nanocomposites, design concept, hardening, ion bombardment, impurities, industrialization, thermal stability, nc-TiN/Si_3N_4, W_2N/Si_3N_4, VN/Si_3N_4, $(Al_{1-x}Ti_x)N/Si_3N_4$, ZrN/Ni, Cr_2N/Ni, ZrN/Cu.

1. INTRODUCTION

Since the publication of our generic concept for the design of novel superhard nanocomposites [1,2,5], many papers have been published and the field is growing fast. Our design concept is based on a strong, thermodynamically-driven and diffusion rate-controlled (spinodal) phase segregation that leads to the formation a stable nanostructure by self-organization [1,2,5]. The hardness shows a maximum of 50 to 60 GPa at a silicon content of about 8-10 at. % (see Fig. 1a) when nanocrystals, e.g. of transmission metal nitrides (or carbides), are covered with about one monolayer of silicon [1,3] or boron nitride (or amorphous carbon) [4]. Thus, it is not the volume content of the Si_3N_4 or BN tissue alone that corresponds to the maximum hardness but its ratio to the specific interface area, which depends on the crystallite size and shape [3,4]. When prepared under

A. A. Voevodin et al. (eds.),
Nanostructured Thin Films and Nanodispersion Strengthened Coatings, 23–34.

conditions that enable the formation of the stable nanostructure during the deposition, the crystallites are of regular shape with a size of about 3-4 nm for the nc-Me_nN/a-Si_3N_4 (Me=Ti, W, V, ($Al_{1-x}Ti_x$)N) systems [1,5,10] and about 20-25 nm for nc-TiN/a-BN free of TiB_2 [4]. Therefore an "optimum" Si-content in coatings that were deposited under conditions of low nitrogen pressure and/or temperature varies in papers of different authors because of different crystallite sizes and shapes. A high thermal stability of the nanostructures and hardness (measured at room temperature after annealing) up to 1100 °C (see Fig. 1b) is a consequence of the strong immiscibility of stoichiometric phases. A high oxidation resistance (e.g. above 800 °C for nc-TiN/a-Si_3N_4, see Fig. 1c [1]) first reported by us and later reproduced by others is a consequence of the strong and dense silicon nitride interface [1].

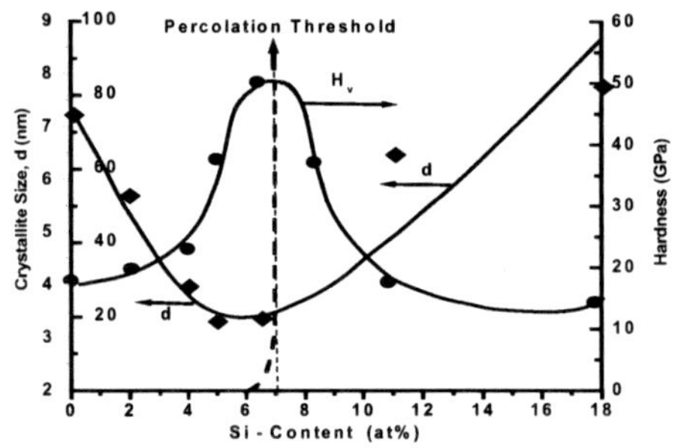

Figure 1a. Dependence of the crystallite size and hardness on the total silicon content (bonded as Si_3N_4) in nc-TiN/a-Si_3N_4 for coatings deposited by plasma chemical vapor deposition (P CVD) upon low-energy ion bombardment [1]. The hardness reaches maximum and the crystallite size minimum at the percolation threshold of the Si_3N_4 phase.

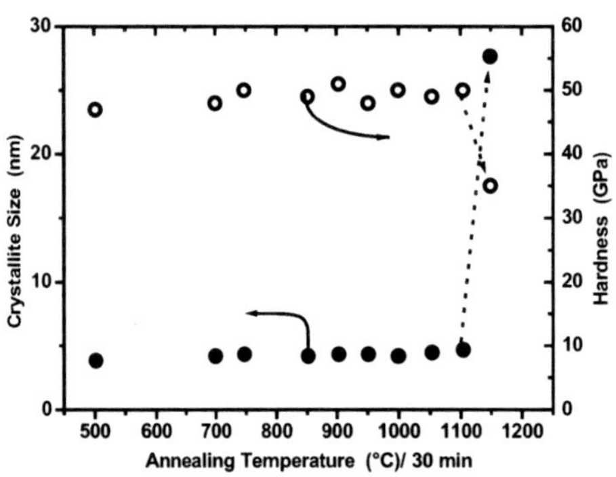

Figure 1b. Example of thermal stability of nc-TiN/a-Si_3N_4 superhard nanocomposites prepared by plasma CVD [5,6].

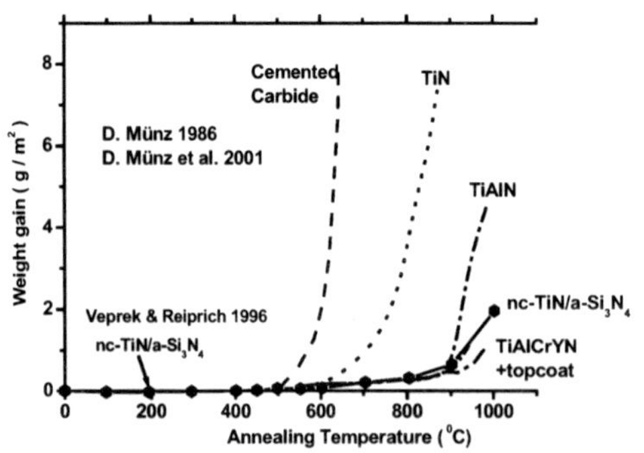

Figure 1c. Example of oxidation resistance of nc-TiN/a-Si_3N_4 nanocomposites wih the optimum composition corresponding to the maximum hardness in Fig. 1a (from [1] with data on the oxidation resistance of other materials from [7, 8].

Already in the first paper [1] we have shown by means of simple thermodynamics calculation that a sufficiently high chemical activity ("nitrogen pressure") is needed for the thermodynamics to drive the phase segregation and a sufficiently high deposition temperature of about 550 °C is required for a fast diffusion in order to complete phase segregation during the deposition. It was shown that under these conditions the crystallites size attains a minimum when the hardness reaches a maximum (Fig. 1a), and the thickness of the amorphous tissue is about 0.3-0.5 nm, i.e. about one monolayer of Si_3N_4 [1]. This was later confirmed by a more detailed analyses for all systems studied in detail so far [3,4]. As predicted in [1], the universality of our design principle was verified by a successful preparation of a variety of superhard nanocomposites based on transition metal nitrides in a combination with covalent, non-metallic silicon or boron nitride, such as nc-W_2N/a-Si_3N_4, nc-VN/a-Si_3N_4, nc-TiN/a-Si_3N_4/a- & nc-$TiSi_2$, nc-TiN/a-BN, nc-TiN/a-BN/a-TiB_2 (see [3,4,5,10] and references therein). Based on the generality of the thermodynamical design principle, we also predicted that it should work in a variety of systems including carbides, borides and others that meet the conditions specified in our first paper [1,5,10]. It was shown in [5] that several researchers indeed prepared such nanocomposites earlier without having realized the generality of the design concept. Superhardness of 50 to 60 GPa was achieved in many binary systems, such as nc-TiN/a-Si_3N_4, etc. and 80 to $\geq$ 100 GPa in the ternary and quaternary nc-TiN/a-Si_3N_4/a- & nc-$TiSi_2$ [3,5,11,12]. Some workers questioned our results [13] but we have shown that our measurement of the hardness was done correctly [14]. Possible artefacts of the measurements of hardness and elastic modulus by means of the automated indentation technique when applied to superhard coatings [15,16] can be excluded in our case because, from the very beginning [1], we reported only the load-independent values of hardness at applied loads of 50 to 200 mN and verified it by means of calibrated scanning electron microscope (SEM) [1,5,15,16]. The unusual combination of the mechanical properties of these nanocomposites can be understood on basis of conventional fracture physic of flaw-free materials [17] whose tensile strength of 10-40 GPa approaches the ideal material strength [18].

1.1. The Earlier Work

Many researchers studied the thermodynamic phase diagrams of the Ti-Si-N system, and the deposition of such films by means of thermal CVD. Because high temperatures are needed for the thermal CVD in this system, no hardness enhancement was reported [19]. Li Shizhi et al. were the first authors who reported on the deposition of superhard "Ti-Si-N" coatings by means of P CVD using chlorides as a source for Ti and Si [20]. Li attributed the hardness enhancement to the precipitation of small Si_3N_4 particles within TiN nanocrystals. A hardness of 60 GPa reported by Li et al. was due to the ternary nc-TiN/a-Si_3N_4/a-$TiSi_2$ nature of these coatings, as found in subsequent joint work [2,3,5,11,12,21]. Together with Li we also found that, after a period of few months, the hardness of these coatings decreased and attributed this degradation to the relatively large chlorine content in these coatings [21].

Recent studies revealed that this degradation is an intrinsic property of these systems due to chemical degradation of the $TiSi_2$ phase by water from air in a similar way as it is known for silica fibres. Furthermore, when we found that the crystallite size is only 3-4 nm it became clear that the original explanation based on a dispersion hardening [20] cannot be correct. This led us to the development of the generic design concept [1,2].

Many researchers (Knotek et al., Andrievski et al., Gissler et al., Hammer et al., Mitterer et al. [5,21]) have studied the deposition of "Ti-B-N" coatings. In the recent review [5] it was pointed out that, based on the immiscibility of the stoichiometric TiN and TiB_2 phases and on a development of the morphology of these coatings as a function of their composition and hardness (reported by Andrievsky, see [5]), these were also nanocomposites. Mayrhofer et al. presented a detailed, convincing evidence for this [22]. To best of our knowledge, Knotek was also the first author who reported the hardness enhancement in the "super-stoichiometric TiC_{1+x}" [23] similar to that found in our nc-TiN/a-Si_3N_4 when the excess carbon content reached about 20 at. %. This and many similar systems were studied in much detail later by others [21,24].

2. SUPERHADR NANOCOMPOSITES AND HARDENING BY ENERGETIC ION BOMBARDMENT

The easiest way to prepare superhard coatings is by applying energetic ion bombardment during the deposition of hard materials, such as stoichiometric TiN, (TiAlV)N, HfB_2 as well as so called "nanocomposites" consisting of a hard transition metal nitride and a soft metal which does not form a stable nitride (ZrN/Ni, ZrN/Cu, Cr_2N/Ni, etc. [25]). The latter coatings are frequently confused with superhard nanocomposites prepared according to our generic principle. Figure 2 shows how the difference can be proved. The hardness that has been enhanced by energetic ion bombardment strongly decreases with temperature to the ordinary bulk value upon annealing to 400-600 °C whereas that of the superhard nanocomposites remains unchanged upon annealing up to 1100 °C (Fig. 2a; for further examples see [4,28]).

Another possibility how to distinguish between the stable, superhard nanocomposites and the ordinary coatings is the dependence of hardness on their composition (Fig. 2b). It is important to notice that superhard nanocomposites prepared according to our design principle [1,5,17,18] show a maximum at the percolation threshold when the surface of the polar transition metal nitrides is covered with about one monolayer of Si_3N_4 [3] or BN [4]. Fig. 2b illustrates the excellent reproducibility of hardness enhancement.

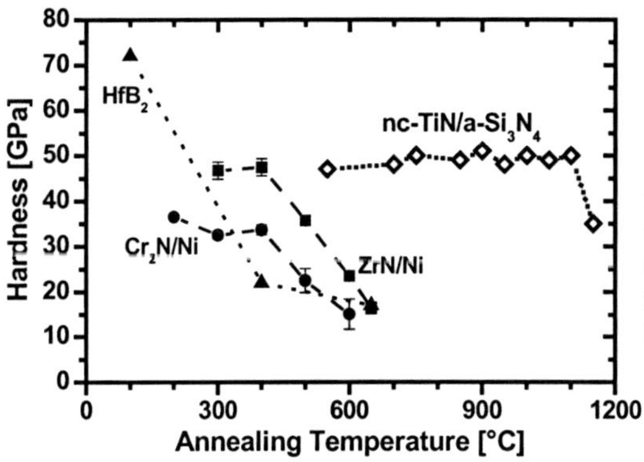

Figure 2a. Dependence of the hardness of HfB_2 [26], Cr_2N/Ni and ZrN/Ni [27] coatings hardened by energetic ion bombardment and the stable superhard nc-TiN/a-Si_3N_4 nanocomposites on the temperature of isochronal annealing in pure nitrogen [5,28]. (The hardness was measured at room temperature after each annealing step.)

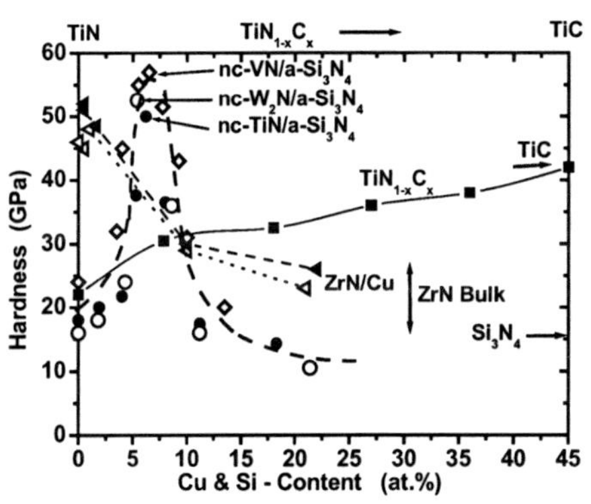

Figure 2b. Dependence of hardness of various coatings on their composition. $TiN_{1-x}C_x$ forms a solid solution and therefore the hardness follows the rule-of-mixtures [29]. In the case of the so called "nanocomposites" consisting of a hard transition metal nitride and ductile metal the maximum hardness is achieved with the pure nitride without that metal (adapted from [30]). The superhard nanocomposites prepared according to our design principle show a maximum hardness at a percolation threshold [1,3,5].

3. REPRODUCIBILITY OF PREPARATION OF SUPERHARD NANOCOMPOSITES

The decisive fact which should to be kept in mind is that the formation of a nanostructure with a high thermal stability (Fig. 1) requires a sufficiently high chemical activity of the system to provide a thermodynamic driving force for phase segregation and a temperature of 500-600 °C to provide a sufficiently fast diffusion that is necessary to complete the segregation during the growth (for details see [1,5,17,18]). When these conditions are not met, the high hardness of 50-60 GPa is difficult to reach in the binary systems, the coatings show self-hardening upon subsequent annealing, and the maximum hardness appears at different total content of silicon. Furthermore, impurities, such as chlorine and oxygen strongly degrade the hardness as reported in Ref. [6,7]. Unfortunately, the majority of published papers on the "Ti-Si-N" and similar coatings do not report these decisive parameters and conditions. Therefore it is difficult to find out what could be the possible reasons for the luck of reproducibility of our data in these works. We shall limit our discussion to few typical examples [19] in Fig. 3, which shows hardness vs. the Si-content for "Ti-Si-N" coatings deposited by plasma CVD (Fig. 3a) and by reactive sputtering (Fig. 3b). As it was shown later [21], the coatings of Li et al. [20] were the ternary nc-TiN/a-Si_3N_4/a-$TiSi_2$ nanocomposites in which hardness of 80 to ≥ 100 GPa was achieved [5,11,12]. As one can see from Fig. 3, the hardness of other coatings just approaches 40 GPa and its maximum appears at different Si-content. Moreover, there is no hardness enhancement found in coatings by reported by Meng et al. [34]. On the other hand, the reproducibility of our nc-MeN/a-Si_3N_4 (Me=Ti, W, V) coatings is quite good (Fig. 2b). The limited space available here does not allow us to discuss the possible reasons in more details. We just state that our coatings were deposited in an intense glow discharge with a sufficiently high pressure of nitrogen of ≥ 0.5 mbar and high deposition temperature of 550 °C as explained in [1]. Furthermore, the

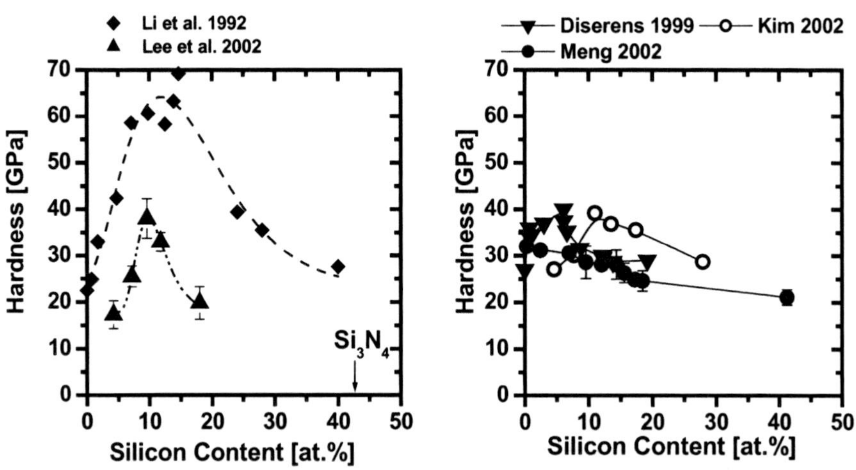

a) b)

Figure 3. a) Hardness vs. Si-content for "Ti-Si-N" coatings deposited by means of plasma CVD by Li et al. [20] and Lee et al. [31]. For our nanocomposites nc-Me_nN/a-Si_3N_4 (Me=Ti, W, V) see Fig. 2b [1,2,29]. b) Hardness vs. Si-content for "Ti-Si-N" coatings deposited by means of reactive sputtering by Disserents et al. [32], Kim et al. [33] and Meng et al. [34].

impurity content was below 0.5 at. % for chlorine and hydrogen and less than 0.05 at. % for oxygen. The poor reproducibility found in many papers (only few examples are shown in Fig. 3) is either due to a choice of a too low nitrogen pressure, or too low temperature, or impurities or a combination of these effects. For example, Meng et al. [34] used a very low partial pressure of nitrogen of 0.198 mTorr (0.00015 mbar = $1.5 \cdot 10^{-7}$ at.) and a too low deposition temperature of about 250 °C. Moreover, in many papers oxygen impurities are likely to be the reason for the low hardness. At typical deposition rates of 0.1 to 0.5 nm/s and a residual gas pressure of 1 to $7 \cdot 10^{-6}$ mbar, the oxygen impurity in the range of several at. % is likely to be incorporated into the coatings during their growth. As it will be shown below, hardness of 40 GPa cannot be reached in such a case.

3.1. The Role of Impurities

Some time ago we reported on the detrimental role of impurities which, when incorporated into the coatings during the growth, made it impossible to achieve the superhardness [6,7]. Recent studies [35] confirmed these finding in a more quantified manner. The hardness decreases when chlorine and

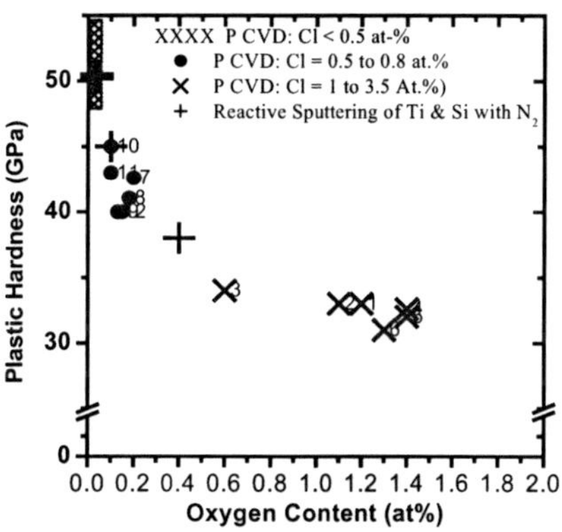

Figure 4. Dependence of the maximum achievable hardness (i.e. for an optimum composition as in Fig. 1a) on the oxygen impurity content for nc-TiN/a-Si_3N_4 coatings deposited by plasma CVD using $TiCl_4$ and SiH_4 and by reactive sputtering (symbols +) [35]. Hardness of ≥ 40 GPa can be achieved only when the oxygen impurity content is ≤ 0.2 at. %. When the oxygen content is above 0.6 at. %, the hardness remains below 35 GPa.

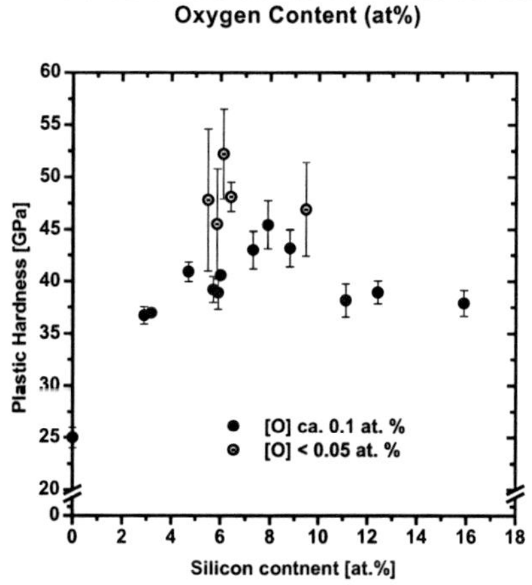

Figure 5. Hardness of the binary nc-TiN/a-Si_3N_4 coatings deposited by magnetron sputtering using only nitrogen as sputtering and reactive gas [36]. Using a high power density of 15 to 17 W/cm^2 a high deposition rate of 1.6 nm/sec was achieved which, together with careful cleaning of the reactor prior to the deposition assured the low oxygen impurity content. In course of a more recent work, the high hardness of > 50 GPa was reproducibly achieved in many coatings (work in progress).

hydrogen impurities of few atomic percent are incorporated. The most dramatic effect provides oxygen as illustrated in Fig. 4. Fig. 5 shows that when the impurity content is sufficiently low the high hardness of more than 50 GPa is obtained by reactive sputtering at a similar Si-content as in the nc-TiN/a-Si_3N_4 nanocomposites deposited by plasma CVD.

The reason of the strong effect of minor oxygen impurities on the hardness is presently not clear. The maximum hardness is achieved when there is about one monolayer of Si_3N_4 acting as a "glue" tissue between the TiN nanocrystals. It could be intuitively understandable why a polar Si-O bond within this tissue may have such a strong effect.

3.2. Conditions Needed to Achieve Hardness of 80 to ≥ 100 GPa

As outlined in our earlier papers [11,12] the ultrahardness was achieved so far only in ternary and quaternary nanocomposites nc-TiN/a-Si_3N_4/a- & nc-$TiSi_2$ when the surface of all nanocrystals was covered with about one monolayer of Si_3N_4 and the oxygen impurity content was 0.01 to 0.05 at. %. There is a very narrow compositional range for the coexistence of all three phases, i.e. TiN, Si_3N_4 and $TiSi_2$, and it is not an easy task to meet all these requirements simultaneously. Nevertheless, we succeeded to deposit more than 15 ultrahard (i.e. H ≥ 80 GPa) coatings. From these 6 reached a hardness of ≥ 100 GPa [14]. In order to avoid possible artefacts that could falsify the hardness measurements we always reported a load-independent hardness, typically in the range of maximum applied load of 30 to 150 mN [1]. Moreover, we verified these values by a calibrated scanning electron microscope and with measurement on single crystal diamond (for details see [11,12,15-18].

4. INDUSTRIAL APPLICATIONS

The industrialization of the nc-$(Al_{1-x}Ti_x)N$/a-Si_3N_4 has been pioneered by Czech company SHM [37] and more recently in a joint venture by SHM and a Swiss company PLATIT [38]. The work included both the development of the coating material and of a new advanced coating technology. The latter uses specially designed **La**teral **r**otating vacuum arc **c**athodes ("**LARC**®") with vacuum arc localized into a narrow track with a fast movement of the cathode spot. This strongly reduces the emission of droplets ("macroparticles") and decreases the surface roughness to ≤ 0.1 μm. The rotation of the cathode results in its uniform erosion and, therefore, a lifetime increases by a factor of 5 – 6 as compared to a conventional planar cathode. Relatively high nitrogen pressure and deposition temperature assure conditions for the formation of stable nanostructure. The presently available coating unit π^{80} is equipped with one cathode made of titanium and second one made of an Al-Si eutectic alloy (11,3 at. % of Si). Both cathodes are mounted within the door of the vacuum coating chamber asymmetrically with respect to a carousel, which holds tools to be coated. This results in a nano-layered modulation of the deposited nanocomposite coatings which increases their toughness. Another advantage of the **LARC®** technology is the possibility to pre-clean the surface of the cathodes by means of a **Virtual Shutter®** **(VS)**, prior to the deposition, that significantly improves the adherence and the lifetime of the coatings. In the **VS**-mode the magnets controlling the movement of the cathode spots are turned backwards and the material of the cathode surface, which was contaminated during the loading of the chamber, is deposited on the wall of the door, whereas the tools are cleaned by means of D.C. glow discharge. After the cleaning the magnets are turned by 180° so that the arc track is facing the carousel [38].

The improvement in the lifetime of the tools coated with recently developed nc-$(Al_{1-x}Ti_x)N$/a-Si_3N_4 nanocomposite coatings is illustrated by Fig. 6 which shows the lifetime of drills made of cemented carbide and coated with the given type of coatings. The comparison of the lifetime of drills coated with "ARC" (conventional planar cathodes) and LARC – TiN shows already a clear improvement. The nano-layered "AlTiN" gradient

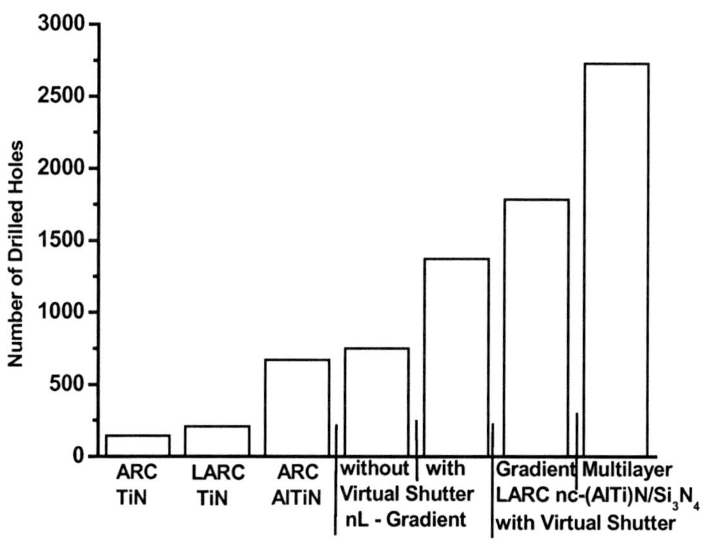

Figure 6. Comparison of the cutting performance of conventional, state-of-the-art TiN and AlTiN coatings with the new coatings prepared by means of the **LARC®** technology and utilizing the **Virtual Shutter®** (see text). Dry drilling of steel was performed with 5 mm diameter drills made of a cemented carbide. The life time is defined by a Corner Wear = 200 μm. In dry drilling the temperature of the coatings reaches high values. The numbers indicate the total number of drilled holes.

coatings ("nL–Gradient") deposited without **Virtual Shutter**® are only slightly better than the conventional "ARC" ones, but their performance increases almost by a factor of 2 when the **Virtual Shutter**® is used (3rd and 5th column). The lifetime is further improved when the nc-(AlTi)N/a-Si_3N_4 coatings are used (c.f. 6th column). Finally, the lifetime of drills coated with the nano-layered multi-layer nc-$(Al_{1-x}Ti_x)N$/a-Si_3N_4 coatings increases by a factor of 4 as compared to the state-of-the art $(Al_{1-x}Ti_x)N$ coatings (c.f. 3rd and 7th column in Fig. 6). A similar improvement of the lifetime and overall cutting performance (e.g. surface finish) was obtained in many other machining operations. More information is available in Ref. [38] and further coming papers.

5. CONCLUSIONS

The superhard nanocomposites prepared according to our generic design principle show high hardness of 40 to ≥ 100 GPa, tensile strength approaching the ideal one, a high thermal stability, and high oxidation resistance. The difficulties in reproduction of the high hardness of the coatings reported in some recently published papers [19,31,32,33,34] can be explained either by too low nitrogen pressure or deposition temperature used in these works or by impurities, as it was explained and shown in our earlier publications [1,5-7,17]. The nanocomposite coatings of nc-$(Al_{1-x}Ti_x)N$/a-Si_3N_4 and a new coating technology were developed and introduced into a large-scale industrial application. The generic nature of our design concepts will allow one to develop even more advanced superhard and stable nanocomposite coatings with different composition and properties tailored to the desirable application.

ACKNOWLEDGMENTS

I would like to thank all my coworkers whose names are listed in the joint publications for their enthusiastic contributions, Prof. Li Shizhi for long-term friendship and many discussions, Prof. A. S. Argon for his crucial contribution to the understanding of the unusual combination of mechanical properties of these new materials and Dr. P. Holubar, Dr. M. Jilek (SHM) and Dr. T. Cselle for interesting collaboration. Financial support of the NATO Science for Peace Programme (SfP Project No. 972379 "Protection Coatings) and of the European Commission in the frame of the 5th RTD Framework Programme (contract No. G5RD-CT-2000-0222 Project “NACODRY”) is acknowledged.

REFERENCES

1. S. Veprek and S. Reiprich, Thin Solid Films 268 (1995) 64.
2. S. Veprek, S. Reiprich and Li Shizhi, Appl. Phys. Lett. 66 (1995) 2640.
3. A. Niederhofer, T. Bolom, P. Nesladek, K.Moto, C. Eggs, D. S. Patil and S. Veprek, Surf. Coat. Technol. 146-147 (2001) 183.
4. P. Karvankova, M. G. J. Veprek-Heijman, O. Zindulka and S. Veprek, Surf. Coatings Technol. 163-164 (2002) 149.
5. S. Veprek, J. Vac. Sci. Technol. A 14 (1999) 2401.
6. S. Veprek, A. Niederhofer, P. Nesladek and F. Glatz, Electrochem. Soc. Proc. 97-25 (1997) 317.
7. S. Veprek, P. Nesladek, A. Niederhofer, F. Glatz, M. Jilek and M. Sima, Surf. Coat. Technol. 108-109 (1998) 138.
8. W.-D. Münz, J. Vac. Sci. Technol. A 4 (1986) 2717.
9. W.-D. Münz, Werkstoffe und Korrosion 41 (1990) 753.
10. S. Veprek, M. Haussmann and S. Reiprich, J. Vac. Sci. Technol. A 14 (1996) 46.
11. S. Veprek, A. Niederhofer, K. Moto, P. Nesladek, H. Männling and T. Bolom, Mater. Res. Soc. Symp. Proc. 581 (2000) 321.
12. S. Veprek, A. Niederhofer, K. Moto, T. Bolom, H.-D. Männling, P. Nesladek, G. Dollinger and A. Bergmaier, Surface Coatings Technol. 133-134 (2000) 152.
13. J. Musil, H. Zeman, H. Kunc and J. Vlcek, Mater. Sci. Eng. A 340 (2003) 281.
14. S. Veprek, S. Mukherjee, H.-D. Männling and Jianli He, Materials Science and Engineering A 340 (2002) 292.
15. S. Veprek, S. Mukherjee, P. Karvankova, H.-D. Männling, J. L. He, K. Moto and S. Argon, Thin Solid Films 436 (2003) 220.
16. S. Veprek, S. Mukherjee, P. Karvankova, H.-D. Männling, J. L. He, J. Xu, J. Prochazka, A. S. Argon, A. S. Li, Q. F. Fang, S. Z. Li, M. H. Manghnani, S, Tkachev, P. Zinin, Mater. Res. Soc. Proc. 750 (2003) 9.
17. S. Veprek and A. S. Argon, J. Vac. Sci. Technol. B 20 (2002) 650-664.
18. S. Veprek, S. Mukherjee, P. Karvankova, H.-D. Männling, J. L. He, K. Moto, J. Prochazka and A. S. Argon, J. Vac. Sci. Technol., A 21 (2003) 532.
19. For lack of space we cannot include a more detailed discussion and all references here.
20. Li Shizhi, Shi Yulong and Peng Hongrui Plasma Chem. Plasma Process. 12 (1992) 287.
21. Li Shizhi, S. Veprek et al., unpublished.
22. P. Mayrhofer et al., these proceedings.
23. O. Knotek et al., Mater. Res. Soc. Fall Meeting, Boston, Dec. 2-5 (1996); See also: O. Knotek, E. Lugscheider, F. Löffler, B. Bosserhooff and S. Schmitz, Mater. Sci. Eng. A 209 (1996) 394.
24. R. Hauert and J. Patscheider, Adv. Eng. Mater. 2 (2000) 247.
25. J. Musil, Surf. Coatings Technol. 125 (2000) 322.
26. W. Herr and E. Broszeit, Surf. Coat. Technol. 97 (1997) 335.
27. P. Karvankova, H. Männling, Ch. Eggs and S. Veprek, Surf. Coatings Technol. 146-147 (2001) 280.
28. H.-D. Männling, D. S. Patil, K. Moto, M. Jilek and S. Veprek, Surf. Coat. Technol. 146-147 (2001) 263.
29. S. Veprek, M. Haussmann, S. Reiprich, Li Shizhi and J. Dian, Surf. Coat. Technol. 86-87 (1996) 394.
30. P. Zeman, R. Cerstvy, P. H. Mayrhofer, C. Mitterer and J. Musil, Mater. Sci. Eng. A 289 (2000)189.
31. E.-A. Lee and K. H. Kim, Thin Solid Films 420-421 (2002) 371.
32. M. Diserens, J. Patscheider and F. Lévy, Surf. Coat. Technol. 120-121 (1999) 158.
33. S. H. Kim, J. K. Kim and K. H. Kim, Thin Solid Films 420-421 (2002) 360.
34. W.J. Meng, X. D. Zhang, B. Shi, R. C. Tittsworth, L. E. Rehn and P. M. Baldo, J. Mater. Res. 17 (2002) 2628.

35. S. Veprek, H.-D. Männling, A. Niederhofer, D. Ma and S. Mukherjee, J. Vac. Sci. Technol. (2003/4) submitted.
36. Jan Procházka, Pavla Karvánková, Maritza G. J. Veprek-Heijman and Stan Veprek, Mater. Sci. Eng. A, submitted.
37. M. Jilek, P. Holubar, M.G.J. Veprek-Heijman and S. Veprek, Mater. Res. Soc. Symp. Proc. 750 (2003) 393.
38. M. Jilek, T. Cselle, P. Holubar, M. Morstein, M. G. J. Veprek-Heijman and Stan Veprek, Plasma Chem. Plasma Processing (2004) submitted.

Chapter 4

NANOSTRUCTURED HARD COATINGS - FROM NANOCOMPOSITES TO NANOMULTILAYERS

J. Patscheider[1], T. Zehnder[2], J. Matthey[2], M. Diserens[1,3]
[1] *EMPA, Section Surface, Coatings and Magnetism, Überlandstr. 129, CH-8600 Dübendorf, Switzerland*
[2] *EIAJ, CH-2400 Le Locle, Switzerland*
[3] *EPFL, SB.CH-1015 Lausanne, Switzerland*

Nanomultilayers and nanocomposites provide a wide range of possibilities to promote hardness, oxidation resistance, improved wear behavior and other properties relevant for protective coatings. Most nanocomposite hard coatings show a maximum of the hardness, as the composition is changed from the pure crystalline phase to compositions dominated by the amorphous phase. At the hardness maximum the grain size is below 10 nm and the amorphous layer separating the nanocrystals is only a few atomic bond lengths thin. The critical dimensions necessary to obtain hardness enhancement in nanomultilayers are of the same order. The reason for this finding in both layer architectures is the absence of dislocation activity due to size restrictions in the single layers or the nanocrystals, respectively. Hindered dislocation movement in nanomultilayers cause enhanced hardness, while deformation of nanocomposites can occur only due to grain boundary sliding which requires more energy than deformation by dislocation movement, thus causing increased hardness.

Keywords: nanomultilayers, nanocomposite, hardness.

1. INTRODUCTION

A study of the Canadian NRC from 1986 estimated the economic loss due to wear and friction to be of the order of 5 billion dollars [1]. This impressive figure corresponds to about two percent of the GNP. The same study estimates that about 70 % of this sum could be saved by improving the surfaces involved in frictional processes. Since environmental costs arising from disposal of liquid lubricants became an economic issue, coatings with improved wear resistance due to increased hardness, toughness and lowered friction due to self-lubrication took an important part in lowering costs improving performance.

The development of new single phase materials for improved resistance against wear is limited by limitations such as increased shear modulus at

A. A. Voevodin et al. (eds.),
Nanostructured Thin Films and Nanodispersion Strengthened Coatings, 35–42.

enhanced hardness. The hardness of a material, being defined as the resistance of a material to plastic deformation, is directly linked to its ability to deform under applied pressure by dislocation movement, by crack initiation or similar mechanisms. By combining materials with different elastic properties these deformation processes can be altered in a composite material. The characteristic size of such a composite - for the case of multilayers the single layer thickness and for isotropic composites the grain size - are key parameters to describe the nanostructure. Tuning of the nanostructure has been shown to be a very effective way to design materials with increased hardness while maintaining comparably low modulus. This is the case for both nanomultilayers and for nanocomposites. In this paper the common aspects linking these two experimental principles for hardness-enhanced thin films shall be outlined.

2. ISOTROPIC NANOCOMPOSITES

The drawback of limited temperature stability of TiN called for improvement which led, among the development of TiAlN [2,3] to the addition of silicon nitride during TiN deposition. Early reports date back to the early eighties, when Hirai et al codeposited Si_3N_4 and TiN [4]. The high deposition temperatures (1050 – 1450°C) lead to partial solution of TiN in Si_3N_4. As it was shown later an important prerequisite for increased hardness, the formation of sharp phase boundaries, was not achieved at these temperatures. Later Li and coworkers used PACVD at temperatures around 550°C to deposit silicon-containing TiN which was identified as nanocrystalline TiN in a silicon nitride matrix; where hardness values of 6000 Vickers were reached [5]. Much work on isotropic nanocomposites, especially on nc-TiN/a-Si_3N_4 where nc stands for "nanocrystalline" and a for "amorphous", has been performed in the group of Veprek [6,7,8]. Extremely high hardness values of nc-TiN/a-Si_3N_4 exceeding 100 GPa were reported by his group [9].They could show that hardness enhancement is also possible for other transition metal nitrides such as W_2N in conjunction with a-Si_3N_4 [10]. An alternative technique to PACVD is Unbalanced Magnetron Sputtering (UBM-PVD) which can be used for the deposition of nc-TiN/a-Si_3N_4 [11,12] and which has the advantage of avoiding hazardous precursors. The nanohardness values reported for these coatings are in the range of 40 GPa [13] to 54 GPa [14]. It is equally possible to deposit other nanocomposite coatings with increased hardness (at a certain composition ratio of the nanocrystalline and the amorphous phase) such as nc-TiC/a-C, prepared by PACVD [15] and by Pulsed Laser Deposition (PLD) [16]. nc-TiC/a-C:H with increased hardness can be deposited by UBM-PVD [17]. All these materials have their hardness maximum at a typical composition of about 80% of the nanocrystalline and 20% of the amorphous phase. All of these coatings have in common that they show increased hardness if three criteria are met: they are composed of at least two immiscible phases, certain nanostructural conditions are fulfilled and the deposition has to be carried out under conditions of sufficiently high plasma density.

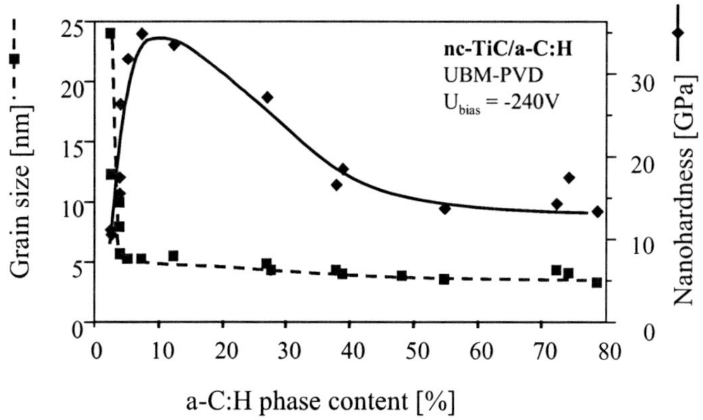

Figure 1: Hardness and mean grain size of nc-TiC/a-C:H as a function of the content of the amorphous phase a-C:H. The hardness maximum occurs at the composition where the grain size is still sufficiently small and the mean thickness of the amorphous phase is thinner than about 2-3 monolayers.

This is exemplified in figure 1 where the hardness of nc-TiC/a-C:H nanocomposite coatings is plotted along with the grain size. The maximum nanohardness is reached, when the crystallites are still of the order of 5 nm. The decay of the hardness at low a-C:H contents is due to the larger grains which deform plastically. At higher contents of the amorphous phase the mean grain separation becomes too large, crack initiation in the amorphous phase is possible and therefore the hardness decays with increasing amorphous phase content.

3. ENHANCEMENT IN NANOSTRUCTURED COATINGS

The resistance to remaining, i.e. plastic deformation is given by the energy barrier of a material that has to be overcome in order to enable deformation by dislocation activity. If dislocation activity can be prevented the hardness of a material is expected to increase. Dislocation propagation can be suppressed when barriers in the form of alternating shear modulus are present. This concept was developed theoretically by Koehler where he showed that the introduction of phase boundaries with alternating elastic properties in a solid material should enhance the strength of such a layered material [18]. The size of a single phase has to be small enough to preclude the operation of Frank-Read dislocation sources . This concept as been experimentally verified on a variety of materials combinations in form of nanomultilayers which were deposited by magnetron sputtering [19,20,21,22]. Most of these multilayer combinations are superlattice coatings with alternating heteroepitaxial layers. Typical hardness enhancement by a factor of two or three could be achieved as long as the layer packet thickness did not exceed the thickness to allow dislocations activity. For most of the nitride materials this value is of the order of ten nanometers. It could be shown that a difference in shear modulus between the two materials is the essential prerequisite to achieve hardness enhancement [23]. Furthermore, coherence stresses originating at the interfaces due to differences in the lattice parameters were identified to be of minor importance to the hardness enhancement [24].

In analogy to superlattice coatings with their nanostructuring in growth direction (atomically sharp phase boundaries after several nanometers), the suppression of dislocation activity is realized also in isotropically nanostructured materials, i.e. in nanocomposite materials, as long as the occurrence frequency of phase boundaries in all directions is the same. This means that a material consisting of isotropical crystallites with grain size below ten nanometers, in conjunction with another phase, gives rise to the same effects. Exactly this is observed for nanocomposite thin films as long as the grain size does not exceed the size where dislocation activity is possible. A simple estimate of the size limit to allow dislocation loops to work in TiC yield a value of 12 nm [25] assuming a contact stress of 2 GPa. An experimentally determined value for the grain size at maximal hardness of CrN yielded 15 nm [26], which is in excellent agreement with this estimate. The only way nanocomposites with grain sizes smaller than these values can be deformed is by grain boundary sliding, a process where intact grains are moved along each other. This requires concerted opening of bonds in the amorphous phase or local debonding of the nanocrystalline and amorphous phase. Both are processes that are energetically less favorable than propagation of dislocations which is synonymous with increased resistance to plastic deformation, i.e. with higher hardness.

Molecular dynamic studies performed by massive parallel computing on nanocrystalline nickel [27] revealed that crystallites smaller than a critical size are virtually free of defects. Below that certain size (around 10 nm for nc-Ni) all crystal imperfections migrate to the grain boundaries and cause a defective amorphous grain boundary zone. Deformation calculations for nanocrystalline copper show that the critical flow stress (strain-independent stress at large strains) increases with shrinking grain size and reaches a maximum at around 10-15 nm [28]. This is correlated by the transition from dislocation-dominated plasticity at large grain size to deformation by grain boundary sliding while the crystallites remain defect-free. Although calculations of nanocomposites are not yet available due to the excessive calculation time required, these nanocrystalline model systems show striking analogies with nanocomposites. They confirm the picture that a) grains under a certain critical size cannot be deformed and b) that grain boundary sliding is the dominant deformation mechanism at small grain sizes.

The mean thickness of the amorphous phase, which is analytically not accessible, can be estimated from stereometric assumptions and amounts for the hardness maximum to only 2 - 3 monolayers, being equivalent to about 0.5 nm. It can be shown that the hardness maximum is given by the common minimum off both grain size and mean thickness of the amorphous phase [29]. Interestingly enough, hardness enhancement was reported for nanomultilayers of TiN/Si_3N_4, when the thickness of the TiN and of the Si_3N_4 layer was 0. 5 nm and 5 nm, respectively [30]. This finding illustrates another close relation between the phenomena found in nanocomposites and nanomultilayer coatings.

4. FORMATION OF NANOSTRUCTURES

When nanocomposites are deposited from the vapor phase the transition metal carbide or nitride will nucleate to form the crystalline phase. In absence of elements leading to the formation of amorphous phases (e.g. silicon), the formation of the crystalline phase follows the known growth modes such as elongated crystals with preferred orientations. The addition of elements leading to amorphous phases, however, drastically influences the growth mode of the crystalline phase. These grains continue to grow until parts of their surface become covered by the codeposited amorphous phase. In the case of a-Si_3N_4 or a-C:H (at sufficient ion energy) these materials are efficient diffusion barriers; the underlying grain thus is no longer exposed to the particle flux from the plasma leading to grain growth. As a consequence newly arriving crystallite-forming species will renucleate and in this way lead to reduced grain size. The higher the fraction of the amorphous phase is, the faster the growing crystallites are covered with the amorphous phase; consequently, the grains remain smaller and their nucleation frequency increases. This behavior can be observed in nc-TiN/a-Si_3N_4 nanocomposites [29] as well as in nc-TiC/a-C:H which is displayed in figure 1. When the grain size drops to small values (e.g. below 10 nm) dislocation activity is precluded and enhanced hardness is measured. Another concept for the formation of nanocomposite coatings describes that the formation of the nanostructure in nanocomposite coatings is the consequence of spinodal decomposition taking place during the deposition [8,31].

While the interfaces in nanocomposites are formed by the immiscibility (or better. extremely low mutual solubility) of the two or more phases, the interface formation in nanomultilayers is simply a consequence of the deposition sequence. The preparation of strong nanomultilayers is done by ion-assisted techniques like magnetron sputtering or PLD in clean environments. Since no thermodynamic requirements (solubility) restrict the formation of nanomultilayers and their interfaces, appropriate measures have to be taken to ensure the formation of atomically sharp and defect-free interfaces. Clean deposition conditions (usually UHV) at elevated temperatures have to be chosen so that the interfaces are free of contaminations and point defects; otherwise debonding at interfaces may occur and no hardness increase is reached [32]. On the other hand, the nanomultilayer concept offers the possibility to form any desired interface independent of the nature of the phases forming the interface.

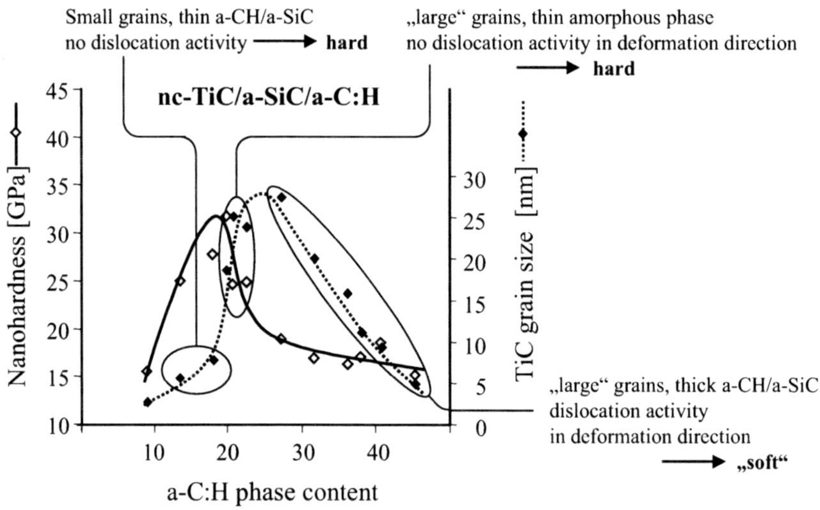

Figure 2: Nanohardness and grain size (in growth direction) of two-dimensionally nanostructured nc-TiC/a-SiC/a-C:H. Not the grain size but the dimension in deformation direction determine the dislocation activity (cf. text and figure 3).

5. TiC/a-SiC/a-C:H - A TWO-DIMENSIONALLY NANOSTRUCTURED COATING

Reports on a phase Ti_3SiC_2, which behaves plastically at low strain rates [33], and reported low friction behavior [34] triggered research in producing this material in thin film form. While single-crystalline films of this material can be grown at 900°C [35], the deposition at low temperatures leads to the deposition of a nanocomposite which consists of nc-TiC, a-SiC and a-C:H [36]. The (111)-oriented TiC crystallites are elongated in growth direction, but small enough to preclude deformation in the preferred slip direction

(110) which lies at an angle of 26.5°. So although large crystallites are present (max 25 nm in growth direction, 5 nm diameter), the size in which the deformation would occur is too small to allow dislocation activity. Only when the mean thickness of the amorphous phase exceeds the values commonly found in hard nanocomposites the material softens due to crack formation in the amorphous phase (see figure 2).

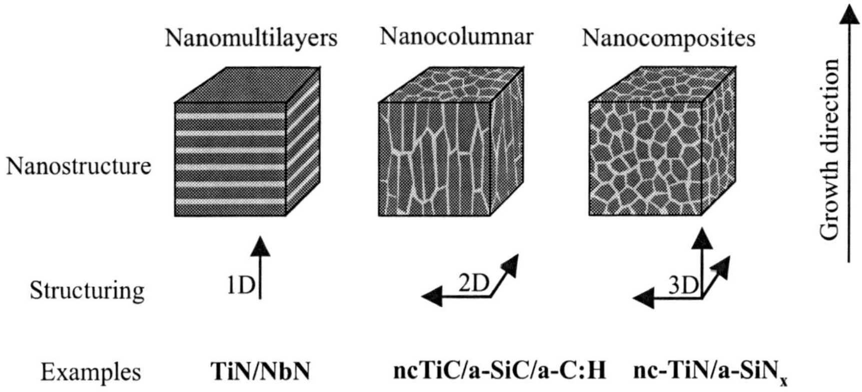

Figure 3: Schematic representation of nanostructuring of coatings with enhanced hardness in one, two and three dimensions.

6. SUMMARY

Although the architecture and preparation methods of nanomultilayers and nanocomposites differ to a large extent, there are many common aspects to the nanostructural requirements for increased hardness. Nanostructuring of thin films can lead to hardness enhancement provided that certain requirements are fulfilled. If the size of the crystalline phases, which are subjected to mechanical load, are below the minimal size for stable dislocations, these phases will not undergo plastic deformation. In nanomultilayers these phases are the single layers, in isotropic nanocomposites the nanocrystals and in two-dimensionally nanostructured coatings the crystallite size in deformation direction. The interfaces need to be atomically sharp in order to provide a barrier for dislocation movement. A further necessary condition for increased hardness in 2D and 3D nanocomposites is a minimal thickness of the amorphous phase.

ACKNOWLEDGMENTS

The financial support by the presidents of EMPA, EPFL and EIAJ as well as by the Swiss Programs PPM (3.3D) and TOP Nano21 (5304.1) is gratefully acknowledged.

REFERENCES

1. R. Melley, P. Wissner, The Real Costs of Lubrication', 99th AGM-CIM (1997), Canada.
2. O. Knotek, M. Böhmer, T. Leyendecker, J. Vac. Sci. Technol. A 4 / 6 (1986) 2695.
3. W.-D. Münz, J. Vac. Sci. Technol. A 4 / 6 (1986) 2717.
4. T. Hirai, S. Hayashi, J. Mater. Sci. 17 (1982) 1320.
5. Li Shizhi, Shi Yulong, Peng Hongrui, Plasma Chem. Plasma Proccss. 21/3 (1992) 287.
6. S. Vepřek, S. Reiprich, Li Shizhi, Appl. Phys. Lett. 66 (20) (1995) 2640.
7. S. Vepřek, J. Vac. Sci. Technol. A 17/5 (1999) 2401.
8. S. Vepřek, S. Reiprich, Thin Solid Films 268 (1995) 64; S. Veprek A. Argon, Surf. Coat. Technol. 146-147 (2001) 175.
9. P. Nesladek , S. Vepřek, Phys. Stat. Sol. (a) 177 (2000) 53.
10. S. Vepřek, M. Haussmann, S. Reiprich, J. Vac. Sci Technol. A 14 (1996) 46.
11. M. Diserens, J. Patscheider, F. Lévy, Surf. Coatings Technol. 108-109 (1998) 241.
12. F. Vaz, L. Rebouta, S. Ramos, A. Cavaleiro, M. F. da Silva and J. C. Soares, Surf. Coatings Technol. 101 (1-3) (1998) 110.
13. M. Diserens, J. Patscheider, F. Lévy, Surf. Coat. Technol. 120-121 (1999) 158.
14. F. Vaz, L. Rebouta, Mater. Sci. Forum 383 (2001) 143.
15. A. Leonhardt, H. Liepack, K. Bartsch, Surf. Coat. Technol. 133-134 (2000) 186.
16. A. A. Voevodin, J. S. Zabinski, Diamond Relat. Mater. 7 (1998) 463.
17. T. Zehnder, J. Patscheider, Surf. Coat. Technol. 133-134 (2000) 138.
18. J. S. Koehler, Phys. Rev. B **2** (2) (1970) p. 547
19. M. Shinn, L. Hultman, S. A. Barnett, J. Mater. Res. 7 / 4 (1992) 901.
20. S. Barnett, A. Madan, Physic World Jan. (1998) 45.
21. M. Setoyama, A. Nakayama, M. Tanaka, N. Kitagawa, T. Nomura, Surf. Coat. Technol. 86-87 (1996) 225.
22. H. Holleck; V. Schier, Surf. Coat. Technol. 76/1-3 (1995) 328.
23. X. Chu , S. A. Barnett, J. Appl. Phys. 77 (9) (1995) 4403.
24. W. D. Sproul, Science 273 (1996) 889.
25. A. A. Voevodin, J. S. Zabinski, Thin Solid Films 370 (2000) 223.
26. P. H. Mayrhofer, G. Tischler, C. Mitterer, Surf. Coat. Technol. 142-144 (2001) 78.
27. H. van Swygenhoven, Science 296 (2002) 66.
28. J. Schiøtz, K. W. Jacobsen, Science 301 (2003) 1357.
29. J. Patscheider, T. Zehnder M. Diserens, Surf. Coat. Technol. 146-147 (2001) 201.
30. Y.-H. Chen, K. W. Lee, W.-A. Chiou, Y.-W. Chung, L. M. Keer, Surf. Coatings Technol. 146-147 (2001) 209.
31. S. Vepřek A. Argon, Surf. Coat. Technol. 146-147 (2001) 175.
32. J. M. Molina-Aldeguia, S. J. Lloyd, M. Odén, T. Joelsson, L. Hultman, W. J. Clegg, Phil. Mag. A 82/10 (2002) 1983.
33. M.-W. Barsoum, M. Radovic, P. Finkel, T. El Raghy, Appl. Phys. Lett. 79 (2001) 479.
34. S. Myhra, J. W. B. Summers, E. H. Kisi, Mater. Lett. 39 (1999) 6.
35. J.-P. Palmquist, U. Jansson, T. Seppanen, P.-O. Persson, J. Birch.; L. Hultman, P. Isberg, Appl. Phys. Lett 81 (2002) 835.
36. T. Zehnder, J. Matthey, P. Schwaller, A. Klein, P.-A. Steinmann, J. Patscheider, Surf. Coat. Technol. 163-164 (2003) 238.

Chapter 5
HARD NANOCOMPOSITE FILMS PREPARED BY REACTIVE MAGNETRON SPUTTERING

J. Musil
Department of Physics, University of West Bohemia, Plzem, Czech Republic
Institute of Physics, Academy of Sciences of the Czech Republic, Praha, Czech Republic

The paper (1) reviews the present state of the art in magnetron sputtering of hard and superhard nanocomposite coatings and (2) reports on recent advances in this field. In the first part of this article, it is shown that nanocomposite coatings represent a new generation of materials, which exhibit new unique properties. A main attention is devoted to the enhanced hardness. It is shown that (1) the hardness H of the nanocomposite film can be more than two times higher than H of its hardest component, (2) there are two groups of hard and superhard nanocomposites: (i) nc-MeN/hard phase and (ii) nc-MeN/soft phase, (3) H correlates well with the film structure and (4) there are three possible origins of the enhanced hardness: (i) dislocation-dominated plastic deformation, (ii) cohesive forces between atoms and (iii) nanostructure of materials. In spite of these findings, there are, however, two fundamental questions: (1) The enhanced hardness of the sputtered film is due to its high intrinsic macrostress σ, induced by the ion bombardment used in their formation, or to its nanostructure? and (2) What is the origin of the hardness in single-phase films? The second part of the paper reports on results of an investigation of last two problems. At the end, trends of next development are outlined.

Keywords: nanocomposite, films, two-phase, single-phase, hardness, physical properties, mechanical properties, magnetron sputtering.

1. PRESENT STATE OF THE KNOWLEDGE

Nanocomposite coatings represent a new generation of materials. They are composed of at least two separated phases with a nanocrystalline and/or amorphous structure. Nanocomposite materials, due to very small (≤10 nm) grains and a more significant role of boundary regions surrounding individual grains, behave in a different manner compared to conventional materials with grains greater than 100 nm and exhibit completely new properties. New unique physical and functional properties of nanocomposite coatings are a main driving force stimulating the development of these materials, see for instance, recent review papers [1-15].

A. A. Voevodin et al. (eds.),
Nanostructured Thin Films and Nanodispersion Strengthened Coatings, 43–56.

At present, the nanocomposite coatings are prepared using different methods, particularly by Plasma Assisted Chemical Vapour Deposition (PACVD) from gas phase, magnetron or ion beam sputtering from a solid target, i.e. by the Physical Vapour Deposition (PVD) process, and combined PVD and PACVD processes, e.g. by cathodic arc evaporation and PACVD [10] or magnetron sputtering and laser ablation, i.e. by the hybrid process called as Magnetron Sputtering and Pulsed-Laser Deposition (MSPLD) [11]. For industrial production of the nanocomposite coatings the most suitable method is the magnetron sputtering.

Main advantages of the magnetron sputtering are the following: (1) the sputtering is a *nonequilibrium process* at an atomic level, which makes it possible to replace a substrate heating (T_s) with the particle bombardment (E_p) and to form films at low T_s even close to the room temperature (RT); here T_s and E_p are the substrate temperature and the energy delivered to the growing film per condensing particle, respectively, (2) no problem in sputtering of *alloys* and their compounds, such as nitrides, carbides, etc., (3) condensing atoms have a high energy (several eV compared with approximately 0.1 eV in evaporation process) what enables (i) to form *high-temperature phases* on unheated substrates due to high cooling rates ($\sim 10^{14}$ K/s) and (ii) to perform *a selective reactive sputtering* of nitride of alloys, i.e. to form a nanocomposite of the type nc-eN/metal, due to a difference in the nitride decomposition temperatures for element A and B of the AB alloy; here nc- denotes the nanocrystalline phase and Me=Ti, Zr, Cr, W, Mo, etc. and (4) *magnetrons can be easily scaled up* into big industrial coating machines.

For sputtering of nanocomposite coatings three basic sputtering systems can be used: (1) one magnetron with an alloyed target, (2) two magnetrons equipped with the targets made of different (i) elements (e.g. Ti, Si), (ii) alloys (e.g. TiAl, CrNi), (iii) compounds (e.g. TiB_2, $TaSi_2$) or their combinations, or (3) pulsed operated dual magnetron, which can easily control individual elements in the alloy film or makes it possible to deposit non-conductive materials at high deposition rates. Hard nanocomposite films are usually prepared by the reactive magnetron sputtering, i.e. the magnetron cathode (target) is sputtered in a mixture of Ar and reactive gas (nitrogen, oxygen, etc.). In the case when only one element is converted into a nitride, this process is called *selective magnetron sputtering.* According to the number of elements in the sputtered alloy, two- or several-phase films can be prepared.

1.1. Classification of Hard Nanocomposite Coatings

According to hardness the nanocomposite coatings can be divided into three groups: (1) hard coatings with hardness H<40 GPa, (2) superhard coatings with H between 40 and 80 GPa and (3) ultrahard coatings with H>80 GPa [13]. At present, two groups of hard and superhard two-phase nanocomposites coatings are known:

1. nc-MeN/ hard phase (e.g. a-Si_3N_4, a-TiB_2, etc.);
2. nc-MeN/soft phase (e.g. Cu, Ni, Y, Ag, etc.);

here nc- and a- denote the nanocrystalline and amorphous phase, respectively, and Me=Ti, Zr, W, Cr, Mo, Al, etc. are the elements forming nitrides.

Hardness H of films from both groups can be continuously varied from low values of about 10 GPa to high values achieving in some cases up to about 70 GPa by varying deposition conditions used in their preparation. Experiments show that nanocomposites with H>70 GPa are the multiphase coatings, for instance, nc-TiN/a-Si_3N_4/a- & nc-$TiSi_2$ [13].

1.2. Mechanical Properties of Hard Nanocomposite Coatings

Mechanical properties of nanocomposite coatings are well characterized by their hardness, H, Youngrs modulus $E^*=E/(1-\nu^2)$ and elastic recovery W_e. These quantities can be evaluated from loading/unloading curves measured by a dynamic microhardness tester such as, for instance, Fischerscope H 100. Measured values of H and E^* permit to calculate the ratio H^3/E^{*2}, which gives an information on a resistance of the material to plastic deformation [16]. The likehood of plastic deformation is reduced in materials with high hardness H and low modulus E^*. In general, a low modulus is desirable as it allows the given load to be distributed over a wider area.

Hard films with H<40 GPa are characterized by low (≤300 GPa) values of E^* and exhibit (i) a low elastic recovery W_e increasing with increasing H up to about 70% for H≈30 GPa and (ii) a *high plastic deformation* increasing with decreasing H up to 70% for H≈10 GPa. On the contrary, *superhard films* with H≥40 GPa are characterized by high (>300 GPa) values of E^* and exhibit (i) a *high elastic recovery* W_e increasing up to 85% with increasing H and (ii) a low plastic deformation increasing only to about 25% for H≈40 GPa with decreasing H. For details see reference [7].

Mechanical properties of nanocomposite coatings strongly depend on (i) elements which form individual phases and (ii) relative content of individual phases in the nanocomposite, see Fig.1. From this figure it can be seen that (1) the film with a given H can have different values of E^*, (2) the value of E^* can be controlled by the chemical composition of nanocomposite and (3) nanocomposite films composed of two hard phases (Ti-Al-N, Ti-Mo-N) and nanostructured hard nitrides of transition metals (e.g.TiN, ZrN) exhibit higher E^* compared to that of the nanocomposites composed of one hard and one soft (metal) phase (e.g. Ti-Al-N, Zr-Cu-N). The Youngrs modulus can be changed not only by the choice of elements forming the nanocomposite but also by the choice of deposition conditions, and mainly by the energy delivered to the growing film by condensing and bombarding particles. This makes it possible to control the resistance of the film to plastic deformation, which is proportional to the ratio H^3/E^{*2}, and to tailor mechanical properties of the nanocomposite for a given application.

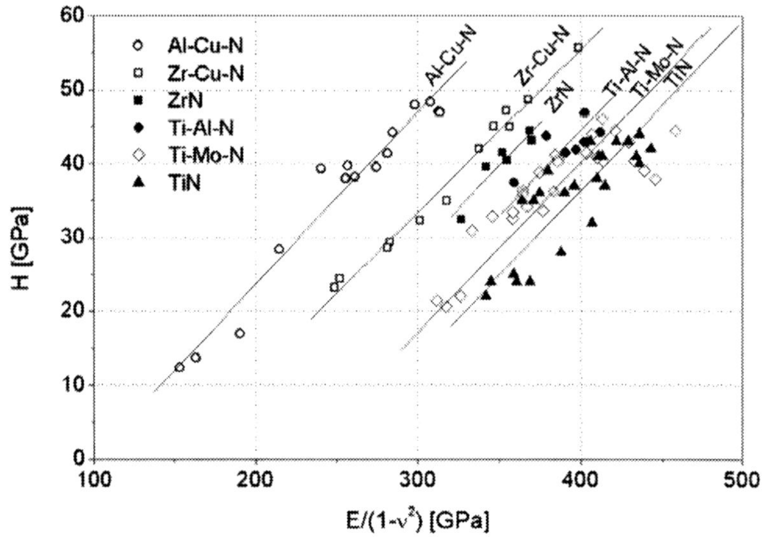

Figure 1. Microhardness of selected hard nanocomposite and binary nitride films magnetron sputtered under different deposition conditions, adapted after reference [9,12].

1.3. Enhanced Hardness

Many experiments clearly demonstrate that nanocrystalline and nanophase materials, which are composed of small (≤10 nm) grains exhibit enhanced properties such as mechanical (hardness), tribological (friction), physical (field emission in ultrananocrystalline diamond, elements solubility, thermal conductivity, photocatalytic effect), optical, magnetic, electrochemical, etc. These enhanced properties occur in the case when the grain boundary regions start to play a comparable or even dominant role over that of grains. Qualitatively new processes, such as grain boundary interaction, grain boundary enhancement or sliding, result in new unique physical and functional properties.

Enhanced properties of nanophase materials do not meet the rule of mixture. For example, hardness of two-phase nanocomposite coating H_n is given by the following inequality

$$H_n > (1/V_n)(H_1V_1+H_2V_2)$$

where H_1, H_2 and V_1, V_2 are the hardness and the volume of the first and second phase, respectively and V_n is the total volume of the nanocomposite coating. A maximum value of the enhanced hardness H_n can be more than two times greater than that of the harder component of the nanocomposite. Main mechanisms, which are responsible for the hardness enhancement, are: (1) dislocation-dominated plastic deformation, (2) cohesive forces between atoms and (3) nanostructure of materials. The magnitude of enhancement depends on processes operating in the material at a given range of the size d of grains, see Fig.4. There is critical value of the grain size $d_c \approx 10$ nm at which a maximum value of H is achieved. A region around the maximum hardness H_{max} at $d=d_c$ corresponds to a continuous transition from the operation of *intragranular processes* at $d>d_c$, dominated by dislocation activity and described by the Hall-Petch low ($H{\sim}d^{-1/2}$) to that of *intergranular processes* at $d<d_c$ dominated by a small-scale sliding in grain boundaries.

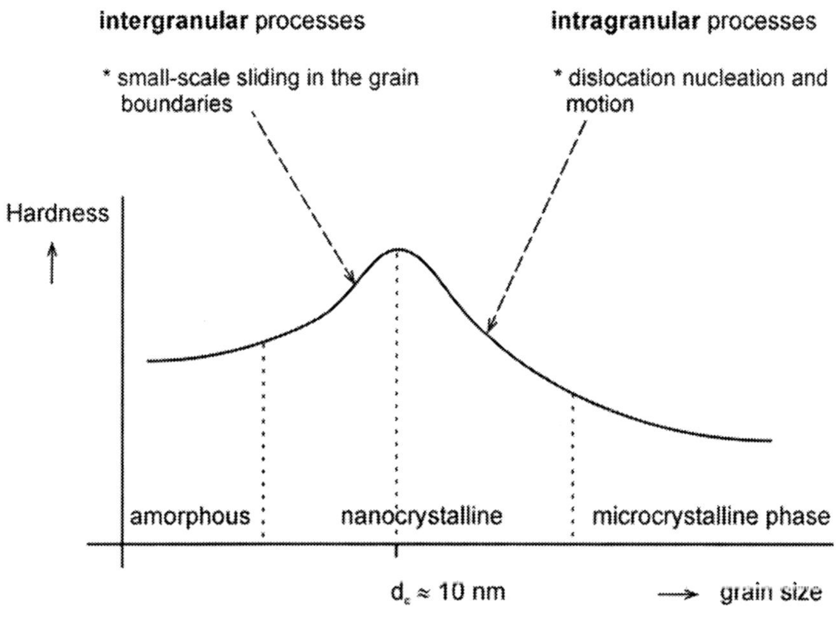

Figure 2. Schematic illustration of development of the hardness in materials with decreasing grain size d, adapted after reference [14,15]

In materials with d≤10 nm (1) the amount of atoms in grain boundary regions is greater than approximately 10% of that in nanograins and (2) dislocations already do not form and so besides chemical bonding namely a nanostructure of materials start to play a dominant role. The properties of these materials strongly depend on the size of grains, their chemical composition, crystallographic orientation and lattice structure. Therefore, the hardness enhancement can be explained by an existence of a *mixture of at least two different kinds of grains*. At present there is only a little knowledge on these materials. It is due mainly to the fact that (i) it is very difficult, and in some cases even almost impossible, to form materials with grains continuously varying in the range between 1 and 10 nm and (ii) relations between the material properties and the size of grains are unknown.

1.4. Open Problems in Formation of Nanocomposite Films and Understanding of Their Unique Properties

In spite of a great progress in the development of hard and superhard films, there is still a little knowledge, which allow us to form nanocomposite films with prescribed properties. This is due to the fact that films are formed in a medium rising in consequence of the action of many process parameters (factors), which are mutually very tightly coupled. In such systems, it is very difficult to change one process parameter, for instance, the flux of ions bombarding the growing film, while keeping the other constant. This may result in jump changes in a structure of the film and its final properties, and/or incorrect explanation of the origin of its enhanced properties. Therefore, a further systematic investigation of the nanocomposite films is highly needed. At present, a considerable attention is concentrated, particularly on the following problems.

1. Macrostress σ generated in sputtered nanocomposite films.
2. Role of the energy in formation of nanocomposite films.
3. Origin of enhanced hardness in single-phase films.
4. Thermal stability of nanocomposite films.
5. Nanocrystallization from amorphous phase.
6. Interrelationships between the energy-structure-size of grains-film properties.

In this paper only the problem 1 and 3 will be discussed in more details.

2. MACROSTRESS IN SPUTTERED FILMS

The main aim of this paragraph is (i) to show that several μm thick hard sputtered films can exhibit a low (<-1 GPa) compressive macrostress σ, when deposition conditions are correctly selected, and (ii) to demonstrate that the enhanced hardness of superhard sputtered films has not to be caused by a high values of σ as some authors incorrectly indicate [6]. At first, we shortly summarize general features of σ, which can be generated in the film during its growth.

The macrostress σ in sputtered films can be controlled by (1) the substrate heating, i.e. by the ratio T_s/T_m, and (2) the energy E_p delivered to the growing film by condensing and bombarding particles; here T_s is the substrate temperature and T_m is the melting temperature of the material of the sputtered film. This possibility is clearly shown Fig.2, where σ as a function of T_s/T_m is given. The total stress σ is composed of two components (1) thermal stress σ_{th} and (2) intrinsic (growth) stress σ_i, i.e. $\sigma = \sigma_i + \sigma_{th}$. The thermal stress σ_{th} is due to the difference in thermal expansion coefficients of the film and substrate. On the contrary, the intrinsic stress σ_i occurs in a consequence of the accumulation of crystallographic defects that are built into the film during its deposition by ion bombardment, i.e. σ_i is controlled by the substrate bias U_s, the substrate ion current density i_s, the deposition rate a_D and total pressure of sputtering gas. σ_{th} dominates at high values of the ratio $T_s/T_m>0.3$, whereas σ_i dominates at low values of $T_s/T_m<0.3$ [17]. Experiments show that the magnitude of σ_i increases with intensity of the ion bombardment and decreases with increasing T_s/T_m. However, the most important finding is the fact that the intrinsic stress σ_i would not occurs at $T_s \geq 0.3\ T_m$ [18]. The last inequality can be shifted to lower values T_s/T_m when compound films, such as for instance TiN, TiO_2, are reactively sputtered and additional energy, released in an exothermic reaction, also contribute to relax σ_i. Because σ_{th} is generally considerably lower than σ_i there are no principal reasons, which prevent to sputter films with low σ.

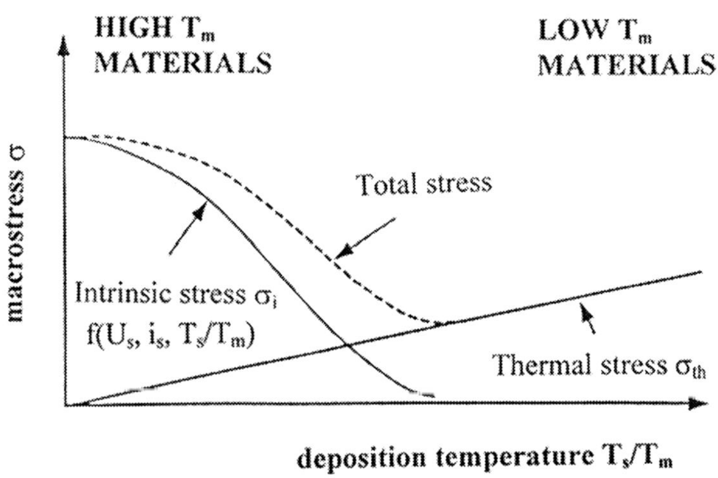

Figure 3. Schematic diagram of the macrostress σ vs. T_s/T_m, adapted after reference [15].

This simple analysis is not only in a good agreement with experiments but also shows under which conditions low (<-1 GPa) stress σ, several μm thick films can be sputtered. Main issues on the macrostress σ can be summarized as follows: (1) Ion bombardment induces in the growing film a compressive stress ($\sigma<0$), (2) σ can be controlled by the energy E_p delivered to the growing film during its growth and (i) increases with increasing negative bias U_s and i_s and (ii) decreases with increasing T_s/T_m; for $T_s/T_m \geq 0.25$ the macrostress $\sigma \leq 0.3$ GPa and (3) thermal expansion coefficient of the film α_f can be controlled by the microstructure of the film [19]. This means that several μm thick, low-stressed, superhard films with H≥40 GPa can be produced.

2.1. Hardness of Sputtered Films

The hardness H of sputtered films can correlate either with increasing macrostress σ or decreasing average size d of grains. This means that there are two hardening mechanisms of the film: (1) macrostress hardening, if high ion energy $E_i \geq 100$ eV is used and $H=f(\sigma)$, and (2) grain size hardening, which follows the Hall-Petch law, if $E_i \leq 30$ eV is used and $H=f(d^{-1/2})$. Experimental evidence of the existence of these two hardening mechanisms is given in [20]. In some sputtered films their hardness can be a result of a combined action of both mechanisms.

Recent experiments indicate that there is a clear correlation between σ and d. The macrostress σ decreases with decreasing d, see for instance Ti-Si-N films [12] or Zr-Si-N films [22]. However, a low σ has not to mean that for the enhanced H only the size d of grains is responsible. The enhanced hardness of low-stressed sputtered films can be also due to a strong chemical bonding as it is in the case of thick (up to 8 μm), low-stressed superhard (H≈77 GPa) $TiB_{x\approx 2.4}$ films [23]. Therefore, a further study is necessary to understand fully the interrelationships not only between d and σ but also between d and physical and functional properties of nanostructured films. For instance, the addition of Si into the film not only contribute to its amorphization but also improves its thermal stability [24]. However, to determine the correlation between d and the film properties a new technological process of preparation of films with grain size d controlled in the interval between 1 and 10 nm is necessary to be developed.

3. FORMATION OF NANOCRYSTALLINE AND/OR X-RAY AMORPHOUS FILMS

The basis for a production of nanocrystalline materials is to find conditions under which very fine grains can be formed. Experiments clearly indicate that nanocrystalline and/or X-ray amorphous films can be formed in a *transition region* between two phases, see Fig.4. From this figure it is seen that in a certain interval x of the $A_{1-x}B_x$ alloy or $A_{1-x}B_xN$ nitride alloy a transition region from the solid solution A(B) to the solution B(A) can exist; here A and B are elements forming alloy and x is the film stoichiometry. In the transition region nanocrystalline and/or X-ray amorphous two-phase films are created. This fact was demonstrated by X-ray measurements of a development of the structure of, for instance, $Ni_{1-x}Ag_x$ alloy film [25], $Ti_{1-x}Al_xN$ [26], Cr_{1-x} AlN [27], $Cr_{1-x}Si_xN$ [28], $Ti_{1-x}Zr_xN$ [29] nitride alloy films with increasing stoichiometry from x=0 to x=1. The transition from one phase to the second phase in the pseudobinary nitrides consisting of the transition metal and AlN can be also predicted by the band parameter method [30]. The agreement of this prediction with experiment is surprisingly good.

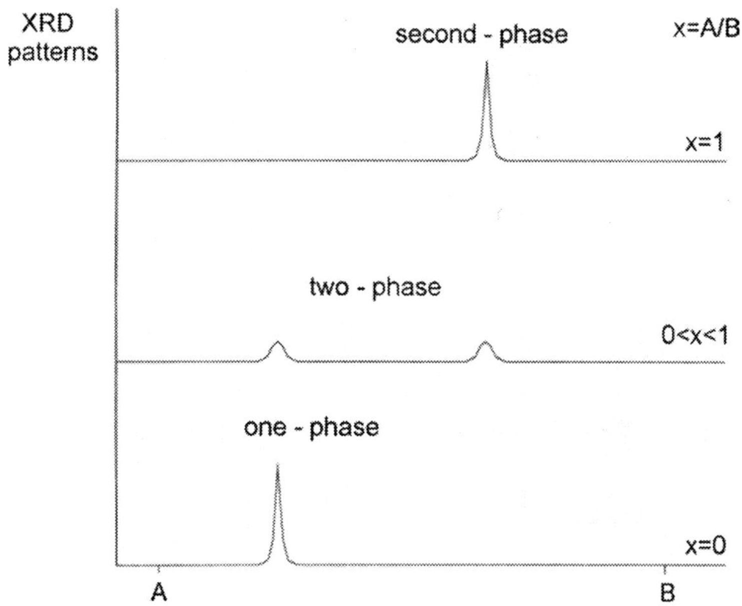

Figure 4. Schematic illustration of the development of XRD patterns from $A_{1-x}B_x$ alloy films and $A_{1-x}B_xN$ alloy nitride films with increasing x.

In the case of both kinds of films, alloys and alloy nitrides, there is a "window", i.e. the interval x, in which films exhibit an X-ray amorphous structure. The width of this window (transition) depends on (i) chemical composition of the film and (ii) deposition conditions under which the film is formed. For instance, the transition region for the $Ti_{1-x}Al_xN$ films is located approximately between 50 and 60 at.% Al. In this concentration range two phases may co-exist and the films composed of a mixture of fine grains with NaCl TiN structure and wurzite AlN structure can be created. The films prepared in the transition region are superhard with H≥40 GPa [31]. This fact simply explains why the $Ti_{1-x}Al_xN$ films with x≈0.5 can not exhibit the enhanced hardness. Recently, the CemeCon AG in Wьrselen, Germany started to produced $Ti_{1-x}Al_xN$ films with x≥0.6 and these films denotes as "supernitrides" because they exhibit the enhanced hardness and fine grained dense structure. In fact, the supernitride films are produced in the transition region between the Ti(Al)N and Al(Ti)N phases.

In the transition region not only films composed of grains of different phases can be created but also those composed of single-phase grains of different crystallographic orientations. Therefore, at present, it is not fully clear what mixture of grains - with (i) different phases or (ii) different orientations- is responsible for the enhanced hardness of binary nitrides. A further study is necessary to separate both these effects. On the other hand, the enhanced hardness observed in the single-phase materials can be caused just by the co-existence of small single-phase grains of different crystallographic orientations.

3.1. Origin of Enhanced Hardness in Single-Phase Films

Similarly as in the case of two- or several-phase materials (nanocomposites), *the enhanced hardness of single-phase materials* can be also explained by an existence of *a mixture of at least two different kinds of grains*. On the contrary to two-phase nanocomposites, which are composed of grains created of materials of different chemical composition (Fig.5a), e.g. nc-TiN/a-Si_3N_4, the single-phase films are composed of a mixture of grains of the same material but with different crystallographic orientation and/or different lattice structure, see Fig.5b. Such single-phase films can be really produced [32,33]. As an example, we present Ti(Fe)N_x films [33].

The Ti(Fe)N_x films with a low (<15 at.%) Fe content is a typical example of the single-phase material. A development of the structure of these films, characterized by XRD patterns, with increasing p_{N2} is given in Fig.6. From this figure it is clearly seen that in the interval x from 0.86 to 1.20 there is a transition region. Films produced in this region are (i) composed of a mixture of fine grains of different crystallographic orientation and (ii) nearly stoichiometric x=N/(Ti+Fe)≈1 and exhibit the enhanced hardness H≥40 GPa.

The rise of the transition region is connected with a jump change in preferred crystallographic orientation of the film from δ-TiN(200) to δ-TiN(220) with increasing p_{N2}. Such a jump change in preferred orientation of grains in sputtered $Ti(Fe)N_x$ films is a typical feature also for other sputtered nitrides containing a small (≤15%) amount of the added element (e.g. Y, Ni) forming a solid solution with the base element, see for instance Zr-Y-N [34], Zr-Ni-N [35], Cr-Ni-N [36]. All these findings indicate that the reactive magnetron sputtering can be a very efficient method to produce nanocomposite films based on nitrides.

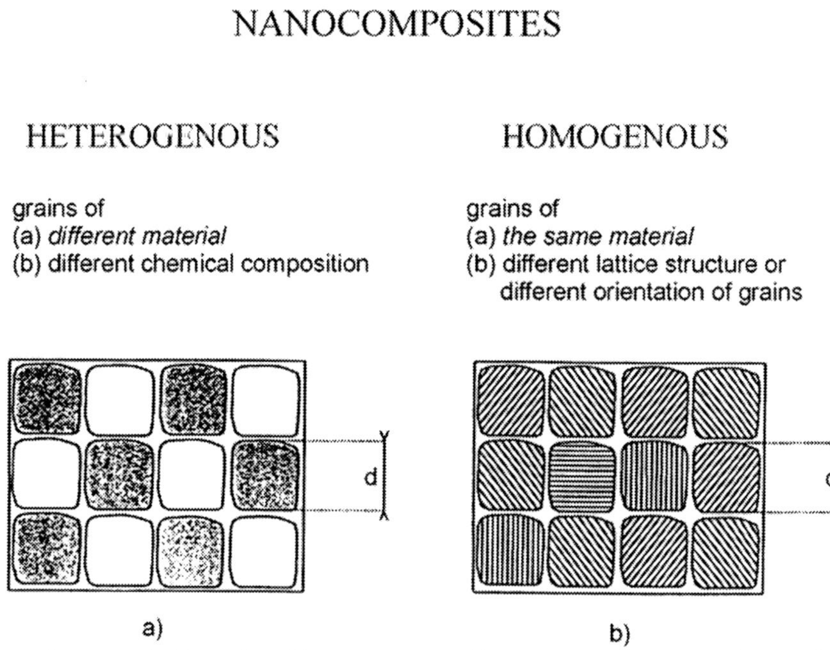

Figure 5. Schematic illustration of heterogeneous and homogeneous nanocomposite materials.

4. TRENDS OF FUTURE DEVELOPMENT

It is well known that there is a strong correlation between the energy E_p delivered to the growing film per condensing particle, the film structure and its physical and functional properties. Recently, it was found that the maximum hardness H_{max} in sputtered films originates from a combined action of physical and chemical processes controlled by the energy E_p and the film stoichiometry x, e.g. x=N/Me for nitrides [33,34]. It was found that two conditions are necessary to be fulfilled to form films with H_{max}: (1) $E_p \geq E_{pmin}$ and (2) the film must have an optimum structure; here $E_{p\,min}$ is the minimum energy needed to form the optimum structure [33]. The optimum structure of films with H_{max} is a very fine grained crystalline structure close to X-ray amorphous. At present, there is no technological process, which can produce films with such a structure in a controlled way. Therefore, one of the basic tasks is to develop a new deposition process, which enables to produce films composed of small (≤10 nm) grains whose size could be continuously varied from approximately 1 to 10 nm. Such a process can be based, for instance, on a nanocrystallization from the amorphous phase.

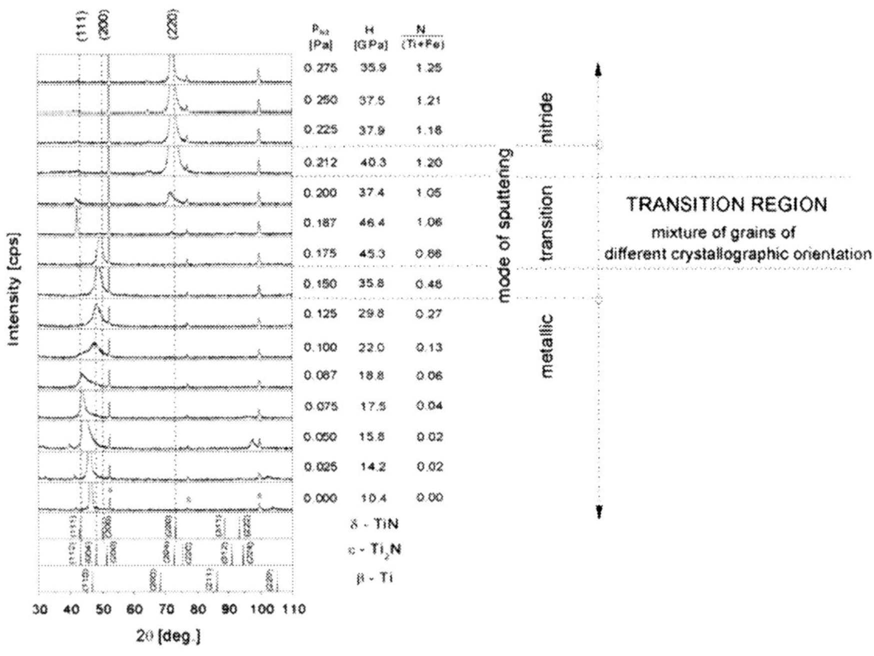

Figure 6. Development of XRD patterns from reactively sputtered Ti(Fe)N_x films with increasing P_{N2}.

As soon as such a process will be developed, a systematic investigation of size dependent phenomena, i.e. dependences of the film properties on the size d of grains from which the film is composed, will start. It will make it possible to develop new nanostructured materials with new unique properties. For instance, in materials with d≤10 nm very important role will play so called electronic effect, i.e. electronic charge transfer which occurs at any interface between two metallic grains with different chemical compositions and different Fermi energies [37,38]. It is expected that utilization of this effect enables to develop materials with new functional properties. Nanocomposite films based on oxides will be developed as well. Their production will require, however, to master fully a high-rate magnetron sputtering of oxides. It is a huge challenge to develop new advanced sputtering sources.

ACKNOWLEDGEMENTS

This work was supported in part by the Ministry of Education of the Czech Republic under Project No. MSM 235200002.

REFERENCES

1. H.Gleiter: Prog.Mater.Sci. 33(1989), 223
2. R.Birringer: Mater.Sci.Eng. A117(1989), 33
3. H.Gleiter: Nanostructured Mater. 6(1996), 3
4. K.Lu: Mater.Sci.Eng. R16(1996), 161
5. J.Musil, J.Vlcek: Materials Chemistry and Physics 54(1998), 116
6. S.Veprek: J.Vac.Sci.Technol. A17(1999), 2401
7. J.Musil: Surf.Coat.Technol. 25(2000), 322
8. A.A.Voevodin, J.S.Zabinski: Thin Solid Films 370(2000), 223
9. J.Musil, J.Vlcek: Surf.Coat.Technol. 142/144(2001), 557
10. S.Veprek, P.Nesladek, A.Niederhofer, F.Glatz, et al.: Surf.Coat.Technol. 108-109(1998), 138
11. A.A.Voevodin, J.S.Zabinski: Diamond and Related Materials 7(1998), 463
12. J.Musil, J.Vlcek, F.Regent, F.Kunc, H.Zeman: Key Engineering Materials 230-232(2002), 613
13. S.Veprek, A.Niederhofer, K.Moto, T.Bolom, et al.: Surf.Coat.Technol. 133-134(2000), 152
14. S.Yip: Nature 391(1998), 532
15. V.V.Brazhkin, A.G.Lyapin, R.J.Hemley: Philosophical Magazine A 82(2)(2002), 231
16. T.Y.Tsui, G.M.Pharr, W.C.Oliver, C.S.Bhatia, et al.: Mater. Res.Soc.Symp. Proc. 383(1995), 447
17. J.A.Thornton, D.W.Hofman: Thin Solid Films 171(1989), 5
18. F.M.drHeurle, J.M.E. Harper: Thin Solid Films 171(1989), 81
19. H.Zeman, J.Musil, J.Vlcek, P.H.Mayrhofer, C.Mitterer: Vacuum 72(2003), 21
20. C.Mitterer, P.H.Mayrhofer, J.Musil: Vacuum 71(2003), 279
21. M.Nose, Y.Deguchi, T.Mae, E.Honbo, T.Nagae, K.Nogi: Surf.Coat.Technol. 174-175(2003), 261
22. J.Musil, R.Daniel, M.Takai, prepared for publication
23. F.Kunc, J.Musil, P.H.Mayrhofer, C.Mitterer: Surf.Coat.Technol. 174-174(2003), 744
24. H.Zeman, J.Musil, P.Zeman: J.Vac.Sci.Technol.A (2003), submitted for publication
25. G.Weigang, H.Hecht, G.von Minnigerode: Z.Phys.B 96(1995), 349
26. U.Wahlstrum, L.Hultman, J.-E.Sundgren, F.Abidi, et al.: Thin Solid Films 235(1993), 62
27. T.Ikeda, H.Satoh: Thin Solid Films 195(1991), 99
28. K.Yamamoto, T.Sato, M.Takeda: Structural analysis of $Cr_{1-x}Si_xN$ coatings and tribological property in water environment, 4th AEPSE, Sept.28-Oct.4,2003, Korea, Paper No. WeA-A04
29. H.Hasegawa, A.Kimura, T.Suzuki: Surf.Coat.Technol. 132(2000), 76
30. Y.Makino: ISIJ International 38(1998), 925
31. J.Musil, H.Hrube: Thin Solid Films 365(2000), 104
32. H.Polakova, J.Musil, J.Vlcek, J.Allart, C.Mitterer: Thin Solid Films 444(2003), 189
33. J.Musil, H.Polakova, J.Juna, J.Vlcek: Surf.Coat.Technol. (2003), in press

34. J.Musil, H.Polakova: Surf.Coat.Technol. 127(2000), 99
35. J.Musil, P.Karvankova, J.Kasl: Surf.Coat.Technol. 139(2001), 101
36. F.Regent, J.Musil: Surf.Coat.Technol. 142-144(2001), 146
37. H.Gleiter, J.Weissmuler, O.Wollersheim, R.Wuschun: Acta Materialia 49(2001), 737
38. H.Gleiter, M.Fichtner: Scripta Materialia 46(2002), 497

Chapter 6
THERMAL STABILITY AND SELF-ARRANGEMENT OF NANOCRYSTALLINE HARD COATINGS

P. H. Mayrhofer
Department of Physical Metallurgy and Materials Testing, University of Leoben, 8700 Leoben, Austria

Nanocrystalline hard coatings have attracted increasing interest in modern development of hard coatings. Their increased volume fraction of interfaces is often responsible for superior properties, but they stimulate also microstructural processes. Whereas for single-phase coatings a remarkable reduction in hardness occurs around 500°C, nanocomposites maybe stable up to 1000°C. In physical vapor deposited (PVD) hard coatings nanostructures can arise during growth or during a post annealing treatment. In non-reactively sputtered overstoichiometric TiB_2 coatings the excess of boron forms a tissue phase surrounding TiB_2 crystals. Thus, a nanocolumnar structure with a diameter of about 5 nm is formed during the deposition process. Another example for segregation driven formation of a nanostructure is TiN-TiB_2, where the coating consists mainly of TiN and TiB_2 nanocrystals of about 3 nm due to their extremely limited solubility. Nanostructures can also arise during annealing of supersaturated coatings. Thermal treatment of $Ti_{1-x}Al_xN$ coatings causes the metastable phase to decompose into its stable constituents TiN and AlN. Initially $Ti_{1-x}Al_xN$ coatings undergo spinodal decomposition generating an increase in hardness at elevated temperatures. The results presented show that next generation's coatings with increased ability for self-arrangement can effectively be prepared by PVD.

Keywords: hardening, nanocomposite, spinodal decomposition, thermal stability, self-arrangement.

1. INTRODUCTION

In the last few years, several hard coating materials with unique properties have been developed or even transferred to industrial applications. Essentially, two different coating material groups can be distinguished. The first one includes materials with inherent unique properties, e.g., diamond-like carbon DLC [1], diamond [2] and cubic boron nitride [3] as inherently hard or even superhard (where the hardness exceeds 40 GPa) materials or aluminum oxide as thermally extremely stable coating [4]. The second group covers heterostructures. Unlike the multilayer coatings like the commercially

A. A. Voevodin et al. (eds.),
Nanostructured Thin Films and Nanodispersion Strengthened Coatings, 57–68.

applied TiC-Al_2O_3-TiN coatings deposited by CVD techniques with layer thicknesses in the order of a few to several tenth of microns [5], nanoscaled heterostructures allow the design of unique properties or property combinations. The first attempts to deposit these structures have been made with the goals (1) to create new superhard materials (see e.g., the review by Vepřek [6]) or to combine properties like high hardness with (2) high toughness [7] or (3) a low friction coefficient [8,9]. The latter task focuses on the design of functional coatings where nanodispersive structures consisting of hard (e.g., TiN) and lubricant phases (e.g., DLC, MoS_2) have been suggested. Here, this topic will not be discussed further. This paper mainly deals with the optimization of mechanical properties represented by hardness and its thermal stability by the formation of heterostructures, which arrange themselves during growth of the coating or during a thermal treatment after deposition.

The science of materials is to a large extend couched in terms of length scales and their interactions. Thus, the mechanical response of materials can be understood from the point of view of the interaction of a characteristic length, which may be the dislocation radius at a given stress, with a size parameter describing the microstructure (e.g., grain size, column diameter, particle size, film thickness) [10]. Hardening of metals is caused by providing obstacles for the dislocation movement, these mechanisms can also be applied to some extent to hard coatings. Hindering of dislocation movement can be achieved by, (1) high density of point and line defects generated by energetic ion bombardment during growth, (2) internal boundaries like grain and column boundaries, (3) second phase particles and (4) solutes. Solutes in a crystal act as obstacles to dislocation motion through their elastic and/or chemical interactions with dislocations. Most solutes are weak hardeners except for the (technologically) important class of interstitial solutes that induce anisotropic distortions of the lattice. Second phase particles are generally the most potent strengthening agent in practical high strength engineering materials, their mechanisms can be divided into coherency, chemical, order, stacking-fault and modulus hardening [11,12].

The different mechanisms are individual in their effectiveness of hardening and especially in their effectiveness during a thermal treatment, which will be discussed in the following chapters.

2. THERMAL STABILITY OF SINGLE-PHASE HARD COATINGS

In single phase hard coatings like TiN a hardness increase can be obtained either by high energetic bombardment during growth of the films causing high compressive stresses or by a reduction of the grain size [13]. The effect of smaller grain sizes on the yield stress is classically described by the Hall-Petch relationship which is based on dislocation pile-up at grain boundaries. The dependence of hardness of single-phase TiN coatings on their compressive biaxial stresses generated by ion bombardment during growth can be seen in Fig. 1a. Using this method even hardness values of about 56 GPa for TiN can be achieved, but their compressive stresses are extremely high [14]. These high stresses are an additional driving force for recovery effects [15] during a thermal treatment of the coatings. Thus, the onset-temperature where annihilation of defects in the coatings occur decreases (cf. Fig. 1b). As soon as the onset-temperature of recovery is reached the hardness of single-phase TiN coatings decreases as point and line defects anneal which determine the hardness of such coatings to a high extent [14,16].

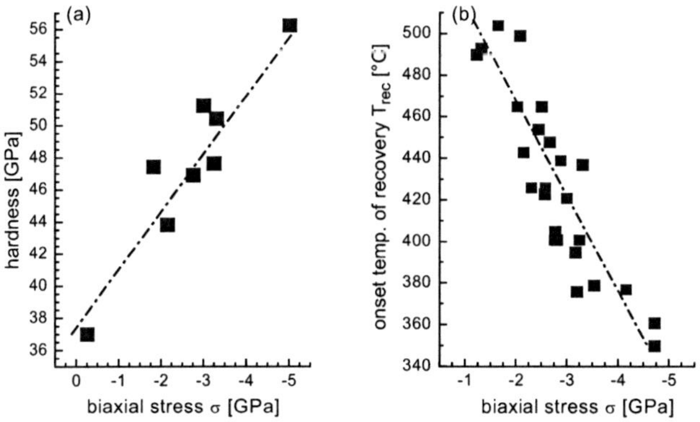

Figure 1. (a) Hardness and (b) onset temperature of recovery as a function of the biaxial compressive stresses in non-reactively sputtered TiN coatings.

Thus, the hardening mechanism for coatings by high-energetic bombardment during growth is not suitable to obtain high hot hardness due to decreasing onset-temperature for recovery [14,17]. Other hardening mechanism like internal boundaries, second phase particles and solutes as mentioned above seem to be much more promising to obtain high hardness which sustains also higher temperatures [17-19]. These mechanisms can be effective due to self-arranging nanostructures either during coating growth an annealing treatment afterwards e.g., during coating operation. The following sections provide some examples to this topic.

3. SELF-ARRANGING NANOSTRUCTURES DURING GROWTH

In 1995, Vepřek and co-workers [18] presented their model on the design of novel superhard nanocomposite coating materials. Using PACVD, they deposited coatings within the system TiN-Si_3N_4 where nanocrystals of TiN due to a segregation driven renucleation are embedded in an amorphous matrix. For these coatings they achieved superhardness values of 80 to 105 GPa [19]. They showed also clearly, that hardness values which are mainly the result of nanostructure sustain higher temperatures than those which are obtained by high energetic bombardment during growth as mentioned above. A very interesting hard coating is TiB_2 [20], which is assumed to be single phase but this coating shows also high hardness even after annealing in vacuum at 800°C. By non-reactive sputtering from a sintered TiB_2 target it was found, that the chemical composition strongly depends on the ion bombardment during growth. Thus, B/Ti ratios between 2.45 and 3.15 could be achieved simply by increasing the ion/atom flux ratio from 1 to 20 at constant ion energy of 30 eV [21]. Figure 2a shows the XRD evolution of such a superstoichiometric TiB_2 coating with annealing temperature. The patterns indicate a well defined (0001) orientation of the TiB_2 phase which remains unchanged for all annealing temperatures. Broadening and intensity of the <0001> peaks indicate almost no structural changes during a thermal treatment up to 800°C, although the peak position shifted gradually to higher diffraction angles. This indicates a reduction of compressive stresses and even the appearance of tensile stresses after the thermal treatment. Due to the different thermal expansion coefficient of coating and steel substrate combined with the stress reduction caused by recovery processes and out-annealing of argon, the highly covalent and thus brittle TiB_2 coating shows crack formation if the annealing temperature exceeds 700°C [21].

Although, there is unambiguously a reduction of compressive stresses during annealing, their hardness remains essentially constant (see Fig. 2b). The results obtained are completely different to single-phase nitride films, where a distinct reduction of their hardness with annealing temperature appears as mentioned above. For comparison data of TiN are added to Fig. 2b. Accompanied with this hardness decrease there is always a pronounced structural change (e.g., defect density, orientation, and grain size) of the coatings caused by recovery and recrystallization processes [14,16]. The investigated overstoichiometric TiB_2 coatings do not show these structural changes during an annealing treatment up to 800°C, excluding the decreasing compressive stresses. This behavior emphasizes, that their hardness is mainly determined by their micro- or nano-structure, and not by compressive stresses. Recent investigations showed, that the excess of boron resulting from non-reactive sputtering forms a tissue phase due to the extremely limited composition range of the TiB_2 phase. In the TiB_2 lattice, a surplus of boron would generate extremely high strain fields. Therefore, the TiB_2 lattice accommodates itself by segregating boron which forms a tissue phase embedded in almost stoichiometric TiB_2. This tissue phase draws through the whole coating and is elongated in the growth direction, thus causes the formation of nanocolumns with an average diameter of 5 nm [21].

The results show that non-reactively sputtered overstoichiometric TiB_2 coatings do not pertain to single-phase coatings. In this way the origin of the extraordinary high hardness of this class of material and their thermal stability can be explained. These findings are in good agreement to previous results about nanocomposite coatings showing that high hardness is obtained when the crystallite size is about 5 nm and a fully percolated amorphous phase separates the crystallites by approximately 0.5 nm [6,22,23].

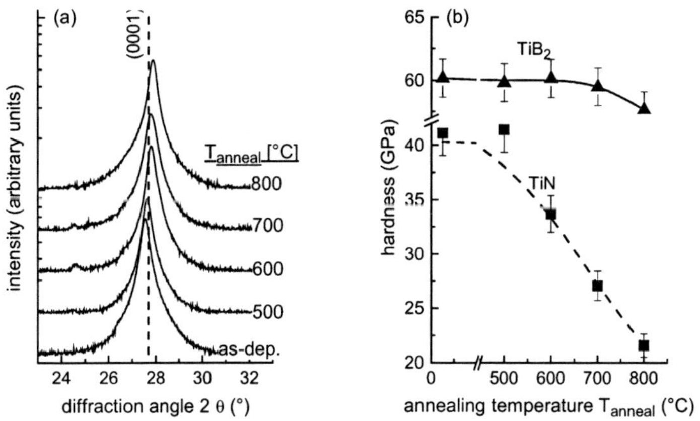

Figure 2. Thermal stability of the TiB_2 coating. (a) XRD patterns in the as-deposited state (as-dep.) and after different annealing temperatures (T_{anneal}). (b) Hardness evolution of TiB_2 as a function of annealing temperature compared to TiN.

Another method to form nanocomposite [24] coatings is co-sputtering of two immiscible phases like TiN and TiB_2 [25,26]. An example of such a coating, which was non-reactively co-sputtered from a segmented TiN-TiB_2 target is given in Fig. 3 which shows the nanostructure of the coating in the as-deposited state with a chemical composition of Ti = 36.5 at-%, B = 36 at-% and N = 27.5 at-% detected by EPMA measurements. In the bright field HRTEM micrograph (Fig. 3a) grains with an average size of about 3 nm can clearly be seen. The diffraction rings in the SAED pattern of the coating shown in Fig. 3b attribute to a possible mixture of TiN and TiB_2 nanocrystals [27]. But due to the limited local resolution or the formation of supersaturated Ti-B-N phases, lattice deformation and amorphous material no clear relation to the individual phases can be found. The z-contrast image (Fig. 3c) shows clearly a random orientation through the whole coating of at least two different phases. The bright dots indicate the appearance of a phase with a higher density than compared to the phase represented by the darker dots. Again the grain size can be estimated to be about 3 nm. Considering the extremely short diffusion paths in 3 nm grains to the nearest sink (i.e., grain and phase boundary) for defects almost 'perfect' crystals should form during the deposition [27]. Thus, the coating is assumed to consist mainly of TiB_2 and TiN crystals with minor defects like incorporated N and B, respectively. In such nanocomposites with an average grain size of about 3 nm also the ‘amorphous’ phase surrounding the crystals plays an important role in determining the coatings properties.

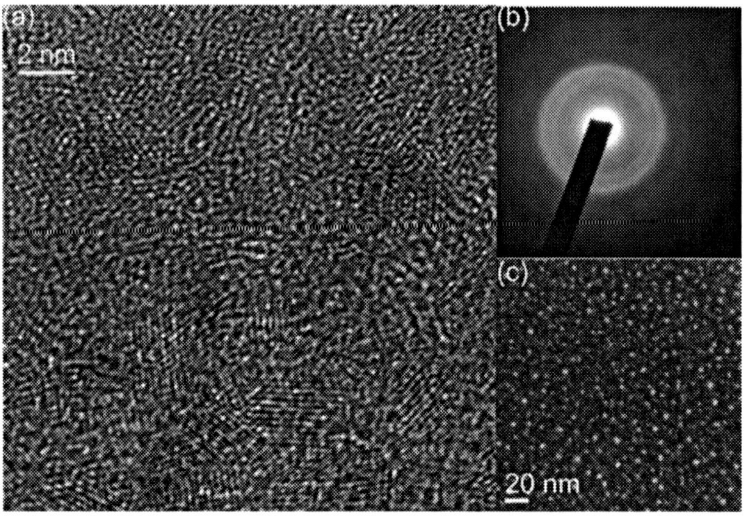

Figure 3. (a) Cross-sectional HRTEM bright field image of $TiB_1N_{0.75}$. (b) SAED pattern with diffraction rings corresponding to a mixture of amourphous phase and TiN and TiB_2 nanocrystals. (c) Z-contrast image indicating a random arrangement of two different phases.

Figure 4 shows a schematic drawing of the nucleation and growth process of the nanocomposite TiN-TiB_2 coating, indicating a self-arranging nanostructure due to segregation driven inhibited growth of the individual nuclei and renucleation, according to oxygen in Al coatings [28,29].

Since there is a high amount of amorphous material in the $TiB_1N_{0.75}$ coating, thermal treatment causes a rearrangement of the nanostructure due to on-crystallization of atoms from this amorphous phase to neighboring TiN and TiB_2 crystals. Thus, the structure and consequently the properties change by thermal treatment, which can occure during operation.

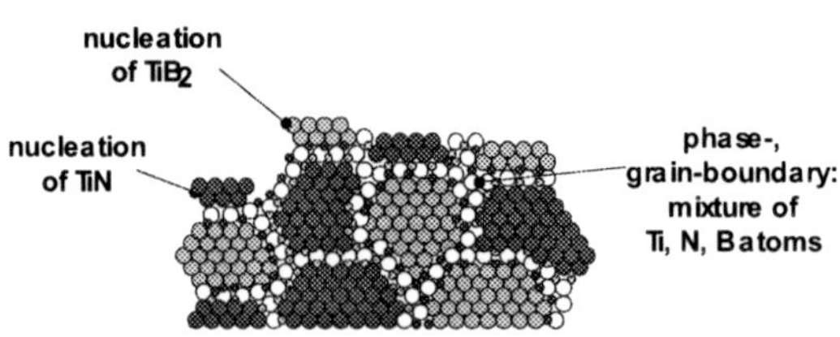

Figure 4. Schematic representation of the nucleation and growth process of $TiB_1N_{0.75}$, indicating segregation driven renucleation of TiN and TiB_2.

4. SELF-ORGANIZED NANOSTRUCTURES DURING THERMAL TREATMENT (OPERATION)

The thermal stability of the TiN-TiB_2 coating is shown in Fig. 5 represented by the XRD and hardness evolution with temperature. In the as-deposited state the broad XRD peaks can be the result of overlapped broad TiN and TiB_2 peaks, indicating a small grain size and a huge amount of amorphous material. After annealing at different temperatures the two phases TiN and TiB_2 become more pronounced indicated by more developed XRD peaks. Especially after an annealing treatment at 900°C a clear separation of the broad XRD peak into TiN and TiB_2 can be seen. After annealing treatment at 1400°C in argon atmosphere, where the coating was removed from the substrate to avoid interdiffusion a fully recrystallized structure containing just TiN and TiB_2 phases can be seen. Corresponding to nanostructural changes within the coating during annealing also the hardness

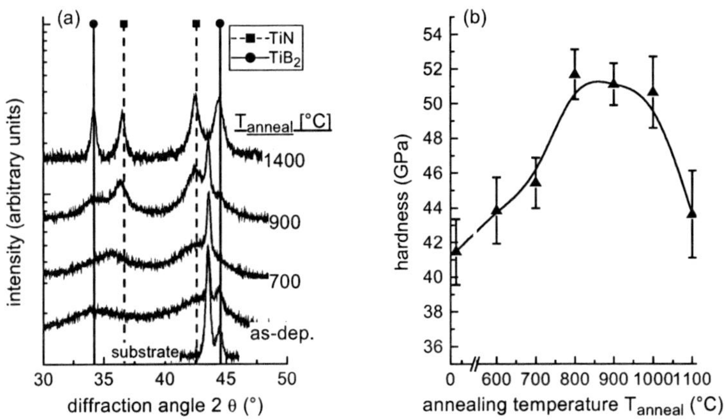

Figure 5. Thermal stability of $TiB_1N_{0.75}$. (a) XRD patterns after different annealing temperatures (T_{anneal}). (b) Hardness evolution with temperature.

changes with temperature (Fig. 5b). An increase from about 41 to about 51 GPa can be obtained by an annealing treatment up to 800 °C. If the annealing temperature exceeds 1000°C the coating hardness decreases due to recrystallization effects [15,27].

Figure 6 shows the nanostructural evolution of the coating for annealing temperatures of 700, 900, 1000, 1100 and 1400°C obtained from HRTEM investigations [27]. After a heat treatment at 700°C slight increase of the average grain size and a reduction of the amorphous phase surrounding the crystals can be observed in comparison to the as-deposited structure. Further increasing grain size and reduction of the amorphous phase is obtained if the temperature is increased to 900 or 1000°C. After annealing at 1000°C the crystals are about 5…7 nm and separated by a thin boundary structure with just a few atoms in thickness. Atoms from the amorphous region crystallize onto the surrounded TiN and TiB_2 crystals. By a reduction of the amorphous phase the cohesive energy between the phases increases and grain boundary sliding is more difficult resulting in an increased hardness of the composite (cf. Fig. 5b). Again the highest hardness is obtained when the crystallite size is about 5 nm and a fully percolated amorphous phase separates the crystallites by approximately 0.5 nm [6,22,23]. At temperatures higher than 1000°C recrystallization and subsequent grain growth occurs resulting in grain sizes above 7 nm [30]. These processes are responsible for the observed hardness decrease if the heat treatment exceeds temperatures of 1000°C.

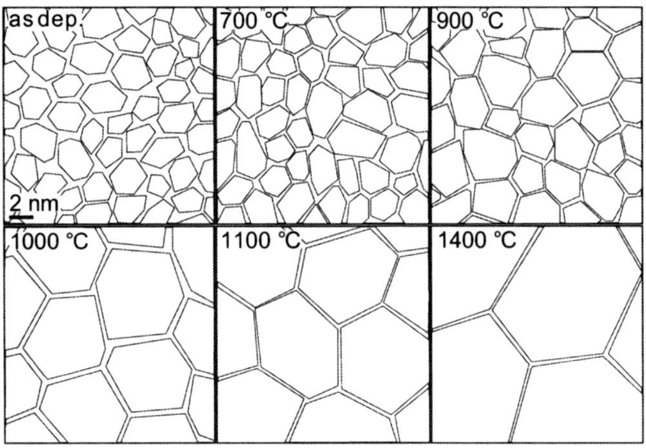

Figure 6. Scheme of the nanostructural evolution of $TiB_1N_{0.75}$ with temperature.

Additional confirmation, that the coating is mainly composed of nanocrystals of the stable phases TiN and TiB_2 surrounded by an amorphous region (mixture of Ti, B and N atoms) is given by DSC measurements since just two main exothermic peaks appear during a thermal treatment in argon atmosphere up to 1450°C (see Fig. 7a). These peaks could be identified to be the result of recovery and grain growth including crystallization effects of the amorphous region onto the surrounded nanocrystals [30]. The latter is represented by a huge asymmetric peak, elucidated in Fig. 7a, following the JMA theory [31]. The two narrow exothermic peaks between 600 and 1000°C could indicate, that the coating has also small amounts of additional supersaturated metastable phases like Ti-B-N which decompose into their stable constituents similar to $Ti_{1-x}Al_xN$ coatings [32]. In the as-deposited state, these coatings show a supersaturated $Ti_{1-x}Al_xN$ solid solution with NaCl-structure [34]. This metastable phase tends to decompose into the stable phases face centered cubic (fcc) TiN and hexagonal close packed (hcp) AlN due to their extremely limited solubility [32-35].

Such structural changes are closely related to changes in total free-energy, that can be detected by DSC. During heating of the $Ti_{1-x}Al_xN$ coatings, four exothermic reactions (recovery, spinodal decomposition [36] to form fcc-AlN and TiN domains and subsequent recrystallization including the transformation of AlN from fcc to hcp) indicated in Fig. 7b were observed, whereas for TiN just one tiny peak appeared. XRD and TEM investigations prior to and after this exothermic peak of TiN indicated recovery processes of deposition-induced lattice point defects that correspond to a compressive residual stress state. Such structural relaxations explain the observed hardness decrease at temperatures higher than 400°C (cf. Fig. 2b) [14,33].

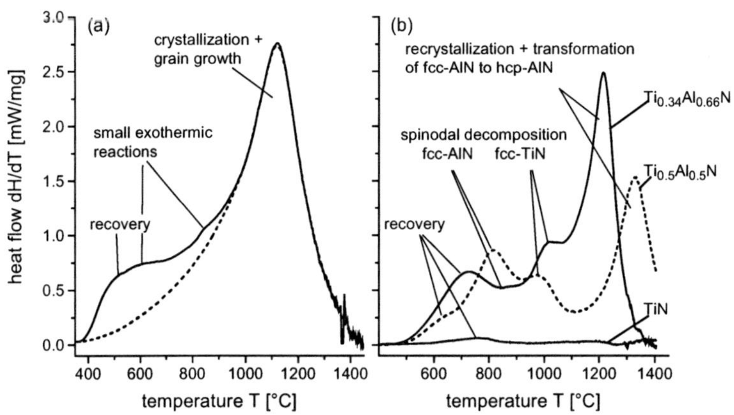

Figure 7. Dynamical DSC measurements of hard coatings. (a) DSC curve of non-reactively sputtered $TiB_1N_{0.75}$. (b) DSC curve of arc-evaporated $Ti_{0.5}Al_{0.5}N$ and $Ti_{0.34}Al_{0.66}N$ coatings compared to arc-evaporated TiN.

After recovery the supersaturated $Ti_{1-x}Al_xN$ coatings initially undergo spinodal decomposition into coherent fcc-AlN and fcc-TiN nanometer-size domains. This transformation generates an increase in hardness at elevated temperatures (age hardening [11,12]), e.g., during cutting. Therefore, the hardness of $Ti_{0.34}Al_{0.66}N$ coatings show a slight hardness increase between 700 and 950°C. The structure can thus be said to self-organize during thermal loading such as in cutting operations. Further decomposition causes a coarsening of the domains and eventual precipitation into the equilibrium fcc-TiN and hcp-AlN phases, and a resulting hardness decrease at temperatures over 950°C for $Ti_{0.34}Al_{0.66}N$ (overaging [11,12]). From detailed investigations of the structural transformations during a thermal treatment it is known, that the coatings start with the formation of Al rich domains ($TiAl_{1+\Delta}N$) thus, in their vicinity the coating shows Al deficiency ($TiAl_{1-\Delta}N$) indicated in Fig. 8. Such segregation amounts to uphill diffusion, i.e.

diffusion in the opposite direction of the concentration gradient (i.e., spinodal decomposition). A schematic representation of the decomposition process of supersaturated $Ti_{1-x}Al_xN$ coatings and the appearance of coherency strains is given in Fig. 8. Differences in the lattice parameter

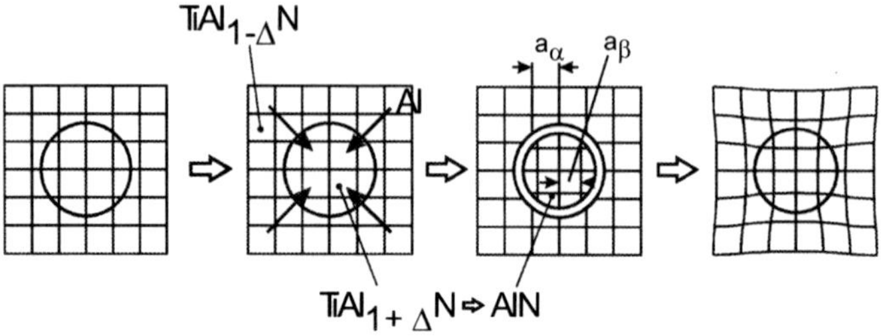

Figure 8. Schematic drawing of the spinodal decomposition process of a supersaturated TiAlN coating and the origin o coherency stresses.

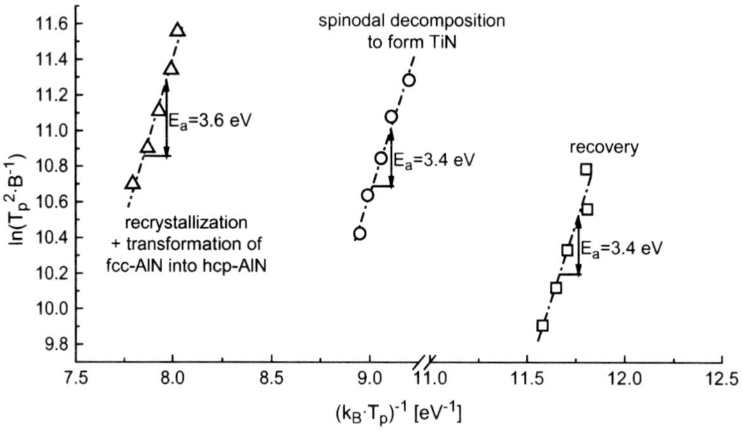

Figure 9. Kissinger analysis of the effective activation energy for recovery, decomposition and recrystallization of a $Ti_{0.34}Al_{0.66}N$ coating, calculated from the change in peak temperature (T_p) with heating rate (B) of the DSC measurement.

between newly formed domain (a_β) and remaining matrix (a_α), give rise to elastic stresses in the vicinity of the domain due to coherency stresses which act as obstacles for dislocations causing a hardening of the material during thermal treatment (i.e., age hardening).

5. CONCLUSION

Activation energies for the exothermic reactions shown in Fig. 7b (i.e., recovery, spinodal decomposition, and recrystallization) in a $Ti_{0.34}Al_{0.66}N$ coating were determined by Kissinger plots (see Fig. 9) using the respective peak temperature for different heating rates during DSC.

The activation energies for recovery and the formation of TiN domains are very similar and yield values of 3.4 eV. Recovery and spinodal decomposition are diffusion-controlled within one phase without nucleation, thus explaining the similar values. For the fcc-hcp transformation of AlN and subsequent recrystallization of the coatings, additional nucleation is needed resulting in a slightly higher activation energy of 3.6 eV [32]. Interestingly, the activation energy for surface diffusion of Ti on TiN was determined to be 3.5 eV [38]. The similar values obtained for diffusional phase transformation in the bulk imply a defect-driven process from, e.g., dislocations, which are present at high densities in the coatings.

Several mechanims of hardening for PVD coatings were investigated in detail. It could be shown, that hardening of coatings by a high density of point and line defects generated by an intense ion bombardment during deposition is not suitable to obtain a high hot hardness. These defects may easily anneal during thermal treatment, e.g. in coating operation.

Hindering of dislocation movement by internal boundaries and second phase particles is much more effective to obtain high hardness which sustains higher temperatures or even increases with temperature. These mechanisms can be effective due to nanosructure self-arranging in growth and/or in thermal treatment (e.g. during coating operation). Examples for self-arranging nanostructures during growth are supersaturated TiB_2, where the surplus of boron causes formation of nanocolumns, and co-sputtered TiN-TiB_2, where a segregation driven nanostructure forms by renucleation. During a thermal treatment of the latter one their nanostructure arranges itself due to the on-crystallization of additional amorphous phase present in the as-deposited state causing an hardness increase with temperature.

Age hardening in hard coatings, as shown for supersaturated $Ti_{1-x}Al_xN$, is very effective to obtain high hardness at increased temperatures. During annealing, this class of materials decomposes into stable constituents, where coherent cubic AlN-rich and TiN-rich domains are formed by spinodal decomposition during early stages of phase-separation. Further aging results in coarsening of these domains to form separated fcc-TiN and fcc- and finally hcp-AlN phases. If dislocations are able to bow between these obstacles, a hardness decrease is observed.

The results presented show, that by choosing the proper elements for hard coatings high hot hardness can be designed using different mechanisms and knowing the underlying materials science.

ACKNOWLEDGMENTS

The Center for Microanalysis of Materials which is partially supported by the U.S. Department of Energy (DOE), at the University of Illinois is highly acknowledged.

REFERENCES

1. A. Grill, Diam. Relat. Mater. 8 (1999) 428.
2. C. Faure, W. Hänni, C. J. Schmutz, M. Gervanoni, Diam. Relat. Mater. 8 (1999) 830.
3. P. B. Mirkarimi, K. F. McCarty, D. L. Medlin, Mater. Sci. Engin. Reports 21 (1997) 47.
4. J. M. Schneider, W. D. Sproul, A. Matthews, Surf. Coat. Technol. 98 (1998) 1473.
5. W. Schintlmeister, W. Wallgram, J. Kanz, K. Gigl, Wear 100 (1984) 153.
6. S. Vepřek, J. Vac. Sci. Technol. A 17 (1999) 2401.
7. H. Holleck, V. Schier, Surf. Coat. Technol. 76-77 (1995) 328.
8. A. A. Voevodin, J. P. O'Neill, J. S. Zabinski, Surf. Coat. Technol. 116-119 (1999) 36.
9. R. Gilmore, M. A. Baker, P. N. Gibson, W. Gissler, M. Stoiber, P. Losbichler, C. Mitterer, Surf. Coat. Technol. 108-109 (1998) 345.
10. E. Arzt, Acta. Mater. 46 (1998) 5611.
11. P. Haasen, Physical Metallurgy, Univ. Press, Cambridge, 1986.
12. D. A. Porter, K. E. Easterling, Phase Transformations in Metals and Alloys, Stanley Thornes, Cornwall, 2000.
13. I. Petrov, L. Hultman, J.-E. Sundgren, J. E. Greene, J. Vac. Sci. Technol. A 10(2) (1992) 265.
14. P. H. Mayrhofer, F. Kunc, J. Musil, C. Mitterer, Thin Solid Films 415 (2002) 151.
15. F. J. Humphreys, M. Hatherly, Recrystallization and Related Annealing Phenomena, Elsevier, Oxford, 1995.
16. L. Hultman, Vacuum 57 (2000) 1.
17. H.-D. Männling, D. S. Patil, K. Moto, M. Jílek, S. Vepřek, Surf. Coat. Technol. 146-147 (2001) 263.
18. S. Vepřek, S. Reiprich., Thin Solid Films 268 (1995) 64.
19. A. Niederhofer, P. Nesládek, H.-D. Männling, K. Moto, S. Vepřek, M. Jílek, Surf. Coat. Technol. 120-121 (1999) 173.
20. C. Mitterer, J. Sol. State Chem. 133 (1997) 279.
21. P. H. Mayrhofer, J. G. Wen, I. Petrov, C. Mitterer, J. E. Greene, Appl. Phys. Lett. submitted.
22. S. Vepřek, A. S. Argon, J. Vac. Sci. Technol. B 20(2) (2002) 650.
23. J. Patscheider, T. Zehnder, M. Diserens, Surf. Coat. Technol. 146-147 (2001) 201.
24. J. Musil, Surf. Coat. Technol. 125 (2000) 322.
25. C. Mitterer, P. Losbichler, F. Hofer, P. Warbichler, P.N. Gibson, W. Gissler, Vacuum 50(3-4) (1998) 313.
26. P. Losbichler, C. Mitterer, P. N. Gibson, W. Gissler, F. Hofer, P. Warbichler, Surf. Coat. Technol. 94-95 (1997) 297.
27. P. H. Mayrhofer, C. Mitterer, J. G. Wen, I. Petrov, J. E. Greene, in preparation.
28. P. B. Barna, M. Adamik, Thin Solid Films 317 (1998) 27.
29. I. Petrov, P. B. Barna, L. Hultman, J. E. Greene, J. Vac. Sci. Technol. A 21(5) (2003) 117.
30. P. H. Mayrhofer, H. Willmann, C. Mitterer, Thin Solid Films 440 (2003) 174.
31. C. Michaelsen, K. Barmak, T. P. Weihs, J. Phys. D: Appl. Phys. 30 (1997) 3167.
32. P. H. Mayrhofer, A. Hörling, L. Karlsson, J. Sjölén, C. Mitterer, L. Hultman, Appl. Phys. Lett. 83 (2003) 2049.
33. A. Hörling, L. Hultman, M. Odén, J. Sjölen, L. Karlsson, J. Vac. Sci. Technol. A 20 (5) (2002) 1815.
34. W.-D. Münz, J. Vac. Sci. Technol. A 4 (6) (1986) 2717.
35. S. PalDey, S. C. Deevi, Mat. Sci. Engin. A 342 (1-2) (2002) 58.
36. J. W. Cahn, Acta Met. 9 (1961) 795.
37. H.E. Kissinger, Anal. Chem., 29 (1957) 1702.
38. S. Kodambaka, V. Petrova, A. Vailionis, P. Desjardins, D. G. Cahill, I. Petrov, J. E. Greene, Thin Solid Films 392 (2001) 164.

Chapter 7
MULTILAYER HARD COATINGS FOR MACHINERY NEEDS

Yu.V. Panfilov
Moscow State Bauman Technical University, Moscow, Russia

Vacuum multi-source coaters for multilayer thin film deposition are shown. Multilayer coating manufacturing process with methods of thin film quality control are developed. Experimental results for multilayer hard coatings are presented.

Key words: multilayer, hard coatings, vacuum coaters, multi-source, ion beam assisted.

1. INTRODUCTION

An important application of thin-film deposition is strengthening of cutting and deformation tools by wear-resistant coatings. The variety of possible thin film materials and the low deposition temperature are the main advantages of PVD methods for increasing wear resistance of cutting and deformation tools. Coating efficiency is ensured by optimum composition, number of layers, and configuration. These coatings are composed of nitride, carbide, and carbonitride compounds based on complex systems of high-melting-point metals (e.g., Ti, Nb, and Cr) [1]. Coatings made of many ultra-thin layers present interest for ultra-hard material preparations. Multilayer coatings provide an opportunity to control thin films adhesion, structure, and density. Layer number and thickness are the main control parameters.

The need of multilayer hard coatings for machinery is caused by extreme field condition: high sliding speed without lubrication, high contact loads and temperatures, oil-free operations, etc. For example, these coatings can be used for wear protection of drills, single-point cutting tools, milling cutter, press tools, drawing machine tools. Multilayer coatings can be most easily obtained in vacuum multi-source coaters because there is no need to transfer substrates between vacuum chambers for deposition of next layers.

2. TWO-SOURCE COATER AND WEAR PROTECTION OF DRILLS

Vacuum ion-plasma coater consists of a vacuum chamber, pumping and gas inlet systems, substrate holder and two plasma source: DC magnetron

A. A. Voevodin et al. (eds.),
Nanostructured Thin Films and Nanodispersion Strengthened Coatings, 69–76.

sputtering and ion-beam (Fig.1). Gas inlet system includes two automatic mass flow controllers and a pressure controller. Gas flows and total pressure are displayed on a process system controller display. The main deposition parameters were: operating gas pressure of 1 – 2 Pa, magnetron target

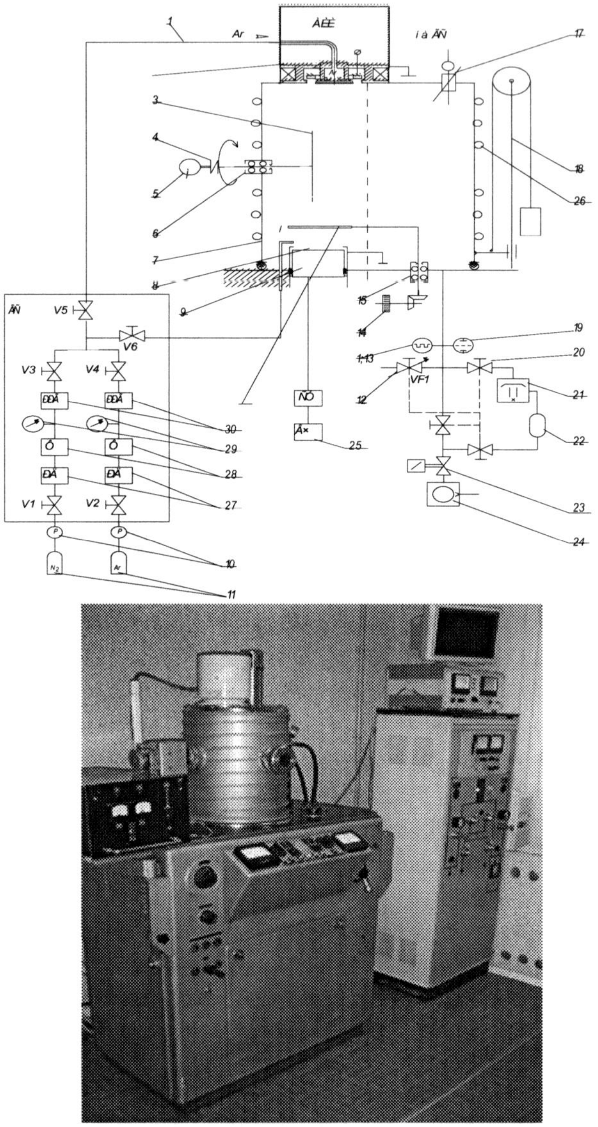

Figure 1. Two-source vacuum coater scheme and photo: 1 – gas inlet line; 2 – ion beam source; 3 – substrate holder; 4 – box joint; 5 – drive; 6, 15 – feedthrough; 7 – vacuum chamber; 8 – target; 9 – magnetron; 10 – pressure redactors; 11 – gas balloons; 12 – inlet valve; 13, 19 - manometers; 14 – drive of shield; 16 – shield; 17 – neutralize cathode; 18 – drive of vacuum chamber; 20, 23 – valves; 21, 24 - vacuum pumps; 23 – vacuum balloon; 25 – power unit; 26 – cooler.

voltage of 500 V, ion current density of 800 A/m^2, reactive gas (nitrogen) concentration of 10 – 22%, deposition time 10 – 60 min. Thickness of the coatings was 400 – 700 nm. One of the process specifics was an ion-beam treatment of substrates before coating deposition. Ion energy of 1.5 keV, residual gas pressure of 0.1 Pa and operating time of 10 – 60 min were used in ion-beam treatment.

Combinations of titanium with silicon, boron and other materials provide an opportunity to nanostructure wear-resistant coatings. Ion-plasma magnetron sputtering from a multi-component target is one method of these coatings preparation. Due to the small grain size, 2 – 5 nm in diameter, the ultra hardness (up to 40 GPa) of these coatings can be achieved [2].

Tool wear resistance was tested by drilling stainless steel samples. Four millimeter drill bits without wear-resistant coating and with a TiBSiN coating were used in the test. The TiBSiN coating increased drill lives of by about 3 – 4 times in comparison to not coated drills. A decrease in friction coefficient was also noted. This was evidenced by a change in dimensions of chip's helix (Fig. 2), comparing drill chips priduced with a not coated drill (small step of helix) and with a coated drill (large step of helix).

3. THREE-SOURCE COATER AND THIN FILM DEPOSITION VERSION

The coating of machinery parts was realized with a three-source vacuum coater (Fig. 3). The coater includes magnetron sputtering system, ion beam source, arc evaporator, substrate rotation and inclination control, substrate heating. The coater vacuum system includes a vacuum chamber with a downstream gas flow control by means of a diaphragm valve. The operation gases inlet system controls flow of three types of operating gases.

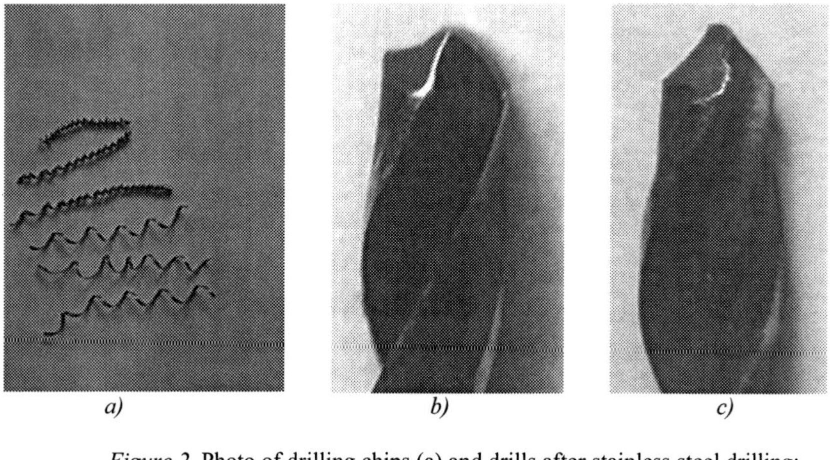

a) *b)* *c)*

Figure 2. Photo of drilling chips (a) and drills after stainless steel drilling: with TiBSiN coating (b) and without coating (c).

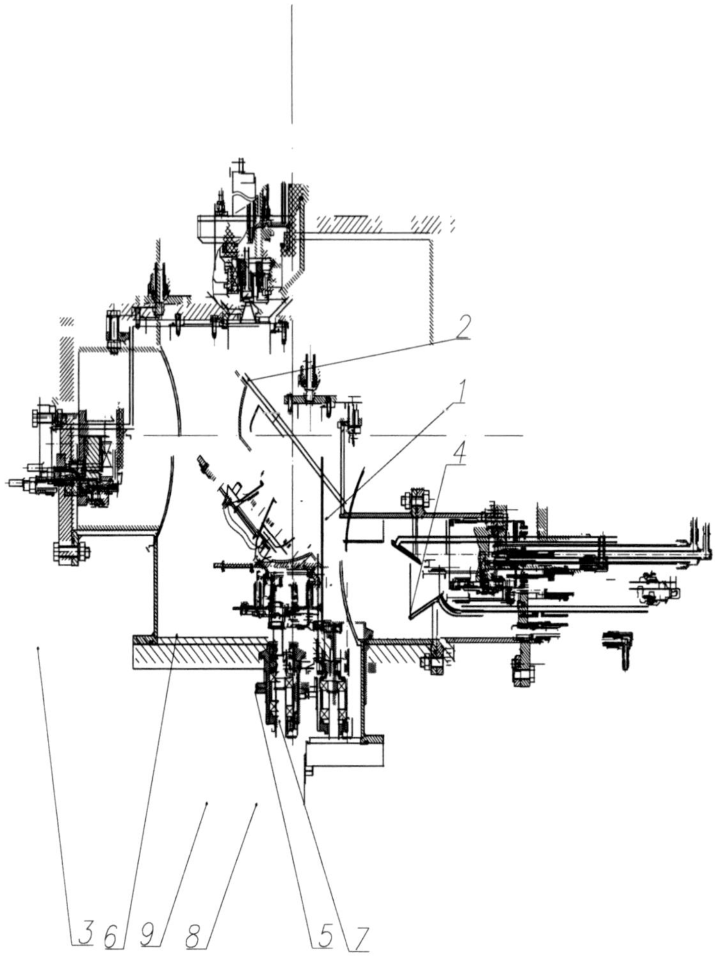

Figure 3. Three-source vacuum coater scheme: 1 – vacuum chamber, 2 – ion-beam source, 3 – DC magnetron spattering system, 4 – arc evaporator, 5 – substrate holder, 6 – substrate heater, 7 – shield, 8 – shield feedthrough, 9 - substrate holder feedthrough.

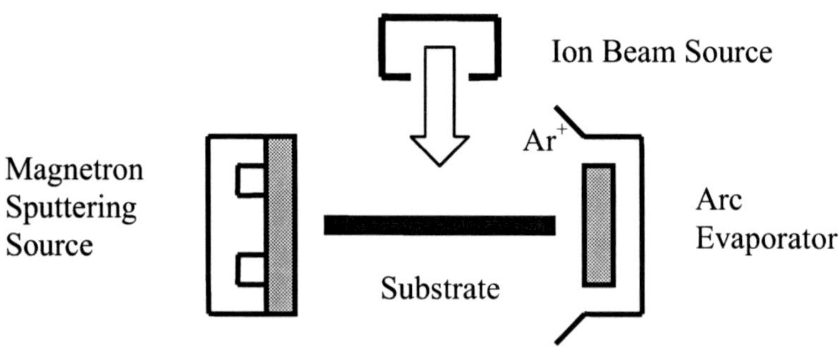

Figure 4. Substrate surface activation before thin film deposition.

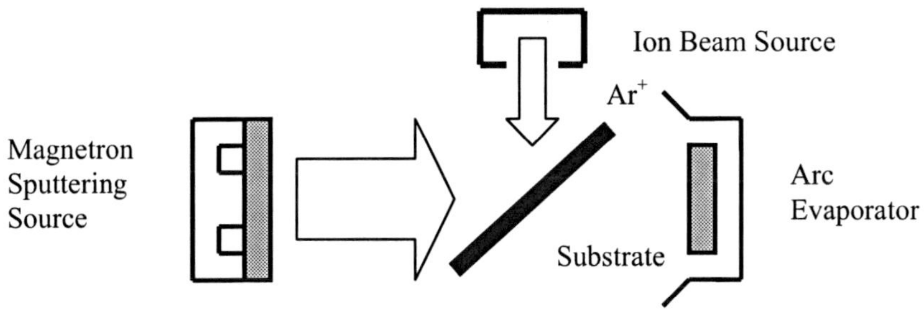

Figure 5. Ion-beam assisted deposition

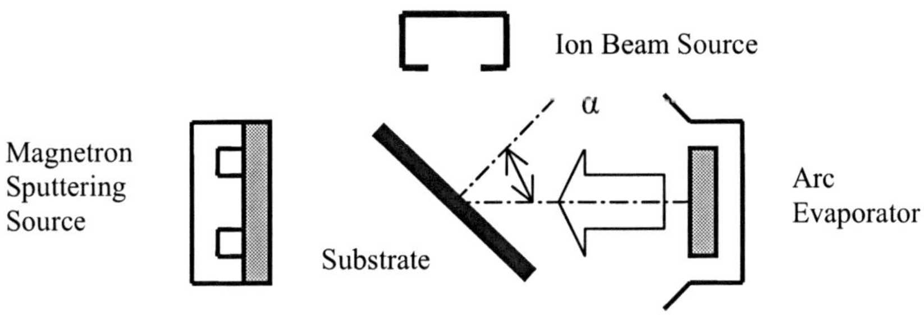

Figure 6. Thin film deposition under angle (α) between film-forming particle flow and substrate surface.

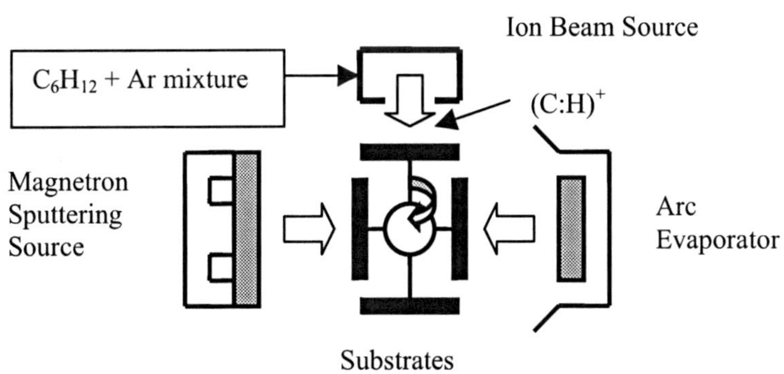

Figure 7. Multilayer thin film deposition.

A multilayer coating manufacturing process and methods for thin film quality control were developed. These methods included the following:

- control of composition, flow and energy of deposited particles;
- control of substrate surface micro-topography and temperature;
- control of composition and flow of residual and operating (e.g. reactive) gases;
- control of the ion beam assisted deposition;
- control of angle between deposited particles flow and substrate surface.

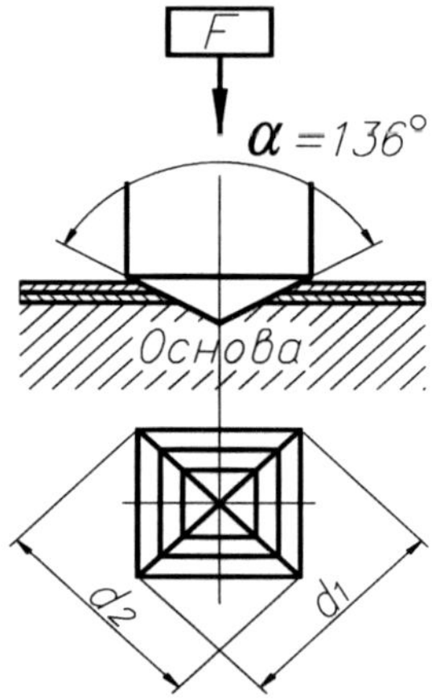

Figure 8. Multilayer thin film coating hardness measurement scheme.

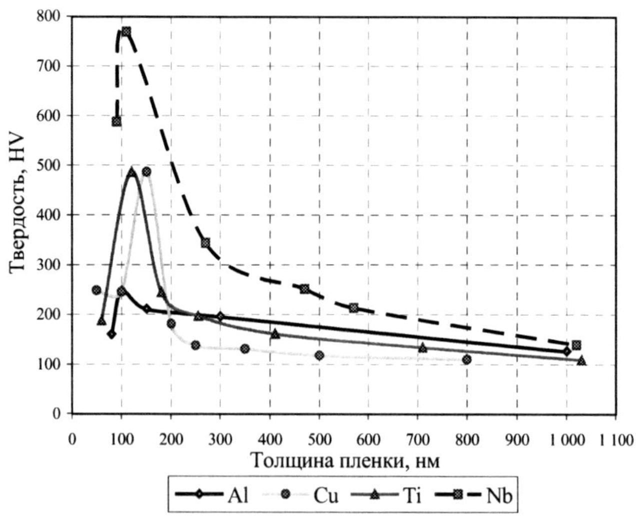

Figure 9. The dependence between coating hardness (HV) and individual layers thickness (nm).

Technological and design parameters can be used to control of thin film nanostructure in a three-source vacuum coater. The main of these parameters are the following:

- thin film deposition to substrate surface with special prepared micro-topography (e.g. epitaxy [3]) or with natural micro-topography;
- substrate surface activation before thin film deposition (Fig. 4);
- ion-beam assisted deposition (Fig. 5);
- thin film deposition with an angle between film-forming particle flow and substrate surface (Fig. 6);

multilayer thin film deposition with α-C:H (carbon diamond like material) layers (Fig. 7).

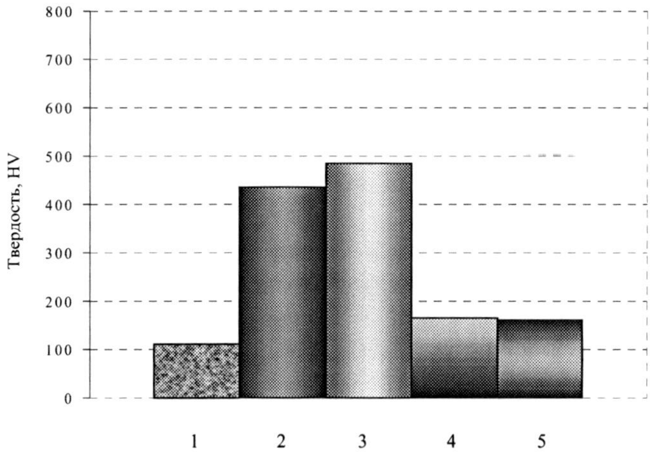

Figure 10. The influence of Ti and Cu thin film thickness on coating hardness (HV): 1 – Al substrate hardness, 2 – Ti film (120 nm) on Al substrate, 3 – Cu film (150 nm) on Al substrate, 4 – 2-layer (120 nm Ti and 150 nm Cu) coating on Al substrate, 5 – 3-layer (120 nm Ti and 150 nm Cu and 120 nm Ti) coating on Al substrate.

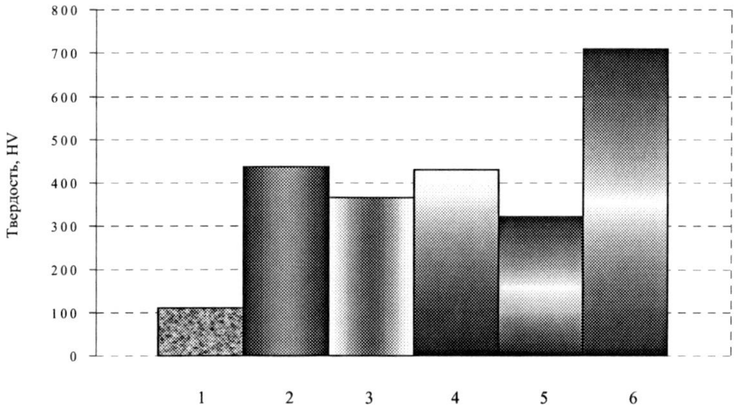

Figure 11. The influence of Ti and α-C:H thin film thickness to coating hardness (HV): 1 – Al substrate hardness, 2 – Ti film (120 nm) on Al substrate, 3 – α-C:H film (100 nm) on Al substrate, 4 – 2-layer (120 nm Ti and 100 nm α-C:H) coating on Al substrate, 5 – 3-layer (120 nm Ti and 100 nm α-C:H and 120 nm Ti) coating on Al substrate, 6 – 3-layer (35 nm Ti and 30 nm α-C:H and 35 nm Ti) coating on Al substrate.

4. EXPERIMENTAL RESULTS

A special system for hardness measurement of multilayer thin film coatings was developed (Fig.8). Al, Cu, Ti, Nb and α-C:H single-layer ultra-thin coatings and Ti/α-C:H, Ti/α-C:H/Ti, Ti/Cu, and Ti/Cu/Ti multilayer coatings were deposited on an Al substrate to demonstrate advantages of multilayer coatings (Fig. 9 - 11). The combination of soft and hard metals and diamond like carbon was used to find a dependence between coating hardness, number of layers, and thickness of individual layers.

As one can see (Fig. 9), hardness of coatings with 100 – 200 nm thin layres is by 2.5 – 7 times higher than that of coatings with about 1 μm layer thickness. Thickness of layers in a multilayer coatings was more important parameters than layer number for hardness increasing (Fig. 10 – 11).

5. CONCLUSIONS

Advanced multi-source vacuum coaters with ion beam assisted deposition were used to produce multi-component and multi-layer coatings. Experimental results compared performance of TiCN, TiBN, TiBSiN and α-C:H multilayer coatings. The TiBSiN coating provided an considerable wear life increase and a reduced friction coefficient in stainless steel drilling. An unusual effect in stainless steel drilling, where drills were worn out of their edges, was revealed when an ion beam surface activation was used before TiBSiN coating deposition. The wear resistant coatings also increased life time of drawing machine drums by above four and more times in comparison with draw drums without hard coatings.

REFERENCES

1. Manufacturing Technologies for Machines of the Future. 21st Century Technologies, Springer-Verlag Berlin Heidelberg 2003, 820 p.
2. New Materials. Moscow State Institute of Steel and Alloys, Moscow 2002, 736 p.
3. Givargizov E.I. Art epitaxy. Science, Moscow 1988, 176 p.

Chapter 8

NANOSTRUCTURED CARBON MATERIALS IN THIN FILM TECHNOLOGY

A.F. Belyanin, M.I. Samoilovich

Central Research Technological Institute "TECHNOMASH", 121108, Ivana Franko str., 4, Moscow, Russia

The present work presents the results of growth of diamond and diamond-like carbon films by plasma methods, including arc and glow discharges, ion beam sputtering of graphite, RF and DC magnetron and diode sputtering, as well as plasma RF and microwave discharges close to the electron-cyclotron resonance. Furthermore, diamond films were synthesized employing a hot filament method. The use of a variety of methods permits producing carbon films with a varying concentration of the diamond phase.

Keywords: diamond, diamond-like carbon, nanostructure.

1. INTRODUCTION

Materials referred to as diamond materials show much promise for electronics applications, and this is due to the properties of diamond itself. Amongst the known materials, diamond offers the highest (i) hardness (~10000 kg/mm^2), which is 4.5 times greater than that of corundum, (ii) thermal conductivity (2000 W/(m·K) at room temperature), which is 5 times larger than that of copper, (iii) velocity of surface and volume acoustic wave propagation (~10 km/s and ~18 km/s, respectively). It also has a high temperature resistance coefficient, displays high stability of electric and semiconductor characteristics (up to ~823 K), and has a high breakdown voltage (up to 10^7 V/cm). Individual samples of natural and synthetic diamond crystals doped with certain materials are wide-band gap semiconductors (band gap width 5.45 eV) with a high mobility of carriers and high chemical stability/tolerance against radiation (by 1-2 orders of magnitude above those of conventional semiconductor materials).

With all great promise that diamond shows for electronics, substrates for different integrated devices, made of single crystals of natural diamonds, may be economically rational with sizes of no more than 5×5 mm. Synthetic diamond crystals grown at high temperatures and pressures must have approximately the same size when used as substrates. Diamond is currently available for wide application in electronics, since by using commercial equipment, it can be grown as films or plates, up to 1 millimeter thick, and

A. A. Voevodin et al. (eds.),
Nanostructured Thin Films and Nanodispersion Strengthened Coatings, 77–90.

its growth space is limited by dimensions of a reaction chamber only. It is expected that the application of diamond and diamond-like carbon films in electronics will permit building radiation resistant semiconductor devices, high-temperature electronic devices, integrated circuits, few sub-microns in size and ultra-high density elements, and ultra-high speed integrated circuits.

2. RESULTS AND DISCUSSION

2.1. Structure and Composition of Polycrystalline Diamond Films Grown by the Arc Discharge Method

Changing deposition parameters makes it possible to change diamond phase concentration in the carbon films over wide ranges (from 0 for diamond-like carbon films to ~99 vol.%). Thus, produced X-ray amorphous films are not just diamond-like carbon, but rather represent a mixture of diamond-like carbon and nanocrystalline diamond, where crystallite regions of X-radiation coherent scattering (RCS) are <1 nm in size. Microhardness of the obtained films is up to 9500 kg/mm^2, thermal conductivity at 303 K is 400-600 W/(m·K), and specific electrical resistance $\rho \approx 10^{13}$ Ohm·m.

Diffraction patterns (graphite monochromator, CuKα-radiation) and electron-diffraction patterns taken from the synthesised films displayed (111), (220), (311) and, more rarely, (400) reflections corresponding to the structure of cubic diamond (space group Fd3m, lattice spacing a = 0.3567 nm) (Fig. 1,a-c). The formation of diamond films on polycrystalline W substrates is always accompanied by formation of tungsten carbides (WC — hexagonal syngony; a = 0.2902 nm, c = 0.2838 nm; W_2C — rhombic syngony; Pbcn; a = 0.4727 nm, b = 0.5990 nm, c = 0.5179 nm) (Fig. 1,b). With Si substrates, the presence of a reflection at the diffraction angle 2θ = 35.6° (interplanar distance d/n = 0.252 nm) indicates formation of a silicon carbide interlayer (α-SiC, hexagonal syngony, $P6_3mc$, a = 0.3073 nm, c = 1.508 nm) in the diamond film-substrate interface area.

Polycrystalline diamond films grown by the arc discharge method revealed no other known carbon phases [1, 2].

A diamond film forms in the presence of atomic hydrogen required for interaction with non-diamond carbon forms, developing on a substrate, and their removal from the growing film surface [3]. The concentration of atomic hydrogen depends on the performance of a gas phase activator responsible for molecular hydrogen dissociation. The rate of formation of a diamond phase dependeds on the gas mixture pressure, methane concentration, crystallization and gas phase activation temperatures, as well as on the availability and concentration of seed diamond microparticles. When seed diamond particles are introduced (Fig. 2,a), the film crystallizes precisely on

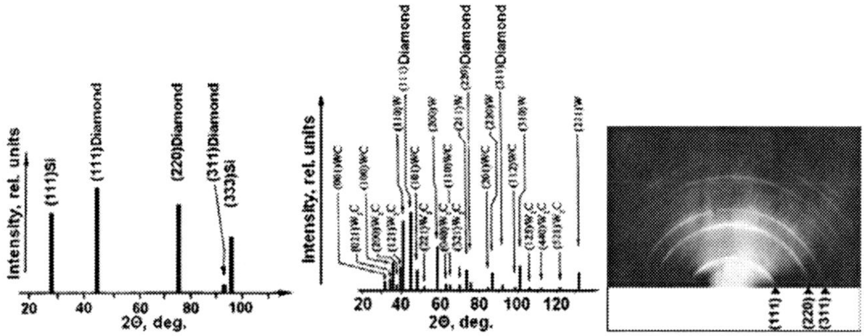

a) b) c)

Figure 1. Diffraction from diamond films grown on {111}-oriented Si substrates and those from polycrystalline W: a,b X-ray diffraction patterns (Cukα-radiation); c) electron-diffraction pattern for reflection (Si/diamond).

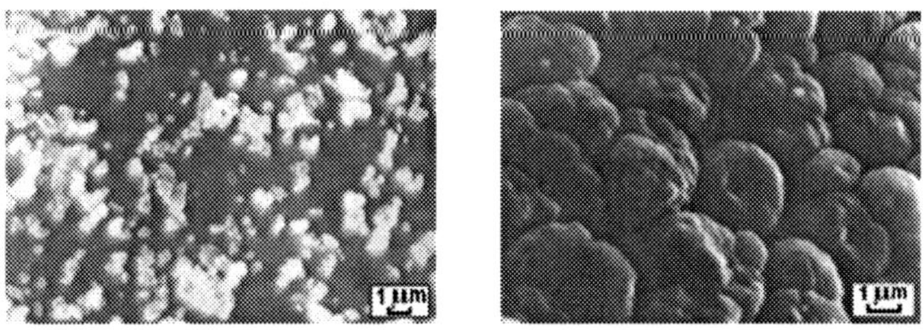

a) b)

Figure 2. a).Si substrate surface coated with seed diamond microparticles. A diamond film was built up for a few minutes. b) Surface of a polycrystalline diamond film in contact with a W substrate coated with seed diamond microparticles (substrate is open).

them, as evidenced by the structure of a diamond film surface at the film-substrate interface (Fig. 2,b).

The temperature dependence of the growth surface morphology and the structure of cross-cut cleavages of diamond layers were investigated with a TESLA BS-350 scanning electron microscope. The sequence of stages involving the growth of films was found to be as follows: I – globular stage; II – formation of {100} facets on globules; III – geometric selection, primary <100> axial texture; IV – secondary <110> and <111> conical textures. The sequence of stages is the same for different temperatures and gas phase activation methods. The stages show differences in development depending on the crystallisation temperature. For example, in the low-temperature region (up to ~1100 K), growth is complete in the globular stage, while in the high-temperature region (>1300 K), the globular stage and the primary <100> axial texture are quickly brought to completion. The formation of diamond as hemispheres (Fig. 3, globular growth stage, temperatures of up to ~1130 K) is distinguished by the small rate (<2 µm/h) and is possible over the entire temperature range of formation of diamond layers (873-1473 K). The thickness of a globular-structured diamond phase may range from a few tens of nanometres to a few tens of microns depending on the production conditions. The globular-structured diamond films are characterized by a high concentration of an X-ray amorphous phase (up to 100 vol.%). The presence of X-ray amorphous phase is typical for the all obtained diamond films. Notably, this phase decreases in abundance with increasing substrate temperature.

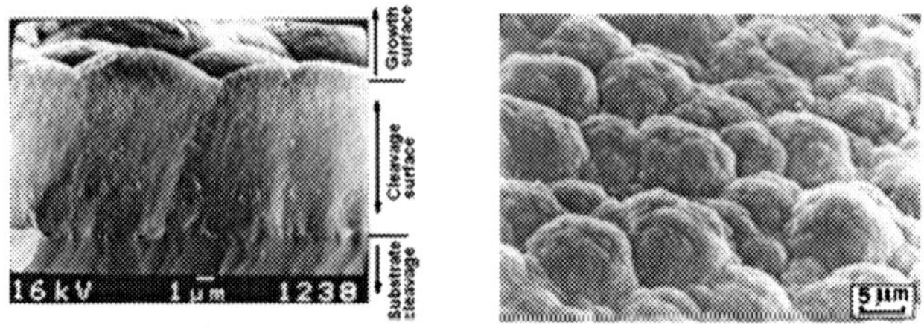

a) *b)*

Figure 3. Structure of the cross-cut cleavage (a) and morphology of the diamond film growth surface (b) at the globular growth stage.

On the globular surface, as the layer thickness increases, a branchy system of reentering angles is formed along the hemisphere boundaries. Such system provides an increase of the rate of formation under the same conditions, as manifested by the coarse structures of globules that are covered with multiple {100} facets (Fig. 4,a,b, temperatures between 1130 and 1180 K).

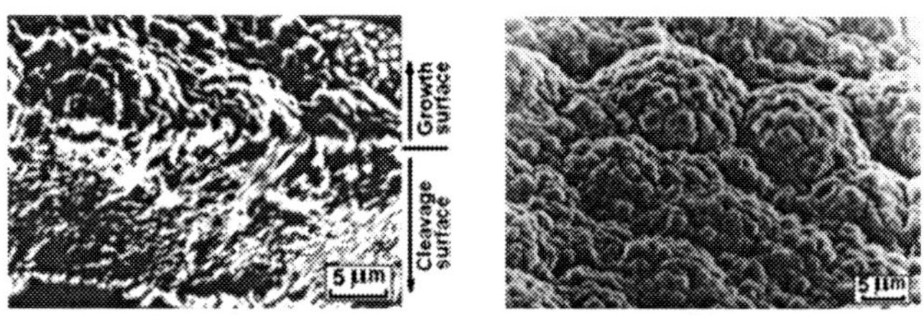

a) *b)*

Figure 4. Structure of the cross-cut cleavage (a) and morphology of the growth surface (b) of diamond films at the stage of development of {100} facets on globules.

The occurrence of {100} plates is a consequence of crystallisation of diamond films by a non-tangential growth mechanism under the conditions of loss of morphological stability in part or in whole (according to Hartman [4], this facet is forbidden in diamond for tangential growth). The formation of multiple minute {100} facets on spherical surfaces inevitably results in the manifestation of the Gross-Moller principle – natural selection and subsequent coarsening of crystallites whose {100} facets are almost parallel to the film surface (Fig. 5,b,c, temperatures between ~1200 and 1300 K). This also provides an explanation for a noticeable texturing of crystallites at subsequent stages of the film growth.

a) *b)* *c)*

Figure 5. Structure of the cross-cut cleavage (a) and morphology of the growth surface (b) of diamond films at the stage of geometric selection and of the primary <100> axial texture. c). Schematic representation of geometric selection during the growth of a film crystalline phase.

With a layer thickness to be dictated by deposition conditions, the formation of a <100> axial texture comes to a close. In this case, numerous reentering angles are formed due to twinning on the {100} surface of plates in order to maintain a high growth rate (Fig. 6,a, temperatures between 1300-1470 K). The twinning on {100} facets under the spinel law (Fig. 6,b) makes it possible to account for the formation of a secondary <110> conical texture on the axial <100> texture by multiple (fourfold) twinning on each {100} facets (Fig. 6,c). Studies of X-ray diffraction patterns of a (220) peak at oscillation revealed formation of a conical <110> texture (semi-apex angle 17.4°) (Fig. 6,d). The $\{110\}_{II}$ plane blunting the $<100>_{II}$ edge of twinned cube II makes an angle of 19.5° with the parent facet of cube I (Fig. 6,b). The angle is almost equal to a semi-apex angle of the <110> conical texture. A second multiple twinning results in the development of a relief, formation of a conical <111> texture, and a box-shaped growth form (Fig. 7, temperature range >1470 K).

The intensity of diamond peaks in X-ray patterns (Fig. 1,a-b and Fig. 8,a) provides no information on the crystalline phase concentration and texture type because of the formation of conical textures, as revealed from X-ray patterns of oscillation (Fig. 6,e). Electron-diffraction patterns and X-ray patterns taken from the grown diamond films show the presence of a texture. Note that the type of the texture under formation was independent of the material and crystallographic orientation of a non-diamond substrate. The texture type depends on the formation conditions and thickness of the grown film.

Increase in the thickness of a diamond film was found to be related to the sequence of development of conical <110> and <111> textures of diamond grown at a substrate temperature of 1323 K. The dependence of the intensities of (220) and (111) peaks (I_{220} and I_{111}) on the film thickness, as constructed from data on diffraction patterns for oscillation (Fig. 8,b,c), shows that the relationship I_{220}/I_{111} remains intact at film thicknesses of 60-200 µm (Fig. 8,d).The numbers of curves (see Fig. 8,a,b), from which the

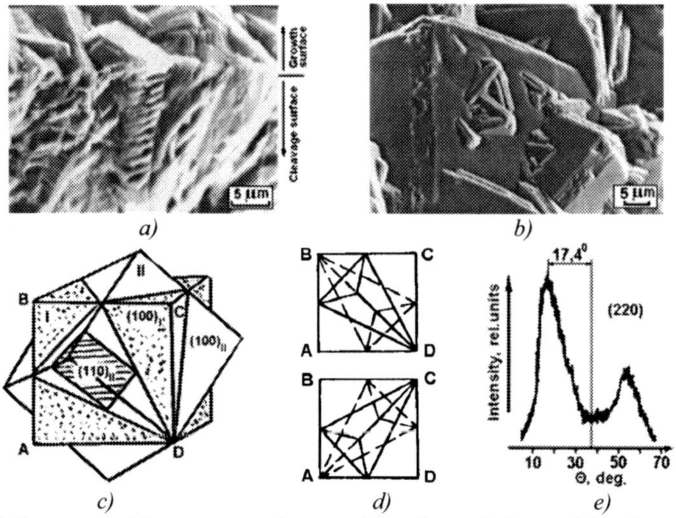

Figure 6. Structure of the cross-cut cleavage (a) and morphology of the diamond layer surface (b) at the stage of formation of conical <110> and <111> textures. c, d) The mechanism of twinning under the spinal law. e) X-ray diffraction pattern of oscillation of a (220) reflection of a 198-µm-thick diamond film.

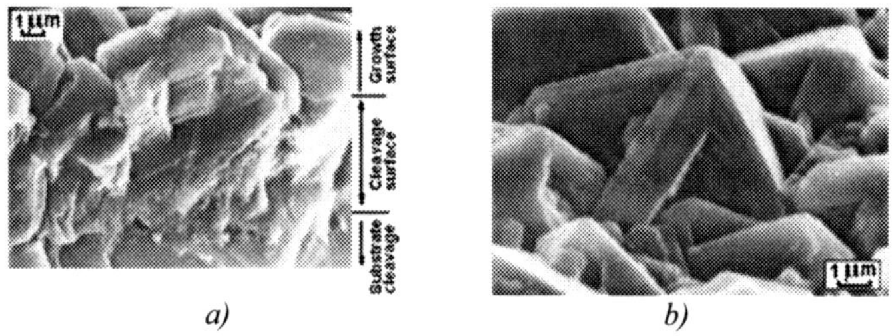

Figure 7. Structure of the cross-cut cleavage (a) and morphology of the diamond film growth surface (b) at the stage of multiple twinning.

dependence of I_{220}/I_{111} on the diamond film thickness was constructed, refer to the following thicknesses: 9, 25, 42, 50, 82, and 198 µm. Thus, at a diamond film thickness >60 µm, two conical textures may be observed at once: a <111> texture (cone semi-apex angle 8.3°) and a <110> one (17.4°). Figure 8,e is a [110] stereographic projection constructed for purposes of identifying the direction of an axial texture as a result of conical <110> and <111> textures.

From these data, it may be concluded that after reaching the stage of mass twinning, grains became oriented to directions close to <553> and <221> perpendicular to the substrate plane (Fig. 8,e). The grains observed on diamond film cleavages are plate-like in structure, regardless of the stages of their formation (formation of facets on globules and mass twinning) (Fig. 9). The RCS size for diamond films, which was calculated from X-ray patterns, fits with the thickness of grain-forming plates observed through an electron microscope.

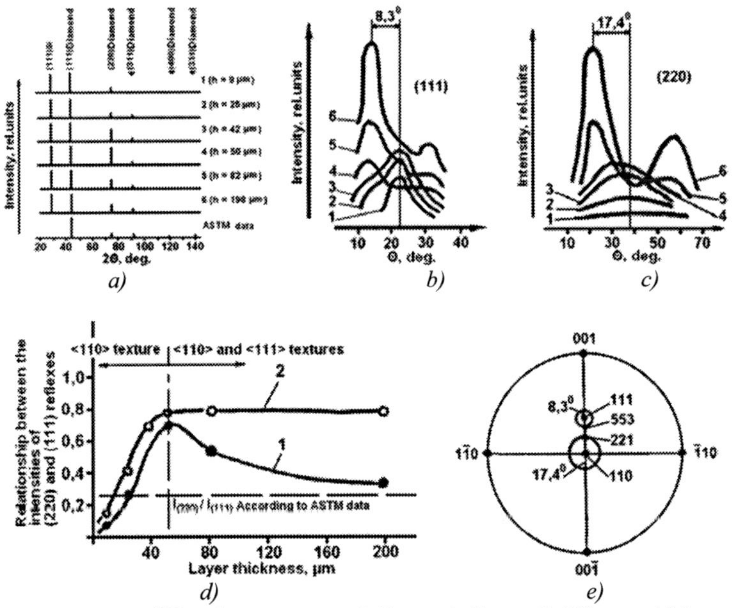

Figure 8. a) X-ray diffraction patterns of diamond films of different thickness. b,c) Diffraction patterns for oscillation of (220) and (111) reflections of diamond films (curve numbers correspond to film thicknesses 9, 25, 42, 50, 82 and 198 μm).d) Relationship between intensities of (220) and (111) reflexes measured from diffraction patterns (curve 1) and from diffraction patterns of oscillation (curve 2).e) Stereographic projection of the conical textures under formation.

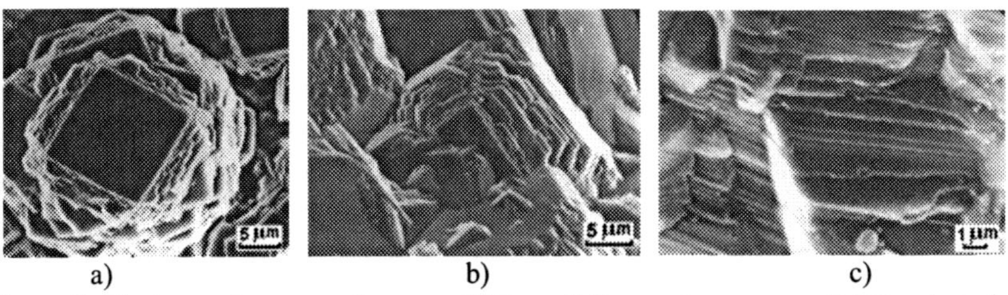

Figure 9. Electron microscope images of films of diamond grown on {111}-oriented Si substrates: a,b) surface at the stages of globular growth and twinning; c) surface of the film cleavage at the stage of twinning.

The lattice spacing of a diamond film grown on a substrate from polycrystalline W was measured from the (220) diffraction. It depended on the film thickness and conditions of its growth and ranged from 0.3566 to 0.3571 nm. The lack of splitting of diamond diffraction doublets indicates a mechanical stress of the grown films. The stress builds up with increase in film thickness, as evidenced by the on-going increase in the lattice spacing. The grain size visible on the growth surface and measured with a scanning electron microscope increases as the film thickness increases (Fig. 10). The RCS somewhat decreases in size as the film thickness increases to become constant thereafter. For diamond films grown at 1323 K, the RCS size becomes invariant at a thickness > 80 μm (Fig. 10).

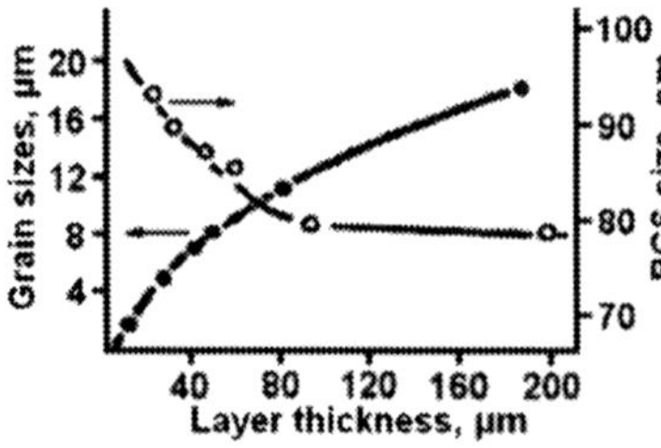

Figure 10. Dependencies of grain sizes and regions of X-radiation coherent scattering on the diamond film thickness.

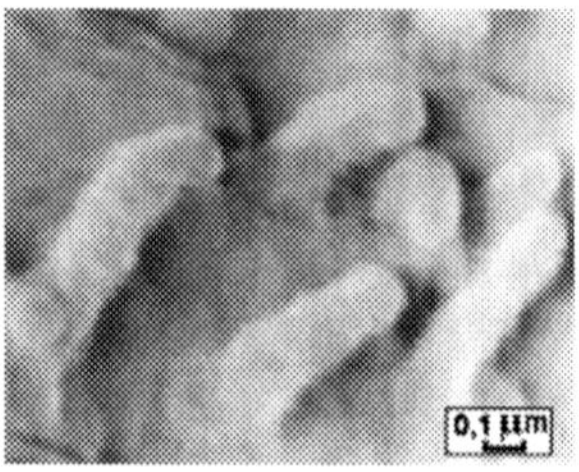

Figure 11. Electron microscope images of filamentary diamond crystals grown on the polycrystalline diamond film surface in contact with a W substrate (substrate is open).

The concentration of the broken bonds C-C (C_{bb}) measured by electron paramagnetic rcsonance (PE-1306 spectrometer, f = 9 GHz) as a function of film thickness and crystallisation temperature of diamond grown on Si and W substrates has no strong tendency to change [5]. It may be asserted that there are no broken bonds along the boundaries of axially disoriented RCS. Similarly, no such bonds are available on the surface of crystalline phase grains in contact with diamond-like carbon filling an intergranular area. Obviously, the change of C_{bb} is due to the change in the concentration and structure of diamond-like carbon in diamond films, not to the RCS size and density of dislocations, as it was believed before [5].

In some cases, filamentary diamond crystals, 300-400 nm in diameter and up to 2 µm in length, (Fig. 11) grow on the surface of a diamond film in contact with a substrate. Similar growths were observed when diamond was grown in a container by the LPSSS method (low pressure solid state source process) [6]. By varying the size and shape of voids on the substrate working surface, one can make up crystalline and non-crystalline ordered structures of filamentary diamond crystals. The latter is an urgent problem because it makes possible to extend applications for classical methods and expand the use of nanoelectronics, a field of electronics little used in practice. The solution to this proble is to combine physical studies and technological developments in order to develop an advanced technique for growing structures of required composition, size and density on real substrates. This will enable a material contribution to the development of nanoelectronics and emission electronics.

2.2. Structure of Diamond Films Produced in a Glow Discharge

Films produced by the glow discharge method were X-ray amorphous (RCS size <1 nm). Films discussed in the present study and produced by the glow discharge method were synthesised explicitly for emission electronics devices; therefore, they were synthesised under the conditions constraining the process at the stage of the globular growth. Electron microscopy

(HITACHI S-405 A scanning electron microscope) indicates that films have a smooth surface (Fig. 12,a; height difference across the layer relief $R_Z \leq 0{,}05$ µm). The growing film may contain embedments of carbon microparticles (Fig. 12,b) depending on the synthesis conditions.

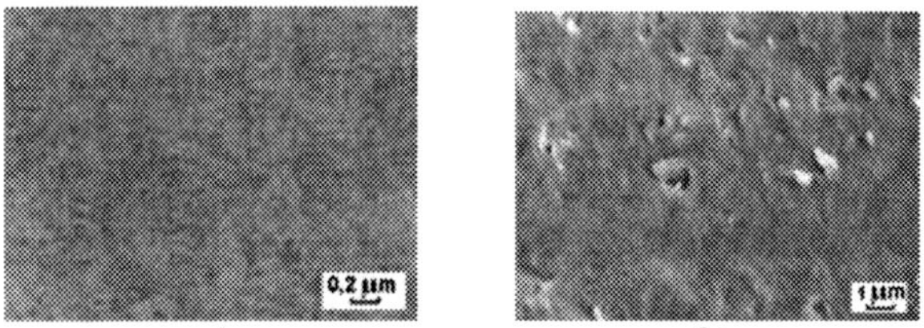

a) *b)*

Figure 12. Surface of diamond-like carbon films synthesised in a glow discharge.

2.3. Structure and Composition of Diamond Films Produced by Hot Filament Procedure

Diamond films produced by hot filament procedure were polycrystalline, their growth surface was granular in structure (visible grain sizes ~10 µm) with a grain cutting along {100} and {111} characteristic of diamond (Fig. 13,a). At phase formation conditions like those used in the paper, the morphology of film surfaces was the same at all thicknesses. X-ray diffraction patterns of the films display (111), (220) and (311) diffractions of diamond. Diamond layers were axially <111> and <110> textured.

The relation between peak intensities in Raman scattering spectra at 1332 cm^{-1} and ~1530 cm^{-1} shows that the degree of crystallinity of carbon films grown by hot filament method is >99 vol.% (Fig. 13,b, U-1000 spectrometer). Presented in the insert of Figure 13,b is a diamond peak shape obtained when working at a high resolution. It is found that at temperatures between 800 and 1300 K, the concentration of carbon amorphous phases decreases with increasing temperature.

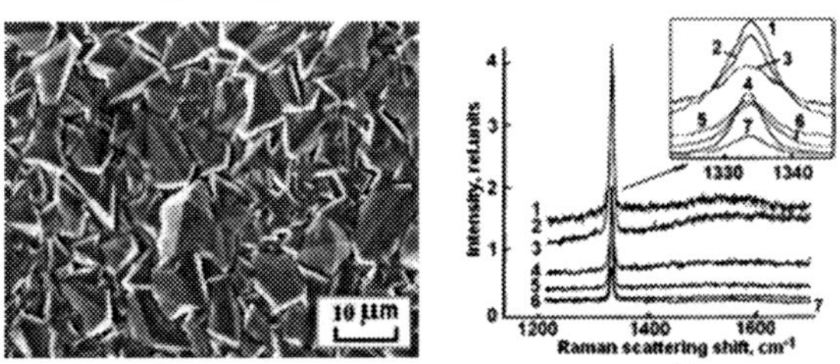

a) *b)*

Figure 13. a) Morphology of the growth surface of the polycrystalline diamond film synthesized by hot filament procedure at the crystallisation temperature T = 873 K (film thickness h = 22,4 µm, growth rate 0,83 µm/h). b) Raman scattering spectra for carbon films grown by hot filament procedure at different temperatures: 1 – T = 1073 K, h = 10,9 µm; 2 – T = 1198 K, h = 15,1 µm; 3 – T = 1183 K, h = 33,7 µm; 4 – T = 1123 K, h = 8,9 µm ; 5 – T = 1108 K, h = 6,4 µm; 6 – T = 1173 K, h = 14,9 µm; 7 – T = 1273 K, h = 5,5 µm.

2.4. Structure, Composition and Properties of Diamond-Like Carbon Films Produced by Plasma RF Discharge Close to the Electron-Cyclotron Resonance

Diamond-like carbon films produced by plasma RF discharge close to the electron-cyclotron resonance were grown on {100}-oriented Si substrates under a variety of formation conditions [7]. The growth surface of diamond-like carbon films, as revealed by means of an electron microscope, consists of protuberances with the size of 40-70 nm. Film surface and cleavages show no cutting elements. An arithmetic mean deviation (R_a) of the growth surface profile for 0.8-μm-thick diamond-like carbon films, as measured from profile images (Fig. 14,a; ALPHA-STEP 200 profilograph-profilometer), does not exceed 5 nm in the 400 μm measurement portion of a 76-mm-diameter Si substrate.

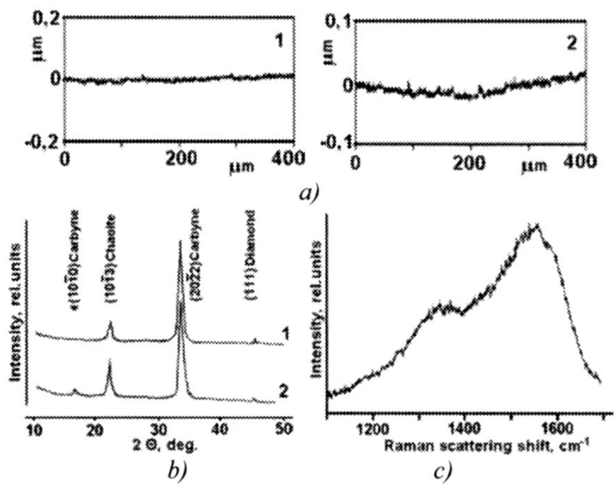

Figure 14. Structure of diamond-like carbon films synthesised on Si substrates via plasma RF discharge close to the electron-cyclotron resonance: a) Profile patterns of the 0,8-μm-thick film growth surface (measurements are made in a perpendicular direction – curves 1 and 2). б) X-ray diffraction patterns (Cukα radiation) of 1-μm-thick films (2 – diamond-like carbon film is doped with zinc). c)Raman scattering spectra for light.

Phase composition and structure of crystalline phases were identified using X-ray diffraction patterns obtained from ~1-μm-thick films. Diffraction patterns (Cukα-radiation) taken from all films displayed diffraction maxima only in the region of the Bragg angles 2θ = 21.85-22.18°, 33.00-33.61° and 44.83-45.27° (Fig. 14,b) matched by the interplanar distances d/n = 0.4066-0.4008 nm, 0.2714-0.2666 nm and 0.2022-0.2003 nm and fitting $(10\bar{1}3)$, $(10\bar{1}1)$ and $(20\bar{2}2)$ diffractions of chaoite, SiC and β-carbyne (superposition of diffractions), respectively, as well as (111)-oriented diamond. Carbyne represents a linear modification to carbon on the basis of *sp* hybridisation of carbon atoms and offers semiconductor properties (band gap width ~1 eV). The SiC layer forms at the substrate-film interface.

Inversion voltamperometry (CBA-1BM voltamperometry system) revealed the presence of Zn in diamond-like carbon films. Attempts to determine the total amount of Zn and its distribution between phases in the film failed because of a lack of a procedure for a complete controllable solution of diamond-like carbon films and selective solution of their individual phases. X-ray patterns of diamond-like carbon films, obtained from Zn-doped samples, also showed a diffraction in the region of $2\theta = 6.50$-16.54°, which fits a $(10\bar{1}0)$ β-carbyne reflection. X-ray diffraction patterns of these samples are characterized by asymmetrical diffuse images of diffraction maxima for $(10\bar{1}3)$ chaoite and $(10\bar{1}0)$ β-carbyne, directed towards larger 2θ angles. The asymmetry of peaks may be due to the formation of substitutional defects Zn_C in crystalline phases of carbon, since the tetrahedral covalent radius of Zn (0.131 nm) is bigger than that of carbon (0.077 nm). The lack of other diffractions in the all films is due to a low concentration of crystalline phases throughout the films and the occurrence of textured grains of crystalline phases. The lattice spacing of diamond incorporated into the film in the amount of up to 3 at% is 0.3469-0.3502 nm. The measured size of the X-ray RCS for crystalline phases formed in films deposited at 400-600 K is: diamond, 4.6-6.4 nm; chaoite, 7.3-12.3 nm; β-carbyne, 9.9-13.7 nm. For other growth conditions, films are are all-X-ray amorphous (RCS size <1 nm). Annealing in air (673 K, 2 h) does not result in any visible change in the X-ray patterns.

A Raman scattering spectrum (S-3000 spectrometer) is common of diamond-like carbon films: its has a maximum near 1580 cm^{-1} and a wide band at 1350 cm^{-1} (Fig. 14,c). From the Raman spectra, it was found that the size of RCS for films is <4 nm.

Thus, diamond-like carbon films produced by plasma RF discharge close to the electron-cyclotron resonance represent a mixture of phases: an X-ray amorphous phase of diamond-like carbon and crystalline phases of diamond, chaoite and β-carbyne. The degree of crystallinity (concentration of crystalline phases throughout the film) reaches 10 vol.%. Some physico-chemical characteristics of diamond-like carbon films are given in Table 1.

2.5. Structure of Diamond-Like Carbon Films Produced by Magnetron Sputtering

Films produced by magnetron sputtering were X-ray amorphous and provided a low hardness. Films that are the closest to diamond in properties were grown in the gas mixture $Ar + H_2$ at an Ar concentration of 5-20 vol.%. With an Ar concentration in the gas mixture >10 vol.%, a graphite phase was produced in films deposited by magnetron sputtering. The specific electrical resistance of films produced over a wide range of parameters was 10^9 Ohm·cm, and their thermal conductivity was ~300 W/(m·K). These characteristics allow diamond-like carbon films formed by magnetron sputtering to be used as insulating and heat-removing layers in fabrication of multichip modules.

Table 1. Properties of diamond-like carbon films produced by plasma RF discharge close to the electron-cyclotron resonance.

Parameter	Value
Microhardness	3000 kg/mm^2
Stability to exposure to aqueous solutions	
(50%) H_3PO_4	>15 min
(10%) KOH	>20 min
Content of diamond phase	≤3 at%
Band gap width	4-5 eV
Electric strength	2,5·10^6 V/cm
Specific surface resistance	>100 Ohm/□
Index of refraction	2,4-2,8
Thermal conductivity	≥200 W/(m·K)

2.6. Structure and Properties of Diamond-Like Carbon Films Produced in a Diode RF Discharge

These films were X-ray amorphous. The relative concentration of diamond type (sp^3) interatomic bonds was 90%; the refractive index was between 1.8-2.6, as measured from Raman scattering data. X-ray phase analysis of diamond-like carbon films grown from an oxygen-containing gas mixture revealed the presence of 13-55 vol.% carbyne in such films. When Ar was used and an electric displacement was passed to a cathode, SiC was always present there. At deposition flux energies of ≥200 eV, diamond-like carbon films with thicknesses >2 μm become opaque.

Diamond-like carbon films with thicknesses >0.2 μm, deposited under steady-state conditions, often develop considerable internal mechanical stresses, which may result in their spontaneous separation. In order to reduce mechanical stresses and improve adhesion, transitional (buffer) carbon layers of few nanometres thickness, were produced at the interface. Such layers compensated for a mismatch of structural and physico-chemical characteristics of film and substrate. Intermediate carbon layers were created by changing concentrations of the gas mixture components.

Mechanical properties of near-surface layers of glass and Si substrates, as well as the same substrates coated with diamond-like carbon films, were investigated by microindentation and microscratching (PMT-3 microhardness tester). The dependence of microhardness on the diamond-like carbon film thickness, as well as the dependence of the crack size on the load value, are given in Fig. 15.

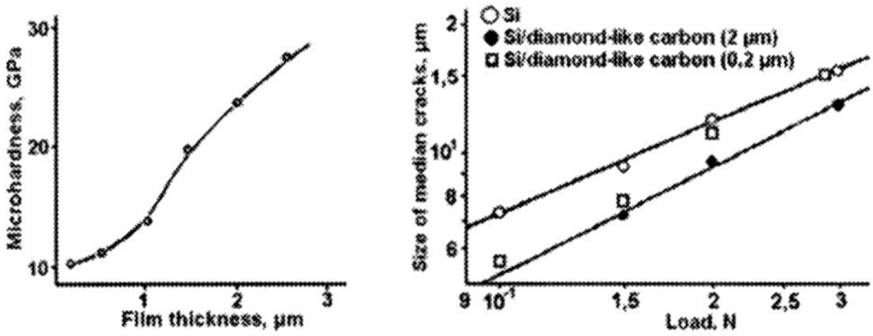

a) *b)*

Figure 15. a) Microindentation hardness as a function of thickness of the diamond-like carbon film formed on a Si substrate under the following conditions: substrate bias 250 V, O_2 concentration 10%, discharge power 300 W. b) Size of median cracks developing in indentation as a function of the indentation load. The cracks are obtained in testing Si single crystals.

Microhardness tests show that gas composition and deposition energy have a dramatic impact on mechanical properties of interface layers.

The presence of diamond-like carbon films on the surface of glass substrates has a pronounced plasticizing effect on the substrate/film system, greatly improving a gradient of mechanical properties across the system. Such a gradient is commonly observed in quartz glass samples on their tests in air. A conversion from glass plasticity to fragility a microsclerometric cocntact tests of glasses coated with micron-and submicron-thick diamond-like carbon films shifts toward lower loads [8]. The presence of a micro and sub-micrometer thick diamond-like carbon films on the surface of a single-crystal Si substrate causes the near-surface layers to harden considerably. This considerably increases limits of elasticity and viscosity of composite system failure under the contact. The adhesive strength of diamond-like carbon coatings applied to silicon single crystals using this technology surpasses the cohesive strength of both the film material and near-surface layers of the single crystals under discussion [9].

2.7. Structure of Diamond-Like Carbon Films Produced by Ion Bean Sputtering of a Graphite Target

Carbon films having a different concentration of a diamond phase were investigated. Films were formed by ion beam sputtering of a graphite target and contained up to 10 vol.% diamond phase.

At certain conditions of an exposure to an electron beam, carbyne crystallized in carbon films. At some sputtering conditions, carbon films were formed, which contain no diamond phase. Such films had a smooth surface ($R_Z < 0.04$ µm). Film diffraction spectra obtained at other parameters of the growth process showed a change in the relative line intensities, width and spectral position. This indicated a change in the phase composition, size of RCS, and extent to which the films become amorphous.

Specially designed equipment and synthesis conditions allowed carbon films growth by sputtering. These films had a fairly high degree of crystallinity (content of a crystalline diamond phase throughout the film) and no impurity atoms. The appearance of non-controllable impurities in the growing films was prevented because ion source elements were fabricated from special materials. By controlling sputtering parameters and conditions of film processing by ion or electron beams, one can control their composition and structure [10].

3. CONCLUSION

The influence of the carbon film synthesis method on film structure is considered. The following methods were described: diamond and diamond-like film deposition from hydrogen-hydrocarbon gas mixtures activated by arc and glow discharges, heated filament, and plasma methods – ionic beam sputtering, RF and DC magnetron and diode sputtering, RF and UHF discharge close to electron-cyclotron resonance. Polycrystalline diamond films with diamond phase content >99 vol. % were obtained by arc discharge and heated filament methods. Films with the various concentration of diamond, carbyne, chaoite and diamond-like carbon were formed by application of plasma RF and UHF discharge close to electron-cyclotron resonance, and also by graphite target sputtering and by ion beam.

REFERENCES

1. ASTM, International Centre for Diffraction Data, 1977.
2. Diamond. Handbook. Ed. V.V. Novikov, Kiev, "Naukova Dumka" Press (1981) 76.
3. L.L. Builov, A.E. Alekseyenko et al., Dokl. Akad. Nauk SSSR 287 (1986) 888 (in Russian).
4. P. Hartman, Crystal Growth., Moscow, "Nauka" Press V, VII (1967) 8 (in Russian).
5. T.A. Karpukhina, A.G. Cherednichenko et al., Communication Facilities Technique, Ser. TPO 4 (1991) 40 (in Rusisan).
6. R. Roy, Advanced in New Diamond Science and Technology, MYU Tokyo (1994) 17.
7. S.M. Naida, Abstract of the Ph.D. Thesis, Moscow 1999, Central Research Technological Institute TECHNOMASH (in Russian).
8. V.I. Savenko, A.F. Belyanin, B.N. Perevozchikov, Physics and Chemistry of Material Processing 5 (1997) 74 (in Russian).
9. V.I. Savenko, A.F. Belyanin, B.N. Perevozchikov, Physics and Chemistry of Material Processing 2 (1997) 59 (in Russian).
10. I.A. Semenova, Abstract of the Ph.D. Thesis, Moscow 2002, Central Research Technological Institute TECHNOMASH (in Russian).

Chapter 9

NANOCOMPOSITE COATING SYSTEMS TAILORED FOR SPECIFIC ENGINEERING APPLICATIONS

D. Zhong[1], K.H. Kim[2], I.-W. Park[2], T. Dennin[1], B. Mishra[1], E. Levashov[3], J.J. Moore[1]

[1] *Advanced Coatings & Surface Engineering Laboratory (ACSEL), Colorado School of Mines, 1500 Illinois St, Golden, CO 80401, USA.*

[2] *National Hard Coatings Laboratory, Pusan national University, San 30, Jangjeon-Dong, Keumjung-Ku, Busan 609-735, South Korea.*

[3] *Center for SHS, Moscow State Institute of Steels and Alloys, Moscow, Russia.*

Nanostructured PVD coatings in the Cr-N, Ni-Al-N, Ti-B-C-N, and Ti-Al-Si-N systems exhibit superior properties and have been tailored for specific engineering applications. These examples demonstrate that ion bombardment (ion energies and relative abundances of plasma species) of the substrate controls the relative amount of constituent phases of a nanocomposite and its crystallite size, and determines the microstructure (including crystal structure, interfacial structure, and grain structure) evolution during film growth. Through engineering the composition and microstructure, the desired relative density, mechanical, physical and chemical properties of the resulting film were obtained.

Keywords: nanocomposite, nanostructure, thin film, functionally graded, Ni-Al-N, Ti-B-C-N, Ti-Al-Si-N, Cr-N.

1. INTRODUCTION

Nanocomposite coating is a most promising novel concept recently developed in the coating community. There exists an increasing amount of on-going research focused on the selection of the constituent phases of a nanocomposite, and the modification of the coating's microstructure, in order to synthesize novel nanocomposite coating systems. On the other hand, most applications have complex requirements for the physical, chemical, and mechanical properties of the coating material, and, in most cases, a single coating cannot satisfy all these requirements. Thus, coating materials with different properties and novel coating concepts can all be engineered into an appropriate coating "architecture" to meet a specific application. To demonstrate coating architecture and microstructural design of

A. A. Voevodin et al. (eds.),
Nanostructured Thin Films and Nanodispersion Strengthened Coatings, 91–102.

nanocomposite coating systems, this paper correlates the process, microstructure and properties of three engineered nanocomposite coating systems for three specific applications: 1) a graded Ni-Al-N/Ti-B-C-N/Ti coating system developed for dies used in high tech, high precision glass molded components; 2) a Ti-Al-Si-N coating system ('working layer') developed for high temperature metal forming tools (high hardness and high toughness); and 3) a graded Cr-N coating system with a specifically engineered interface developed for providing a hard, wear resistant surface on a precipitation hardened aluminum alloy (relatively soft) substrate.

2. EXPERIMENTAL METHODS

Coatings in Ni-Al-N and Ti-B-C-N systems were produced in different argon-nitrogen atmospheres using RF magnetron sputtering from a NiAl compound target and a TiB_2-TiC composite target (40 at.% TiB_2 and 60 at.% TiC), respectively. The targets were synthesized using a one step SHS (self-propagating high-temperature synthesis) consolidation method [1,2]. If sputter cleaning of the substrate was performed prior to the deposition, the substrate bias was set to -500 volts, producing a glow discharge sputter etching of the substrate. When a titanium adhesion layer was produced prior to the deposition, a 50-nm thick titanium thin film was sputtered from a titanium target using 3 W/cm^2 power from a 40 kHz AC plasma source. Detailed deposition conditions and characterization procedures are well documented elsewhere [3-6].

Ti–Al–Si–N films were deposited on WC–Co substrates by a hybrid deposition system that consisted of cathodic arc evaporation (CAE) from a Ti_3Al target and DC magnetron sputtering from a silicon target. A typical deposition was conducted at a substrate temperature of 300 ^{o}C and in a nitrogen-argon (10:1) atmosphere of 6.7×10^{-2} Pa using an Arc current of 60 A, sputter currents of 0-6 A, and a substrate bias of –25 V. Detailed procedures for deposition and instrumental analyses can be found elsewhere [7].

The effects of high-energy metal ion bombardment to the substrate/coating interface region were investigated by comparison of Cr-N coatings deposited by CAE on 7075-T6 aluminum at low temperatures (<120°C) with or without the assistance of a high energy (up to 30 keV) microsecond pulsing of the substrate using a Hyper-Ion. The optimal Hyper-Ion conditions used for this investigation were determined to be 10 kV, 1 μs and 1 kHz. These parameters were restricted by the amount of thermal energy generated by varying the high-voltage and the pulse frequency to avoid over-aging of the aluminum substrate. The Cr metal adhesion layer was deposited with an arc current of 72 amperes and a substrate bias of –100 volts, and without gas. The Cr-N graded coatings were deposited using an arc current of 62 amperes and a substrate bias of −100 volts, and with the nitrogen pressure systematically altered from 3.5 to 22.0 mTorr.

3. RESULTS

3.1. Graded Ni-Al-N/Ti-B-C-N/Ti Coating System

Nanostructured NiAl and Ni-Al-N thin films were RF magnetron sputtered from a NiAl compound target [1,3,5]. Equatomic NiAl thin films were deposited in an argon sputtering atmosphere. Introducing 10% nitrogen into the sputtering gas resulted in the formation of AlN in the Ni-Al-N films, with a relative atomic concentration of nitrogen of 21.4% in the film.

The microstructure of thin films based on the ternary Ni-Al-N system was tailored through biasing the substrate. As seen in Fig. 1, applying a substrate bias of −50 volts during deposition resulted in larger NiAl grains and crystallization of AlN particles (about 2 nm in size). The NiAl particle sizes in the NiAl and Ni-Al-N films increased from approximately 12 and 2 nm for unbiased films to about 15 and 4nm for biased films, respectively.

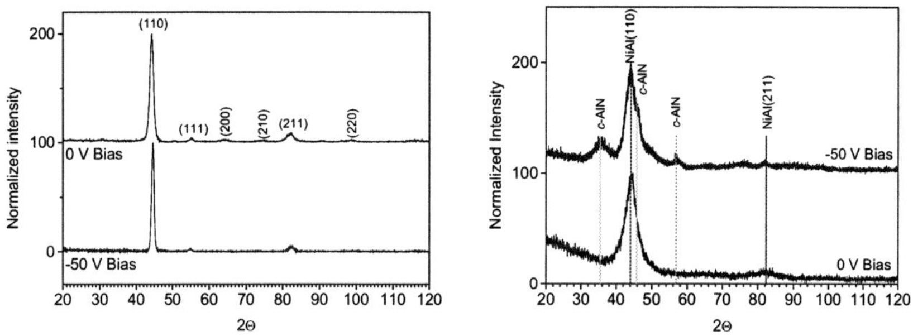

(a) *(b)*

Figure 1. XRD pattern comparison of (a) NiAl and (b) Ni-Al-N films deposited on AISI 304 stainless steel substrates without substrate bias or with −50 V DC substrate bias, respectively.

The NiAl thin films exhibited a dense columnar microstructure [3,5]. The microstructure of the Ni-Al-N film deposited without substrate bias exhibited a homogeneous microstructure with nano-scale NiAl (110) grains distributed in an amorphous matrix (see Fig. 2a). When a -50 volt substrate bias was applied during deposition, the Ni-Al-N coatings exhibited a dense, homogeneous microstructure with particle size on the order of 4 nm (see Fig. 2b). The electron diffraction pattern indicates the presence of more than one crystallographic phase or orientation, which agrees with the XRD pattern.

Nanostructured Ti-B-C and Ti-B-C-N composite thin films were deposited using RF magnetron sputtering [4,6]. On introducing nitrogen into the sputtering gas (5 to 20%), the resultant atomic concentration of nitrogen in the films was fixed at approximately 30%, and the XPS C_{1s} spectra shifted from the typical position for carbide to graphite [6]. The films in the Ti-B-C-N system deposited without a substrate bias exhibited a completely amorphous structure. Applying a substrate bias of -50 volts during the deposition process resulted in quasi-amorphous, nanocomposite films with ultra fine grain sizes on the order of 2.5-3 nm.

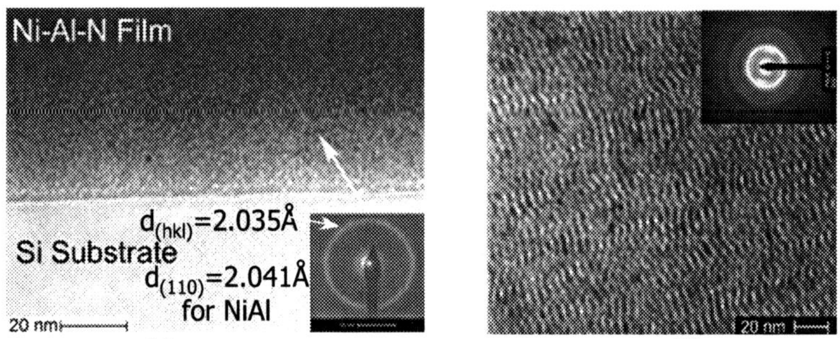

(a) *(b)*

Figure 2. Cross-sectional STEM photomicrographs and electron diffraction patterns of the Ni-Al-N thin films deposited on Si<100> wafer in an Ar-N_2 deposition atmosphere of 10% N_2, (a) without substrate bias, and (b) with -50 V DC substrate bias.

The unique microstructures of these coatings resulted in different physical, chemical and mechanical properties (see Table 1). The best mechanical properties and tribological performance of these coatings were achieved by depositing the film with a titanium interlayer and using a –50V substrate bias. Overall, fundamental experiments identified NiAl to be the most promising working surface layer in terms of non-sticking (with molten glass) [8] and oxidation resistance (in air), while the Ti-B-C film and the Ti-B-C-N film (deposited in 10% nitrogen) exhibited the best adhesion and best tribological performance [4,5]. On the other hand, glass molding dies and forming tools operate in air at elevated temperature (ideally up to 1200 K), and thus the contacting surface ideally needs to be non-sticking (non-wetting by molten glass), oxidation resistant, and wear resistant. Therefore, an optimized coating architecture must be built to enhance the performance and reliability of the coating system for glass forming applications. Based on finite element analysis [9], an optimized graded coating architecture of Ni-Al-N/Ti-B-C-N/Ti was developed. Compared to the bilayer coatings consisting of a titanium adhesion layer and a thin film based on the Ni-Al-N or Ti-B-C-N systems, this graded coating system not only provided excellent oxidation resistance and non-sticking by molten glass, but also gave better adhesion, better wear resistance, and an intermediate value of residual stress.

Table 1. Properties and performance of coatings in the Ni-Al-N and Ti-B-C-N systems. [1,3-5,8]

Film	Substrate Bias (V)	Ti Adhesion Layer	Micro-Structure[1]	Residual Stress (GPa)	Critical Load L_C (N)	Wear Volume ($10^5 \mu m^3$)	H/E[2] (GPa)	Friction Coeffi-cient	T_{ox}[3] (°C)	Q_{ox}[4] (eV)	T_C of Wetting[5] (°C)
Ti-B-C	No	No	(a)	-5.65	17	10.7	-	0.5	-	-	
Ti-B-C	-50	No	(nc)	-2.38	42	2.6	24/285	-	650	1.64	1027
Ti-B-C	-50	Yes	(nc)	-1.41	46	2.1	25/289	0.32	-	-	
Ti-B-C-N	No	No	(a)	-5.83	8	9.3	-	0.39	-	-	
Ti-B-C-N	-50	No	(nc)	-2.53	43	2.3	16/212	-	650	1.33	1050
Ti-B-C-N	-50	Yes	(nc)	-1.36	48	2.2	22/217	0.37	-	-	
Ni-Al	No	No	(nc)	-2.68	16	84.6	-	0.95	-	-	
Ni-Al	-50	No	(nc)	-0.57	35	9.0	14/200	-	1010	3.57	1078
Ni-Al	-50	Yes	(nc)	-0.16	40	8.0	14/201	0.54	-	-	
Ni-Al-N	No	No	(nc)	-3.00	10	68.1	-	0.79	-	-	
Ni-Al-N	-50	No	(nc)	-0.83	31	7.5	12/176	-	975	2.57	1064
Ni-Al-N	-50	Yes	(nc)	-0.38	38	6.6	18/201	0.45			

[1] (a) for amorphous, (nc) for nanocrystalline/amorphous; [2] H for nanohardness, E for Youngs Modulus;
[3] T_{ox} is the minimum temperature of oxidation at which oxidation products were first observed in STA;
[4] Q_{ox} is the activation energy for oxidation; [5] T_C of wetting is the critical temperature above which the film can be wetted by glass.

3.2. Ti-Al-Si-N Coating System

By means of embedding nano-crystallites in an amorphous matrix, superhard nanocomposite coatings with high apparent fracture toughness and enhanced thermal stability may be produced [10]. Quaternary Ti-Al-Si-N Nanocomposite films have been explored in this work as a high hardness and high toughness candidate surface "working layer" for high temperature metal forming tools.

Without Si addition, a crystalline $Ti_{0.76}Al_{0.24}N$ film was synthesized [7]. As silicon was incorporated into the Ti-Al-N, the Ti to Al atomic ratio was maintained at around 3 (Ti_3Al target), and the Si content could be up to 30% [7]. Increasing the Si addition, the intensities of X-ray diffraction peaks from the Ti-Al-Si-N films gradually decreased and became amorphous for films with a Si content above 19%. XRD peaks corresponding to crystalline Si_3N_4 or titanium silicides were not observed.

Fig. 3 shows XPS high-resolution spectra near the binding energy of Si 2*p* with various Si contents. The peaks appeared at 101.8 eV, which is in good agreement with that of a Si_3N_4 compound [11]. The peak intensity gradually increased with an increase in Si content. In addition, the peaks corresponding to free Si (at 99.28 eV) and $TiSi_2$ (at 98.8 eV) were not detected. The absence of free Si in these samples is considered to be due to the very high ionization rate of nitrogen gas in the arc plasma. XPS analyses indicated that Si existed mainly as a phase of the Si_3N_4 compound in the Ti–Al–Si–N film, which is believed to be amorphous from XRD results.

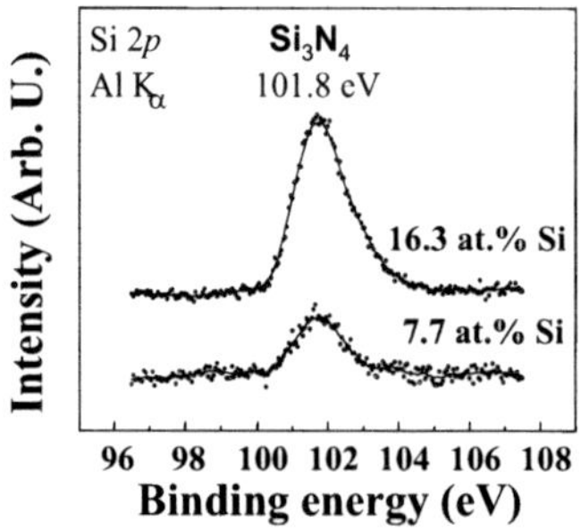

Figure 3. High-resolution Si 2*p* XPS spectra of Ti–Al–Si–N films with various Si contents.

The cross-sectional HRTEM images and corresponding diffraction patterns (see Fig. 4) clearly show that the Ti–Al–Si–N films were composites of nanocrystallites embedded in an amorphous phase which were distinguished from each other by lattice fringe contrast. HRTEM, XPS, and XRD analyses revealed that the synthesized Ti–Al–Si–N films had composite microstructures consisting of solid-solution (Ti,Al,Si)N crystallites and amorphous Si_3N_4. For the film with 4 at.% Si, the crystallites showed irregular and ellipsoidal shapes, and the sizes ranged from 10 to 20 nm. As the Si content increased to 9 at.%, the nanocomposite film exhibited much finer crystallites with approximately 8 nm size distributed uniformly in the amorphous matrix. The average grain size of crystallites, calculated from XRD and observed from HRTEM, decreases as the silicon incorporation increases (see Fig. 5a). As the Si content increased (see Fig. 5b), the hardness and Young's modulus values of the Ti–Al–Si–N films steeply increased and reached maximum values of approximately 55 and 650 GPa, respectively, at a Si content of 9 at.%. The hardness and modulus decreased again with a further increase in Si content. This is believed to be due to the reduced grain-boundary sliding and high cohesive energy in interphase boundaries which were simultaneously caused by the percolation of *a*-Si_3N_4 into a Ti–Al–N film. This nanocrystallite structure was possible only using the higher ion bombardment energies in CAE; lower energy sputtering of the Ti_3Al target did not provide nanocrystallites.

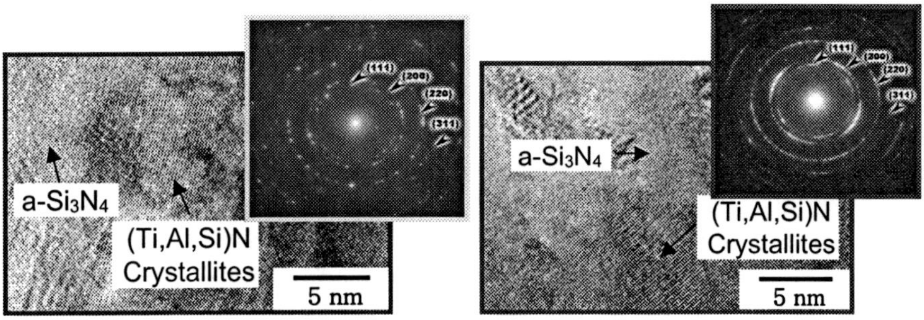

Figure 4. Cross-sectional HRTEM images and their electron diffraction patterns of Ti–Al–Si–N films containing (a) 4 and (b) 9 at. % Si, respectively.

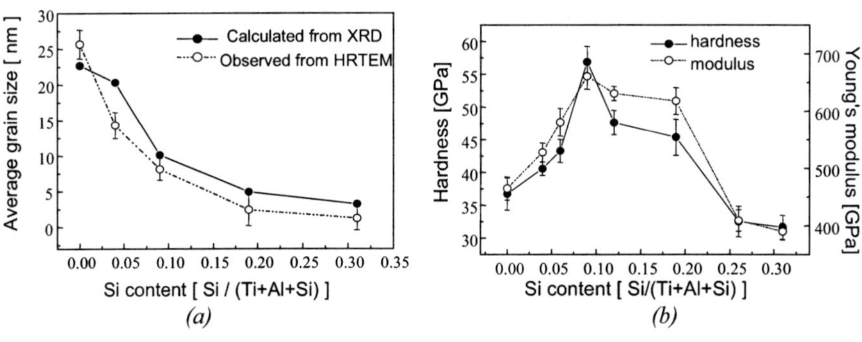

Figure 5. (a) average grain size of (Ti,Al)N crystallites, and (b) nanohardness and Young's modulus values as a function of the Si content in the Ti-Al-Si-N films.

3.3. Graded Cr-N Coating System

It is well known that ion bombardment energy can be used to control adhesion and coating microstructure, and thus to alter mechanical properties, wear resistance and corrosion behavior of the coating system. Cathodic arc evaporation (CAE) of Cr-N coatings on 7075-T6 Al substrate was used as one of the model systems to explore this effect.

Cathodic arc evaporation was used to deposit the Cr-N graded coating system because of its advantages of high deposition rate, high ionization of species (90-100 %) and high ion kinetic energy (50-150 eV) compared with other PVD methods to deposit wear resistant thin films.

The stoichiometry of the Cr-N coatings was determined by secondary neutral mass spectrometry (SNMS). As thc amount of nitrogen pressure increased from 3.5, 10.5 to 22.0 mTorr, the overall film stoichiometry changed from approximately 66.5, 57 to 56 at.% Cr respectively with the high energy pulsed bias. XRD showed that all coatings consisted of varying amounts of Cr, Cr_2N, and CrN phases. This made compositionally grading of the coating system possible through controlling the nitrogen pressure.

From TEM observations, the coating produced without the aid of high-energy ions showed a microstructure with a defined interface at the substrate surface, while the coating produced with the aid of high-energy ions exhibited an "affected region" which broadens the interface. This "affected region" increased from 50 to 225 nm with increasing nitrogen pressures from 3.5 to 22.0 mTorr. SNMS and TEM coupled with electron diffraction have been used to identify specific regions of the coating (i.e. Cr-N region, Cr rich region, affected region, and substrate). Fig. 6 shows TEM micrographs and associated electron diffraction patterns from all regions of the Cr-N graded coating on the aluminum substrate. The TEM results illustrate that the coatings are dense, nano-culumnar zone T structures of approximately 20 nm in width. It can be seen that the "affected region" is polycrystalline and has nanocrystalline precipitates, resulting in a decrease in the amount of mismatch between the Cr-rich interlayer that has been traditionally linked to improved adhesion. The build-up of precipitates within the aluminum substrate demarks the affected region from that of the unaffected aluminum substrate. Fig. 7 shows the SNMS depth profiling of the Cr-N coating deposited with the aid of the Hyper-Ion (10kV, 1ms, 1kHz) and at 22.0 mTorr N_2. The "affected region" is clearly shown in the depth profile. For the coatings deposited with the aid of high-energy ions, the nitrogen concentration shows a decrease in concentration through the Cr rich region and then an increase in the affected region for a distance of 50 nm. Once into the unaffected substrate, the maximum nitrogen implantation depth is approximately 600 nm. It can be seen that the nitrogen penetration depths are greater than the chromium penetration depths due to the large differences in atomic size for chromium and nitrogen atoms. The TEM results coupled with XRD data and SNMS depth profiles allow for a schematic representation of the Cr-N coating system deposited with the aid of high-energy Cr ions on to 7075-T6 aluminum (see Fig. 8).

Compared to the non-Hyper-Ion aided CAE coatings at the same varying nitrogen pressures, the Hyper-Ion assisted CAE Cr-N coatings exhibited increased adhesion and wear resistance (by approximately 100%), and reduced residual stress (by at least 30%), while maintaining at least the same fracture toughness.

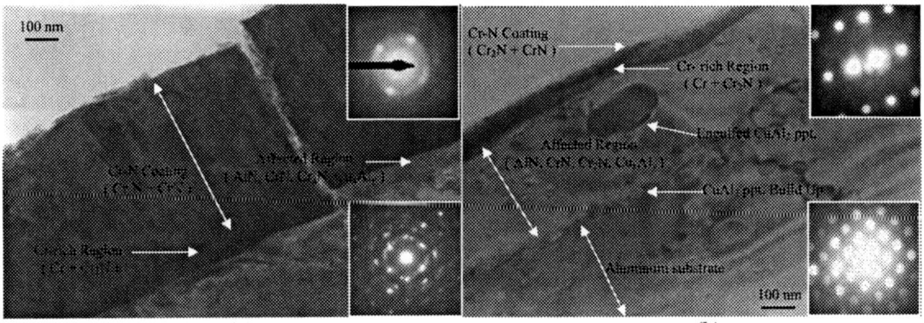

(a) *(b)*

Figure 6. Cross-sectional TEM photomicrographs and electron diffraction patterns from high energy pulsed bias-assisted CAE Cr-N coatings deposited at (a) 10.5 mTorr and (b) 22.0 mTorr nitrogen pressure. Electron diffraction pattern identifying (a) the Cr, Cr_2N phase for the Cr-rich region and (b) the Cr_2N, CrN, AlN, and Cu_xAl_y phases in the affected region.

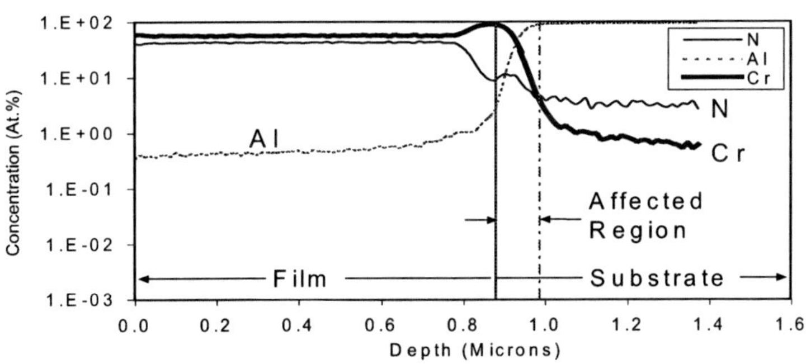

Figure 7. SNMS depth profiling for the Cr-N coating deposited at 22.0 mTorr N_2 with the aid of high-energy ions produced by the Hyper-Ion.

4. DISCUSSION

These examples show that film growth and microstructure and thus film properties can be tailored through controlled ion bombardment. Through controlling the momentum, energy and relative abundance of plasma species, the resultant film composition and microstructure (crystal, defect, grain and interfacial structures) were modified, which in turn, ultimately determined the properties of these coatings. In addition, functionally graded coating architectures have been used successfully in the coating systems in order to meet the application requirements.

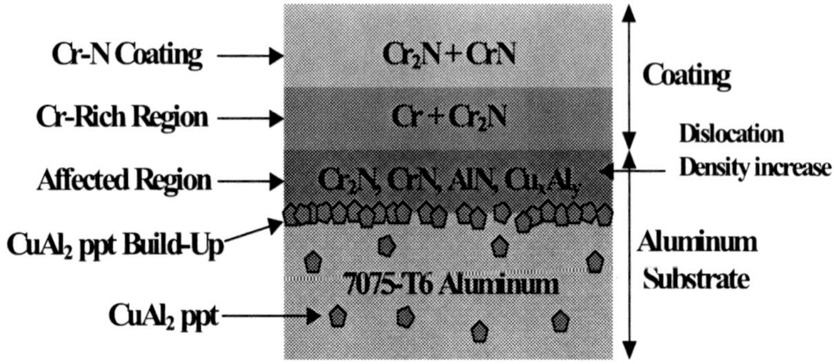

Figure 8. Schematic representation of the Cr-N coating produced by high-energy pulsed bias-assisted cathodic arc evaporation at nitrogen pressures between 3.5 – 22 mTorr.

Nanocrystalline-amorphous composite thin films (e.g. Ti-Al-Si-N, Ti-B-C-N, and Ni-Al-N) appear to offer superior properties. In view of the complexity of these systems, any attempt to elaborate in more detail on the possible contributions of the different potentially "hardening", "toughening", and "stabilizing" mechanisms would be, at this stage, too speculative. Most probably, various mechanisms of strengthening and possibly also toughening are active simultaneously which results in a multiplication of strengthening and toughening effects. When nano-size crystallites are glued together by a few tenths of a nanometer thin amorphous layer with a high adherence to the surface of the crystallites, the dislocation activity (e.g. formation and multiplication) in such a nanostructure is absent and the stress concentration factor on the tip of a nanocrack is very small, resulting in a "superhardness", and blocking crack propagation in the surrounding thin amorphous tissue [12]. In nanocrystalline-amorphous nanocomposites, the chemistry operating in these systems not only determines the constituent phases but also stabilizes their nanostructure and makes it fairly free of flaws. Thus, there are critical values of crystallite size and relative amount of nanocrystalline phases that control the interaction between the nanocrystallites and amorphous phase and thus determine the properties and performance of the nanocomposite. It was shown in this paper that the crystallite size was controlled by the composition of the films and/or by the ion bombardment during deposition.

Pulsed plasma operation has been found to manipulate particle energies and plasma composition in both PVD [13] and CVD [14] processes. Besides the conventional PVD process parameters, such as target materials, power to target, bias to substrate, gas pressure and composition, and substrate temperature, pulsing offers a new set of parameters (pulse frequency, pulse duration, and pulse magnitude) to control ion bombardment (ion energies and relative abundances of plasma species) and thus film growth and resultant film properties. The future trend of ACSEL Research on nanocomposite coatings will be based on the Ti-Si-Al-C-B-N system using pulsed-plasma in a closed-field unbalanced magnetron sputtering system (CFUBMS), coupled to high energy substrate bias pulsing. Plasma diagnostics using a Hiden Analytical Ltd., electrostatic quadrupole plasma analyzer (EQP) and a Langmuir Probe will be employed to fundamentally study the chemistry and physics of this complex plasma.

5. CONCLUSIONS

Nanostructured thin films in the Cr-N, Ni-Al-N, Ti-B-C-N, and Ti-Al-Si-N systems have been synthesized and exhibit superior properties. Controlling the ion bombardment (ion energies and relative abundances of plasma species) of substrate is a very effective approach to modify the microstructure (including crystal structure, interfacial structure, and grain structure) and composition of film growth, and, in turn, to obtain desired relative density, mechanical, physical and chemical properties of the resulting film. For example, as using controlled ion bombardment via substrate biasing, the Ni-Al-N and Ti-B-C-N films became denser and nanostructured; superhard Ti-Al-Si-N nanocomposites were synthesized; and the interface structure between Cr-N coatings and 7075-T6 Al substrate was graded, allowing superior adhesion and wear resistance.

These three example coating systems demonstrate that the microstructure and properties of each coating system can be tailored to provide an optimum combination of functionality to meet specific engineering application requirements. Thorough process diagnostics and high resolution microscopy of resultant films will facilitate correlation of plasma properties with structure, properties and performance of these nanocomposite coating systems, and merit further investigations.

REFERENCES

1. D. Zhong, J. J. Moore, J. Disam, S. Thiel, and I. Dahan, Surf. & Coat. Technol. 120–121 (1999) 22.
2. E. A. Levashov, V. I. Kosayanin, L. M. Krukova, J. J. Moore, and D. L. Olson, Surf. & Coat. Technol. 92 (1997) 34.
3. D. Zhong, J. J. Moore, T. R. Ohno, J. Disam, S. Thiel, and I. Dahan, Surf. & Coat. Technol. 130 (2000) 33.
4. D. Zhong, E. Sutter, J. J. Moore, G.G.W. Mustoe, E. A. Levashov and J. Disam, Thin Solid Films 398-399 (2001) 320.

5. D. Zhong, J. J. Moore, G.G.W. Mustoe, E. Sutter, B. Mishra, and J. Disam, Structure and mechanical properties of magnetron sputtered coatings in the Ni-Al-N ternary system, in: Surface Modification Technologies XV, ASM International, Materials Park, OH 44073, USA, March 2002, pp. 53.
6. D. Zhong, J. J. Moore, B. M. Mishra, T. Ohno, E. A. Levashov, and J. Disam, Surf. & Coat. Technol. 163-164 (2003) 50.
7. In-Wook Park, Sung Ryong Choi, Mi-Hye Lee, and Kwang Ho Kim, J. Vac. Sci. Technol. A 21 (2003) 895.
8. D. Zhong, E. Mateeva, I. Dahan, J.J. Moore, G.G.W. Mustoe, T. Ohno, J. Disam, and S. Thiel, Surf. & Coat. Technol. 133-134 (2000) 8.
9. D. Zhong, G.G.W. Mustoe, J.J. Moore and J. Disam, Surf. & Coat. Technol. 146-147 (2001) 312.
10. P. Nesládek and S. Veprek, Phys. Status Solidi A 177 (2000) 53.
11. J. F. Moulder, W. F. Stickle, P. E. Sobol, and K. D. Bomben, Handbook of X-ray Photoelectron Spectroscopy, Physical Electronics, Eden Prairie, MN, 1995, pp. 238.
12. S. Veprek and Ali S. Argon, J. Vac. Sci. Technol. B 20 (2002) 650.
13. C. Muratore, J.J. Moore, and J.A. Rees, Surf. & Coat. Technol. 163 –164 (2003) 12.
14. S. Morrison, Ujjwal Das and Arun Madan, Solar Energy Materials and Solar Cells 76 (2003) 281.

Chapter 10
HYBRID PLASMA DEPOSITION METHODS FOR SYNTHESIS OF NANOSTRUCTURED MATERIALS

A. A. Voevodin and J. S. Zabinski
Air Force Research Laboratory, Materials and Manufacturing Directorate
Wright Patterson Air Force Base, OH 45433-7750, USA

Hybrid deposition techniques facilitate a high degree of structural control at low substrate temperatures, permitting growth of nanostructured materials, which are not possible to produce by other methods. Several hybrid processes were studied, where pulsed plasma plumes from an excimer laser ablation were combined with continuous plasma generated by non-pulsed sources. One was a hybrid of laser ablation and ion beam deposition, where short-lived interactions between two plasma sources were explored to produce AlON films. The process was used as a base for the development Al_2O_3/MoS_2 nanocomposite coatings. Another hybrid process was a combination of laser ablation with magnetron sputtering (MSPLD), where highly energetic plasma plumes from laser ablation were intersected with a magnetron-generated plasma containing sputtered metal atoms. MSPLD was used to prepare a number of nanostructured coatings, including WC/DLC, $WC/DLC/WS_2$, YSZ/Au, and $YSZ/Au/MoS_2/DLC$ nano-composites for wear protection of engineering components. Recently we initiated exploration of a new hybrid process, using filtered vacuum arc and magnetron sputtering, as an aletranative to MSPLD. Initial studies were performed in preparing Ti-TiC-DLC and Ti-TiC-CN_x functionally gradient coatings.

Keywords: plasma deposition, hybrid processes, nanostructured materials, coating, tribology, laser ablation, sputtering, vacuum arc, ion beam.

1. INTRODUCTION

Over the years we had explored various plasma based methods for synthesis nanostructured materials for tribological applications. Among these methods, pulsed laser deposition (PLD) received a special attention as a versatile tool of new material synthesis with a broad applicability in the growth of various metallic, oxide, nitride, and carbide films [1]. PLD is most frequently performed by using excimer laser beams focused on a target surface to deliver an instantaneous power density of 10^8-10^9 W/cm^2. The use of short 10-20 ns pulses of ultraviolet 193 nm, 248 nm, or 308 nm irradiation causes the domination of photon sputtering processes over

A. A. Voevodin et al. (eds.),
Nanostructured Thin Films and Nanodispersion Strengthened Coatings, 103–111.

melting/evaporation processes, which greatly facilitates flexibility in target material selection and film compositional control. PLD has a number of unique characteristics, which are not available using other deposition methods. For example, it can be used to produce plasma plumes from high melting point target materials, such as graphite or metal oxides. Plasma plumes produced by PLD are typically a few microseconds in duration, but very dense and provide conditions for the growth of correspondingly dense and impurity-free films. Furthermore, the kinetic energy of the generated plasma is on the order of 100-1000 eV, since laser ablated plumes explosively expand from the target surface into the chamber vacuum, reaching supersonic speeds. Such high kinetic energy of ablated atoms can be used to produce coatings with crystalline structures at low substrate temperature.

The major drawback of PLD in comparison to other deposition techniques, *e.g.* magnetron sputtering or vacuum arc, is a relatively low deposition rate. This is aggravated by the high initial investment expense for laser equipment, which results in high coating cost. Evaluation of process economics under those circumstances limits PLD to mainly microelectronic applications, where precision in composition control and film purity may override higher cost. However, the cost of laser energy has significantly decreased in recent years, and the low deposition rate has become the predominant factor restricting broader commercial use of PLD. This is especially true for film growth using laser ablation of carbides, nitrides, some oxides, and graphite. A possible solution to the problem is combining PLD with other deposition methods to increase deposition rate. The additional plasma sources may be selected to overcome specific limitations of laser deposition. For example, deposition of many oxides and DLC is quite effective using PLD compared to magnetron deposition, whereas metal deposition is usually more effective via magnetron. Hybridization of PLD with other techniques can also facilitate a high degree of structural control at low substrate temperatures, permitting growth of materials that are not possible to produce by other methods. Development of such hybrid technologies is discussed in this paper, including also a use of filtered vacuum arc as an alternative to laser ablation in hybrid processes.

2. HYBRID OF MAGNETRON SPUTTERING AND PULSED LASER DEPOSITION

A hybrid of magnetron sputtering and pulsed laser deposition (MSPLD) was initially developed for preparation of functionally gradient transition metal carbides and multilayer structures in the TiC and diamond-like carbon (DLC) material systems [2,3]. MSPLD was then expanded toward WC/DLC, WC/DLC/WS_2, YSZ/Au, and YSZ/Au/MoS_2/DLC nanocomposites for wear protection of engineering components [4-7]. The technique was adapted by other researchers for the synthesis of TiC/Ag nanocomposite coatings with advanced mechanical characteristics [8,9].

A schematic of MSPLD is shown in Fig. 1. A pulsed laser beam is used to generate a carbon flux, while a magnetron generates a flux of titanium, tungsten, or tungsten+sulfur atoms. The target orientations define the point of intersection of the fluxes from the different sources. Magnetrons are placed into an unbalanced configuration to sputter metal (Ti, W, Au, Ag, etc.) and/or dichalcogenide (WS_2, MoS_2) materials with kinetic energies of several electron volts in the direction of the substrate. For sputtering, an argon pressure of 0.2 Pa is typically maintained in the deposition chamber, which is low enough to prevent thermalization of the ablated plumes. The main task of magnetron sputtering is to deliver a high flux of metal and/or sulfur atoms to the condensation surface and thus provide a high growth rate. Carbon plumes from the laser ablation of graphite or Y-Zr-O plumes from YSZ targets are also directed to the condensation surface, intersecting with the metal atom flux from magnetron sputtering.

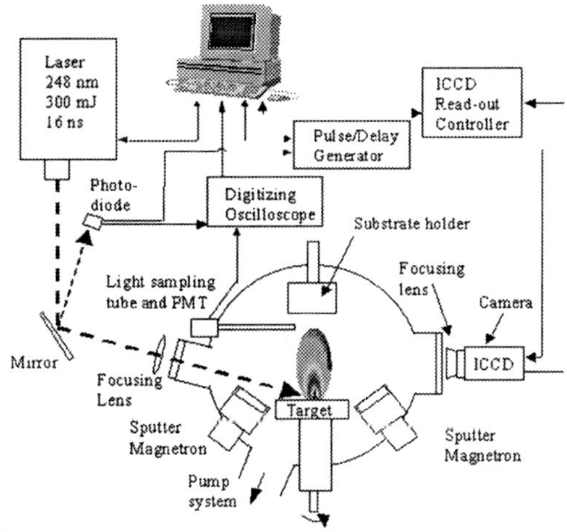

Figure1. Schematic of MSPLD arrangement. Magnetron sputter sources provide W, Ti , Au and/or WS_2 fluxes, while laser ablation provides the carbon plume.

In the case of carbide growth, the independent operation of metal and C sources provided a means to control composition and was based on the relationship between film stoichiometry and the deposition rates of each element. In particular, a direct dependence of the Ti_xC_y and W_xC_y film stoichiometry on the laser repetition rate (R) and a linear relationship between R and the ablation rate of carbon was found, which made composition control easy. When an *y/x* ratio of stoichiometric coefficients is used, the film composition can be predicted with the following equation:

$$y/x = \alpha \beta R,$$

where α is the ratio of the product of mass and density for the two elements $\alpha=(M_{Ti}\rho_C)/(M_C\rho_{Ti})$, and β is the ratio of the sputtering rate of Ti or W to the ablation rate of C. The energetic carbon atoms arriving at the condensation surface provide sufficient energy for the growth of nanocrystalline carbide phases at low substrate temperatures.

With MSPLD, crystalline TiC, WC, and ZrO_2 coatings with grain sizes regulated between 10 and 100 nm were grown at near room temperature and the deposition rates were similar to that in magnetron sputtering. MSPLD has a significant advantage over regular magnetron sputtering, where substrate temperatures of at least 350-400ºC are required for carbides and 500-600 °C for ZrO_2. Furthermore, MSPLD opens the possibility of growing "next generation" coatings of unique composite materials made of amorphous DLC, MoS_2, WS_2 and crystalline carbide or oxide phases. The task is not trivial, which is evident from the fact that DLC is typically grown at temperatures below 200 ºC to escape graphitization during growth, while carbides requirc 350-500 ºC temperatures to form crystalline phases. Similarly, both MoS_2 and WS_2 dissociate at the typical growth temperatures of crystalline ZrO_2. MSPLD solved these problems by compensating for a low substrate temperature with a high kinetic energy of laser ablated carbon or zirconium atoms, providing crystalline carbide and oxide formation without raising substrate temperature. Growth rates of about 1 μm/hour were maintained, which were close to that of magnetron sputtering. While local growth rates on par with magnetron sputtering have been achieved, additional work is required to coat numerous parts at once – in other words, scale-up is the next issue.

3. HYBRID OF ION BEAM AND PULSED LASER DEPOSITION

A hybrid of ion beam plasma and PLD (Fig. 2) is sometimes referred as ion-beam assisted pulsed laser deposition (IAPLD). IAPLD was found to be very successful in structure control of Y_2O_3-ZrO_2 and CeO_2 buffer layers for subsequent epitaxial growth of superconductive oxides [10]. Another application of IAPLD was the production of β-C_3N_4 [11] and cubic-BN [12] hard films for surface wear protection. In both cases, high kinetic energy and activation of gas

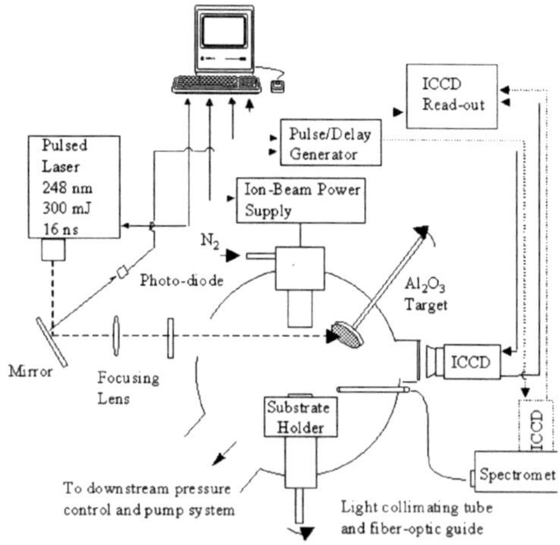

Figure 2. Schematic of IAPLD arrangement configured for an AlON coating growth. An ion-beam source above the substrate provides the N^+ plasma, while laser ablation is used to generate plasma plumes from an alumina target.

species in the ion beam plasma were key factors for film synthesis.

A generally accepted model of IAPLD growth considers separately the processes of film deposition during short (5-10 μs) pulses of ablated plasma and structure and chemistry adjustment by ion-bombardment during relatively long (10-100 ms) pauses between laser pulses. Until recently, little attention was paid to a synergistic interaction between the plasma generated by the ion beam and laser. However, such interaction could lead to new chemical reactions in the plasma as well as the generation of additional excitation states important for control of film growth.

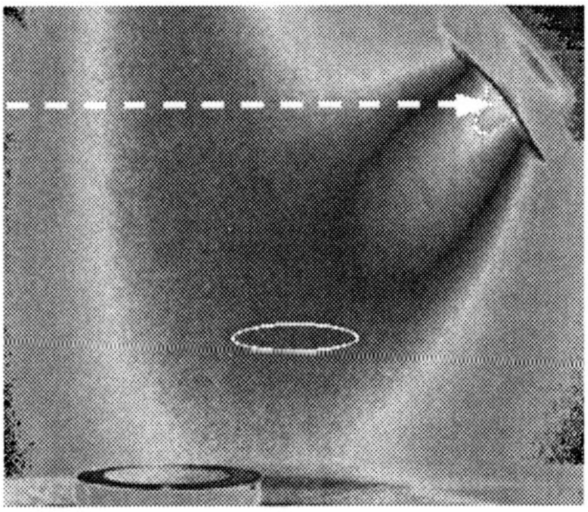

Figure 3. Photograph of the IAPLD process obtained at 10 μs ICCD exposure, showing a plasma interaction channel between the ion-beam and laser ablation sources. The white circle indicates the substrate position, where AlON films with a high nitrogen content were produced.

Some of these interactions were explored while growing Al-O-N films [13], which have enhanced thermal, tribological, and dielectric properties. Plasma plumes from laser ablation of Al_2O_3 and a nitrogen ion beam were intersected in the vicinity of the substrate (Fig. 2). A KrF excimer laser (Lambda Physik, COMPex 205) produced pulses of 248 nm wavelength and 300 mJ energy with 20 Hz frequency. The laser beam was directed onto a rotating polycrystalline Al_2O_3 (alumina) target under a 45° angle and focused to an instantaneous energy density of about 1.5 GW/cm^2 . A 3 cm diameter Kaufman ion beam source (Commonwealth Scientific Corp.) was operated using high-purity nitrogen gas at beam a voltage of 300 V. Temporal and spatial plasma distributions of the plasma generated by laser ablation and the ion-beam were investigated with a gated image intensified charge-coupled detector (ICCD) camera at nitrogen background pressures from 0.1 to 3.9 Pa.

An interaction effect was discovered, revealing plasma channels connecting the ion beam plasma and laser-ablated plume (Fig. 3). These plasma channels existed only during the short period of the ablated plume development and their spatial location was dependent on the nitrogen background pressure in the chamber. Increasing the pressure resulted in interaction channel locations closer to the plasma sources and away from the substrate. The plasma connection between the ion-beam and laser ablation sources resulted in an instantaneous 40-60% increase of the current extracted from the ion-beam source during laser plume development. Furthermore, plasma spectroscopic analyses indicated the presence of exited NO radicals during this short 5-10 μs interaction event. These molecules were identified from rotation/vibration $A^2\Sigma^+ \rightarrow X^2\Pi$ and $B^2\Pi \rightarrow X^2\Pi$ excitation transitions. There were also additional excitation and ionization states for atomic nitrogen and oxygen occurring within plasma interaction channels. The ionization level of atomic aluminum was not dependent on the ion-beam source operation, and the emission bands of AlO or AlN molecules were not detected.

From the main chemically active elements in the plasma, Al, N, O and NO radicals, it was reasonable to suggest that nitrogen and oxygen predominately react with each other in the gas phase to form molecular NO. The reaction may occur only during short intervals of the interaction between the ion-beam, supplying nitrogen, and laser ablated plumes, supplying oxygen. The growth of Al-O-N films is then occurring through a heterogeneous reaction on the sample surface between arriving atomic Al and molecular NO species.

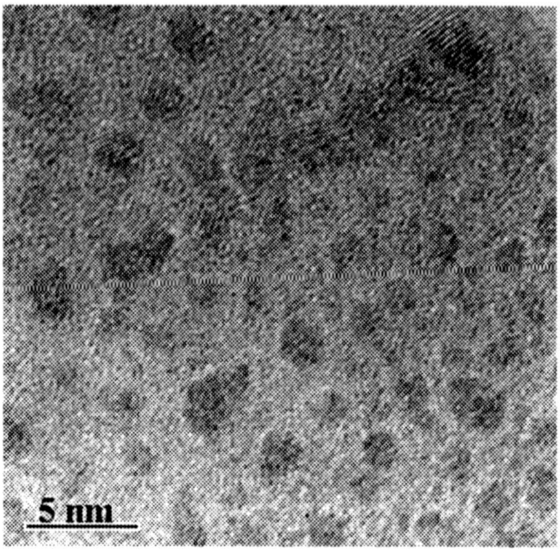

Figure 4. Nanocomposite tribological coating made of a hard Al_2O_3 matrix with sealed MoS_2 solid lubricant grains (dark spots).

The synergistic effect of plasma enhancement in IAPLD growth was used to produce highly nitrogenated (20-30 at. % N) AlON films by moving substrates into the brightest part of the interaction channel (Fig. 3), where the strongest emission signal from excited NO radicals was detected. This sample location was off axis of the nitrogen ion-beam source, and thus would not be considered as an optimum location to maximize nitrogen content, when considering ion-beam source operation separately from laser ablation. However, this location provided a factor of two increase in film nitrogen content as compared to the film grown on a substrate located at the centerline of the ion-beam source.

Furthermore, the increase in plasma excitation of both atomic and molecular species helped to produce nanocrystalline AlON films at about 300°C substrate temperature. Such moderate temperature was very favorable in comparison to 600-800°C temperatures required by other deposition techniques to achieve film crystallinity. This was used to prepare advanced tribological coatings, where MoS_2 nanoparticles were sealed inside AlON and AlO matrices (Fig. 4). These nanocomposite coatings provided an excellent tribological performance in dry environment with friction coefficients as low as 0.02-0.04. They also had an extended life in humid environments due to the preservation of encapsulated MoS_2 from an oxidation in contact with moist environment. This success of IAPLD technology confirmed once more that hybrids of PLD with other deposition techniques have great potential for the discovery and production of unique materials.

4. HYBRID OF FILTERED VACUUM ARC AND MAGNETRON SPUTTERING

Most recently research was initiated to explore a possibility of replacing the PLD technique with filtered arc deposition (FAD). The approach explores the similarity in energetic parameters of plasmas produced by laser ablation and by cathodic vacuum arc evaporation with ion component filtering in a strong magnetic field. A hybrid system incorporating magnetron sputtering and filtered arc deposition (MSFAD) was assembled to compare coating produced by MSPLD and MSFAD. The system schematic is shown in Fig.5. The filtered cathodic arc assembly was manufactured by the Kharkov Institute of Technology, following original reports by Aksenov et al. [14,15].

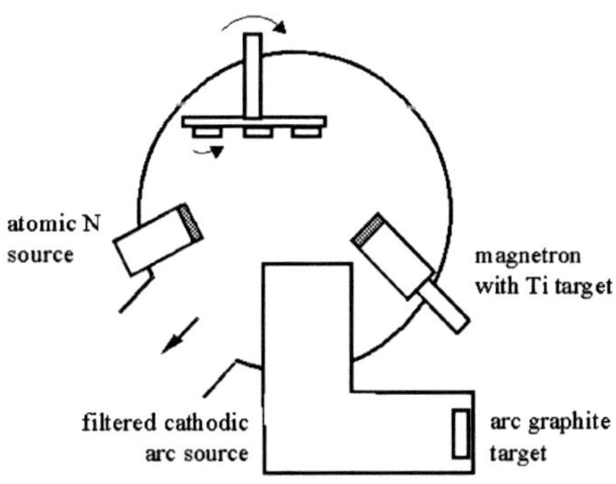

Figure 5. Schematic of MSFAD arrangement configured for a growth of Ti-TiC-DLC and Ti-TiC-CN_x gradient coatings. Magnetron sputter sources provide Ti plasma, which is intersected with carbon plasma from filtered vacuum arc source and a nitrogen flux from an atomic gas source (for CN_x).

The system was used to produce Ti-TiC-DLC and Ti-TiC-CN_x functionally gradient coatings and compare with a similar composition coatings produced by MSPLD [2,3,16]. In MSFAD, a graphite cathode was evaporated with a 125 A arc current, generating a carbon plasma. Carbon ions were guided by an electromagnetic field to the substrate location with a 90° turn, while droplet emission from arc spots was prevented to reach the substrate. Design and operation of filtered cathodic arc plasma sources was discussed in available reviews [17-20]. For the Ti-TiC-CN_x coating, an atomic nitrogen source was used in combination with MSFAD (Fig.5). An HD25 atomic gas source from Oxford Applied Research, Inc. was operated by leaking nitrogen through a heated ceramic tube (chamber pressure was within 10^{-4}-10^{-5} Torr), inside which a 500 W r.f. discharge was supported to dissociate nitrogen. The exit of the source was blocking ion component and releasing atomic nitrogen in the substrate direction. Coatings were produced with about 15 nm/min growth rate.

A comparison of Ti-TiC-DLC and Ti-TiC-CN_x coatings produced by MSPLD and MSFAD had indicated that both methods provided coatings with very similar chemistry, structure and properties. Independent of the growth technique, Ti-TiC-DLC coatings had hardness within 52-57 GPa and elastic modulus within 490-560 GPa. The Ti-TiC-CN_x coatings offered a reasonably high hardness of 28-30 GPa, while their elastic modulus was as low as 160 GPa. The results indicated a possible interchangeability of MSFAD and MSPLD, which opens an avenue for commercialization of advanced tribological coatings, e.g. "chameleon" and super-tough nanocomposites developed with MSPLD [6,7,21,22]. Vacuum arc is a relatively easy to scale-up for an industrial coating growth and many production type arc deposition units are available at the engineering coating market today.

5. CONCLUSIONS

The advantage of PLD in the production of a high energy and short duration plasma was used beneficially in hybrid processes, where plasma plumes from excimer laser ablation were combined with continuous plasma from non-pulsed sources, providing a much faster deposition rate. In the MSPLD process, intersection of a highly energetic carbon plasma from laser ablation with a magnetron sputtered plasma was critical for the formation nano-crystalline carbide and amorphous DLC phases in the same film. This was used to produce nano-composite materials made of nanocrystalline TiC, WC, YSZ, MoS_2, WS_2 and amorphous DLC with layered and grain/matrix nanostructures for wear protection of engineering components. In the IAPLD process, short-duration interaction channels between laser and ion beam plasma sources were discovered. They were found critical for AlON film growth by laser ablation of Al_2O_3 in the presence of a nitrogen ion beam. In particular, formation of NO molecules was found, providing a new growth mechanism for low temperature synthesis of AlON na d Al_2O_3. This was used to prepare nanostructured tribological coatings made of alumina matrix with nanosized inclusions of MoS_2 grains. It was demonstrated that MSPLD and IAPLD hybrid processes have high potential for laser ablation assisted technology applications. At the same time, combination of magnetron sputtering and filtered vacuum arc in the MSFAD technique can be a way to commercialize advanced engineering coatings developed with MSPLD. The first experiments with functionally gradient Ti-TiC-DLC and Ti-TiC-CN_x coatings indicated that the two techniques may be interchangeable and yet compliment each other. MSPLD characterized by an easy and independent control of the coating composition and structure, which is most important for a new coating development stage, while MSFAD has a better potential for a scale-up stage, making a more easy transition of the developed coatings into applications.

REFERENCES

1. M. S. Donley, J. S. Zabinski, Tribological Coatings, in: D. B. Chrisey, G. K. Hubler (Eds.), Pulsed Laser Deposition of Thin Films, John Wiley and Sons, New York, 1994, pp. 431.
2. A. A. Voevodin, M. A. Capano, A. J. Safriet, M. S. Donley, J. S. Zabinski, Appl.Phys.Lett. 69 (1996) 188.
3. A. A. Voevodin, M. A. Capano, S. J. P. Laube, M. S. Donley, J. S. Zabinski, Thin Solid Films 298 (1997) 107.
4. A. A. Voevodin, J. P. O'Neill, J. S. Zabinski, Tribol.Lett. 6 (1999) 75.
5. A. A. Voevodin, J. P. O'Neill, S. V. Prasad, J. S. Zabinski, J.Vac.Sci.Technol.A 17 (1999) 986.
6. A. A. Voevodin, J. G. Jones, J. J. Hu, T. A. Fitz, J. S. Zabinski, Thin Solid Films 401 (2001) 187.
7. A. A. Voevodin, J. J. Hu, T. A. Fitz, J. S. Zabinski, J.Vac.Sci.Technol.A 20 (2002) 1434.
8. J. L. Endrino, J. J. Nainaparampil, J. E. Krzanowski, Surf.Coat.Technol. 157 (2002) 95.
9. A. R. Phani, J. J. Nainaparampil, J. E. Krzanowski, Mater. Res. Soc. Symp. Proc. 697 (2002) 409.
10. R. P. Reade, S. R. Church, R. E. Russo, Rev.Sci.Instrum. 66 (1995) 3610.
11. C. Niu, Y. Z. Lu, C. M. Lieber, Science 261 (1993) 334.
12. T. Friedmann, P. B. Mirkarimi, D. L. Medlin, K. F. McCarty, E. J. Klaus, D. R. Boehme, H. A. Johnsen, M. J. Mills, D. K. Ottesen, J. C. Barbour, J.Appl.Phys. 76 (1994) 3088.
13. Voevodin, A. A., Zabinski, J. S. Advances in laser ablation assisted processes. SPIE Proceedings 4157, 269-274. 2001.
14. I. I. Aksenov, V. A. Belous, V. G. Padalka, V. M. Khoroshikh, Instruments and Experimental Techniques 21 (1978) 1416.
15. I. I. Aksenov, V. A. Belous, V. G. Padalka, V. M. Khoroshikh, Soviet Journal of Plasma Physics 4 (1978) 425.
16. A. A. Voevodin, J. G. Jones, J. S. Zabinski, Zs. Czigany, L. Hultman, J.Appl.Phys. 92 (2002) 4980.
17. A. Anders, Surf.Coat.Technol. 120-121 (1999) 319.
18. R. L. Boxman, V. Zhitomirsky, B. Alterkop, E. Gidalevich, I. I. Beilis, M. Keidar, S. Goldsmith, Surf.Coat.Technol. 86-87 (1996) 243.
19. D. A. Karpov, Surf.Coat.Technol. 96 (1997) 22.
20. P. J. Martin, A. Bendavid, Thin Solid Films 394 (2001) 1.
21. A. A. Voevodin, J. S. Zabinski, J.Mater.Sci. 33 (1998) 319.
22. A. A. Voevodin, J. S. Zabinski, Thin Solid Films 370 (2000) 223.

Chapter 11
RECENT DEVELOPMENTS ON IONIZED PHYSICAL VAPOUR DEPOSITION: CONCEPTS, DETERMINATION OF THE IONISATION EFFICIENCY AND IMPROVEMENT OF DEPOSITED FILMS

J. Bretagne, C. Boisse-Laporte, L.de Poucques, G. Gousset, M.C. Hugon, J.C. Imbert, O. Leroy, L.Teulé-Gay, M. Touzeau, P. Vašina[1], O. Voldoire

Laboratoire de Physique des Gaz et des Plasmas, Bât.210, Université Paris-Sud, 91405 Orsay Cedex, France

[1] *Department of Physical Electronics, Kotlářská 2, 61137, Brno, The Czech Republics*

While most of the IPVD reactors use radio-frequency (RF) coils to create additional ionization, we developed an alternative technique consisting of a home made magnetron sputtering device in which the ionization of the emitted sputtered vapor is achieved by two microwave antennas. Langmuir probe measurements were used to determine electronic density and temperature. Emission optical spectroscopy has been performed and argon and titanium line intensities have been measured, showing an increase of Ti^{+*} to Ti^{*} line intensity ratio. Optical absorption spectroscopy using a titanium hollow cathode lamp powered with a pulsed power supply has also been performed to determine the ionized fraction of the sputtered vapor. Preliminary results are also given for a conventional IPVD system (with RF loops) used for the deposition of Ti-based biomaterials.

Keywords: Ionized PVD, magnetron sputtering, absorption spectroscopy.

1. INTRODUCTION

In usual magnetron devices, thermalization of energetic sputtered neutral atoms on the buffer gas limits the energy available on the substrate and increases the loss of material by diffusion. Ionized physical vapor deposition (IPVD) [1-4] is actively studied to improve the quality of deposited films obtained on the basis of conventional PVD magnetron techniques.

The concept of IPVD is to ionize a significant fraction of the metal vapor, which is emitted by the cathode sputtering as neutral species, and to collimate the obtained ion particle fluxes by applying an electric potential

A. A. Voevodin et al. (eds.),
Nanostructured Thin Films and Nanodispersion Strengthened Coatings, 113–122.

bias on the substrate. Most of the IPVD reactors use radio-frequency (RF) coils to create additional ionization [1-7]. Helicon discharges have also be considered [8]. In Orsay, we developed an alternative technique consisting in a home made magnetron sputtering device in which the ionization of the sputtered vapor is achieved by two microwave antennas [9,10] working with surface wave plasma excitation. The potential advantage of using microwave is the efficiency in terms of electron density and the possibility to work with rectangular cathode for continuous linear processes such as deposition on glass.

The present work describes the plasma study by using optical spectroscopy for quantitative data about the ionized fraction of the vapor, which is ionized. Optical Emission Spectroscopy (OES) is a technique easy to implement but Optical Absorption Spectroscopy (OAS) is, by far, more reliable to obtain quantitative results for the determination of the ionized fraction.

Analysis of deposited films has also been made in order to compare their properties in terms of composition and density for the various experimental conditions that we used (injected microwave power, buffer gas pressure, and bias voltage). In order to avoid the problem encountered in the deposition of conductive material with the microwave IPVD technique, we recently developed an IPVD reactor using a classical RF loop arrangement. Preliminary results are given in the case of Ti-based biomaterials deposition.

2. MICROWAVE ASSISTED IPVD

2.1. Microwave Assisted IPVD Reactor

The reactor represented on Fig. 1 consists of a planar rectangular titanium cathode (22 cm x 9 cm) and of two coaxial-type microwave applicators located perpendicularly to the substrate-magnetron axis, on both sides of the sputtered vapor flow.

The cathode is supplied by a DC generator (0.1 to 3 A and 250 to 330 V) and each coaxial microwave applicator by a generator (up to 1 kW). This reactor can operate on a wide pressure range: 0.2 to 60 Pa. Buffer gas is argon. The aspect of the discharge viewed from the top of the reactor is represented on Fig. 2. To avoid the deposition of metallic films on the quartz tubes, we placed a metallic protection around each of them, which limits the exposition of the quartz to the sputtered vapor issued from the magnetron cathode.

Optical emission spectroscopy has been performed using collimators. Argon and titanium lines intensities have been measured with a spatial resolution better than 5 mm. Spatially resolved measurements (between the target and the substrate) of Ti and Ti+ densities and corresponding temperatures have also been performed by absorption spectroscopy using a hollow cathode lamp powered in a pulsed regime.

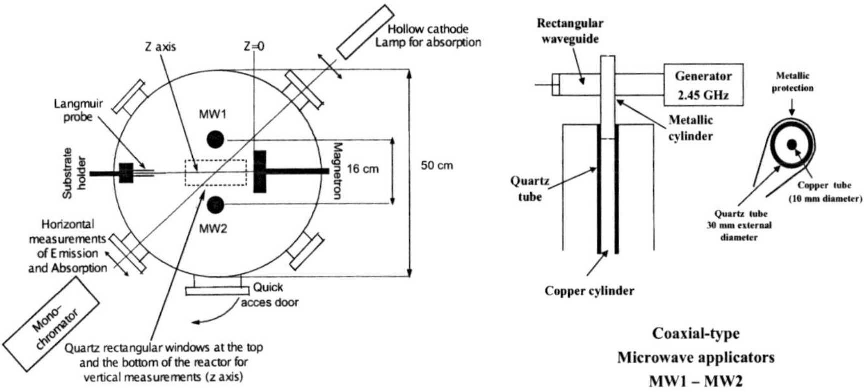

Figure 1. Schema of the IPVD reactor.

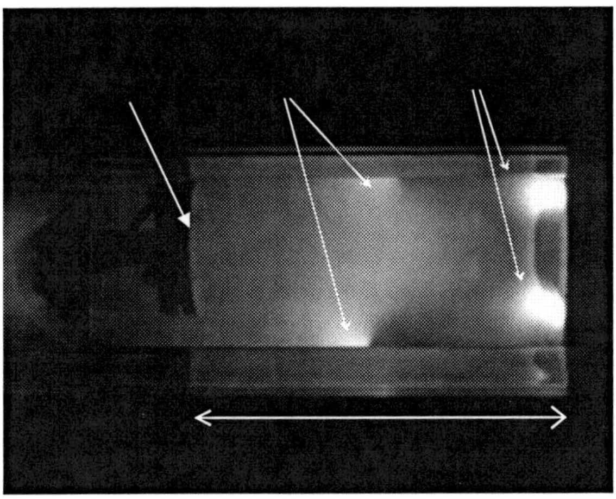

Figure 2. Top view of the discharge.

In order to optimize the antennas system and to study the dependence of the ionization on parameters such as injected microwave power on each antenna and gas pressure, we systematically determined the electron density and temperature for various experimental conditions by using Langmuir probe techniques. Results obtained for a pressure of 5 Pa are reported on Fig. 3. We observe that the electron temperature T_e is about 1.8 eV for the most part of the discharge and the electron density n_e reaches a value of about 1.3 x 10^{11} cm^{-3} in a region which faces the aperture of the antenna protection (about 9 cm from the cathode). Note that n_e remains at a value of about 9 x 10^{10} cm^{-3} near the substrate (located at 12cm).

The values of T_e and n_e let us expect a good efficiency of the microwave plasma for the ionization of sputtered metal atoms for which the ionization potential is generally in the 6 – 8 eV range while the effect on the argon ionization can be expected to be low.

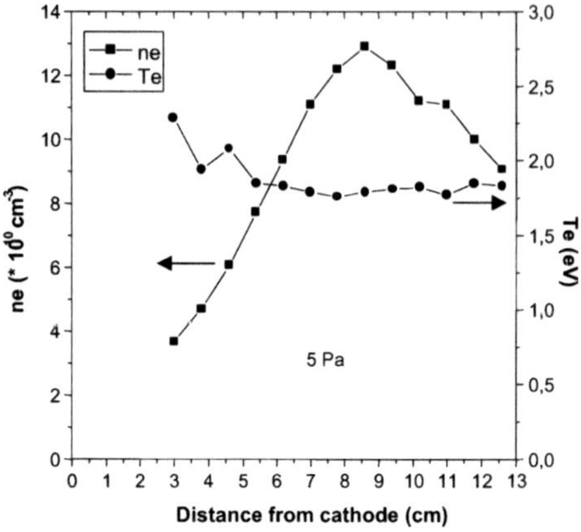

Figure 3. Electron density and temperature as a function of the distance to the cathode. A microwave power of 800 W is injected on each antenna. The pressure was 5 Pa.

2.2. Influence of Microwave Excitation on the Ti Ionization

The OES technique is very useful for the analysis of plasma excited species and in the present situation it made available the trends on the influence of the additional plasma source on the ionization of the sputtered vapor. So, it was used in order to compare the intensity of lines emitted by the titanium vapor. The variations of the intensity ratio of lines emitted by $Ti^{+}*$ and Ti* reflect qualitatively the Ti^{+} to Ti density ratio. Typical results for ion to neutral line ratio as a function of the microwave power injected on each antenna are reported on Fig. 4. We observe an increase of this line intensity ratio for increased distance from the cathode. This indicates, if we assume that ion lines are excited from the ion ground state that the ion ground state density increases more rapidly than that of the neutral ground state does or to detriment of it. In particular, the rapid increase of the intensity ratio for distances at which microwaves are effective may characterize the role of microwave excitation. But it is difficult to draw further conclusions due to the fact that we do not master the ways in which respective emitting states are formed (direct or stepwise excitation, energy dependence of excitation cross-sections).

In order to get free of the effect of excitation we have developed absorption techniques. The absorption measurements are made with a pulsed titanium hollow cathode lamp (100 μs pulse, 100 Hz frequency). The absorption A is determined experimentally by $A = 1 - I_t / I_0$, with I_0 being the intensity emitted by the lamp and I_t the transmitted intensity measured after the crossing of the plasma by the light beam.

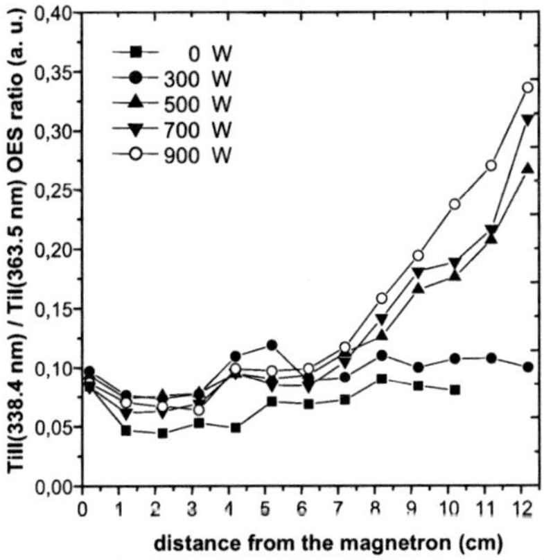

Figure 4. Ti^{+*} (TiII) to Ti* (TiI) line intensity ratio as a function of the distance to the magnetron target for various microwave powers injected on each antenna. The DC magnetron current was 0.5 A and the pressure 30 mtorr.

The absorption of four transitions were measured: three titanium lines involving the three ground state sublevels at 3635.5 Å ($^3G^0_{J=3}$ – $^3F^0_{J=2}$), 3642.7 Å ($^3G^0_{J=4}$ – $^3F^0_{J=3}$) and 3653.5 Å ($^3G^0_{J=5}$ – $^3F^0_{J=4}$) and a line with the same lower level but a different oscillator strength at3981.8 Å ($^3F_{J=2}$ – $^3F^0_{J=2}$) [11, 12]. The absorption is a function of the absorption coefficient k_0, the absorption length L and the parameter α defined as $\alpha = (T_l / T)^{1/2}$ where T_l is the emission temperature of the lamp and T the Ti atom temperature. The numerical procedure to determine α and the Ti ground state sublevels densities was described in a previous paper [11]. This procedure assumes that the Boltmann temperature which is deduced from the densities of the ground state sublevels corresponds to neutral gas temperature (argon gas); this means that the thermalization of sputtered atoms on the argon gas is achieved and that the pressure times the distance from the target is high enough (a few tenths of cm x mTorr); moreover, this assumes that the effect of electron collision on the state population is negligible compared to collisions with neutrals. The use of a pulsed regime for the hollow cathode lamp made also possible the determination of the population of the Ti $^5F_{J=5}$ metastable [Ti^{nm}] neutral state (energy level at 0.848 eV) from absorption on $^5F_{J=5} \rightarrow {}^5G^0_{J=6}$ transition at 4981.7 Å and determination of the population [Ti^{+g}] of the ion ground state from the absorption of the 3383.8 Å line ($^4F_{J=3/2}$ $\rightarrow {}^4G^0_{J=5/2}$ transition).

Results for the variation of neutral ground state Ti density (summed over the three sublevels) and temperature are reported as a function of the position from the cathode, for various microwave power P_{MW} on the antennas, on Figs 5 and 6. We observe a decrease of the ground state Ti density versus P_{MW}. This decrease cannot be directly interpreted to be due to an increase of the ionization of the sputtered vapor. For the interpretation of the absorption measurements we must take into account the modification of the gas and the plasma parameters induced by the microwave source. In particular, we have to consider the changes induced on the titanium ground state density due to the coupling of the two plasmas (the DC magnetron plasma and the additional microwave one), which modifies the discharge voltage and, consequently, the sputtering yield. Moreover, the existence of neutral titanium metastable states closely coupled to the ground state must be taken into account in the total neutral Ti density [9,12,13].

The spatial variations of densities of Ti ground and metastable neutrals and of Ti^+ ions are reported on Fig. 7 and that of the ionized fraction defined as the ratio of local Ti neutral atom density to the ion density on Fig. 8. We observe that we achieve an ionization fraction of about 5 %. This value could seem quite low but figures change when we consider the respective fluxes of neutrals and ions on the substrate. We evaluated the neutral flux to the substrate from the densities determined by OAS and the mean velocity deduced from gas temperature while the ion flux was calculated from the Bohm flux = $[Ti^+] \times V_{Bohm}$. For a pressure of 30 mTorr, a DC magnetron current of 2 A and a microwave power of 2 x 800W, we respectively got $\phi(Ti^n) = ([Ti^{ng}] + [Ti^{nm}]) \times \langle V_{Ti}{}^n \rangle /4 = 1.9\text{x}10^{15}$ cm^{-2} s^{-1} for the neutral flux and $\phi(Ti^+) = [Ti^{+g}] \times V_{Bohm} = 0.6\text{x}10^{15}$ cm^{-2} s^{-1}resulting in a ratio $\phi(Ti^+)$ / $\phi(Ti^+) = 0.3$. In these evaluation we neglected ionic metastable states and other excited states. We can conclude that by applying a bias voltage, the energy provided by ions can be much higher than that provided to the substrate without additional ionization. This conclusion was confirmed by the analysis of thin films.

2.3. Influence of Microwaves on Ti Thin Films

The composition, density and morphology of deposited films were analyzed. The composition of the titanium thin films deposited during 4 minutes was performed ex-situ by NRA (Nuclear Reaction Analysis) for [O] and RBS (Rutherford Backscattering Spectroscopy) for [Ti]. The sample thickness was determined by X-ray reflectrometry. The presence of oxygen in the films is due to the contamination when opening the reactor.

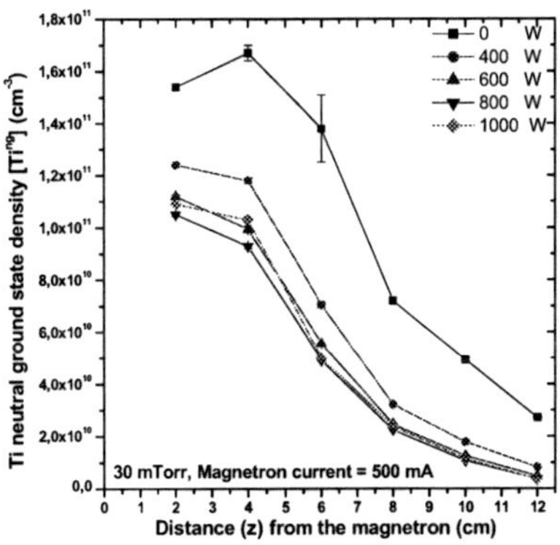

Figure 5. Spatial variation of neutral Ti density for various microwave power on each antenna. The pressure is 30 mTorr.

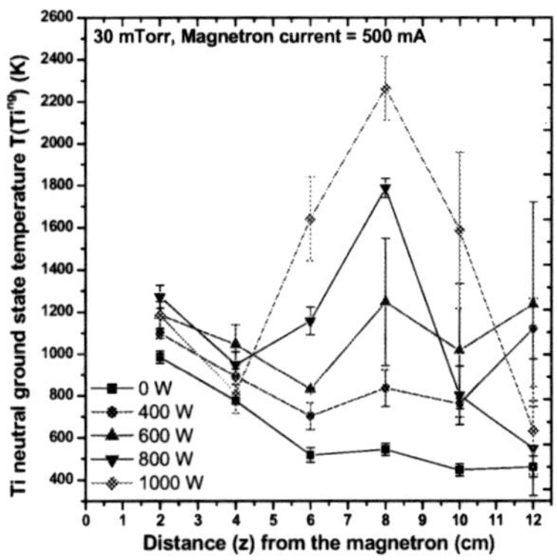

Figure 6. Spatial variation of neutral Ti temperature. Same conditions as Fig. 6. The pressure was 4 Pa.

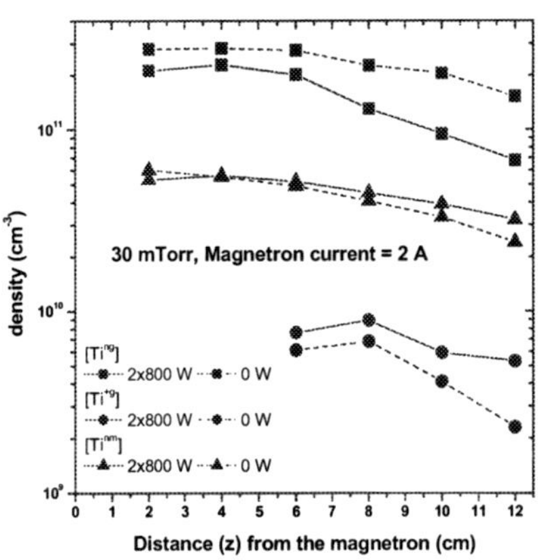

Figure 7. Spatial distributions of densities of ground and metastable neutrals and ions with and without microwave excitation.

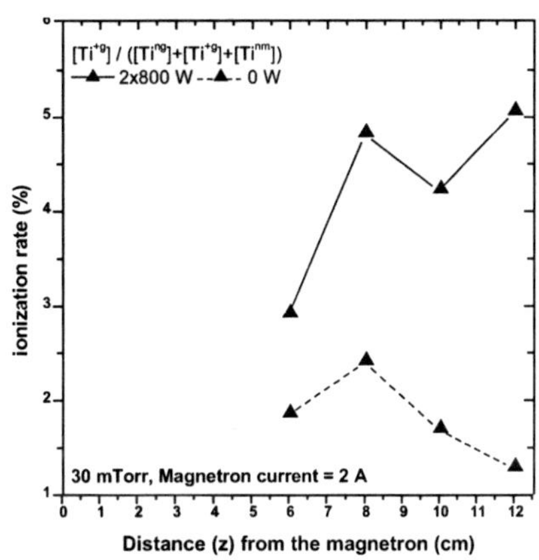

Figure 8. Spatial distribution of the ionization rate with and without microwave excitation.

Results for various deposition conditions with and without microwave excitation and substrate bias are reported in Table 1. These results show that the lower oxygen contamination [O]/[Ti] is obtained with microwave excitation and bias; it is also for these conditions that the density is the highest.

The latter results are confirmed by SEM (Scanning Electron Microscopy). The results reported on Fig. 9 for films deposited on a pin clearly show a more homogeneous structure and a lower roughness of the film deposited with the additional microwave system.

Table 1. Analysis of films deposited at 4 Pa for the various mentioned conditions (see text).

Experimental condition	Thickness (nm)	O density ($*10^{15}$ at. cm^{-2})	Ti density ($*10^{15}$ at. cm^{-2})	[O] / [Ti]	Density (g cm^{-3})
Magnetron (2 A)	73	177	251	0.71	3.38
Magnetron (2A) + Microwaves (2*800W)	82	217	233	0.93	2.96
Magnetron (2A) + Bias (-40V)	70	88	254	0.35	3.20
Magnetron (2A) + Microwaves (2*800W) + Bias (-40V)	39	47	185	0.26	4.08

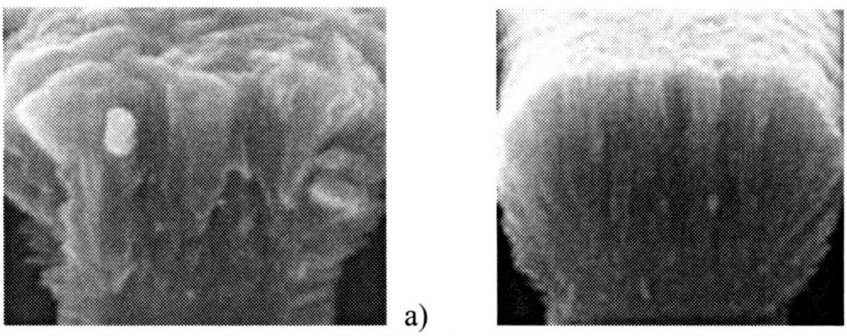

a) b)

Figure 9. Comparison between thin films deposited on a pin without (a) and with (b) additional microwave excitation. The following conditions were used: gas pressure, 2 Pa, magnetron current, 1.5 A, DC magnetron power, 500 W, microwave power, 2 x 800 W, substrate bias – 40V. The deposition time was 1 hour.

3. RF ASSISTED IPVD REACTOR

The previous results lead to interesting conclusions about the efficiency of the microwave systems for ionizing the sputtered vapor. Unfortunately, the deposition that we observed, even with a protection around the microwave applicators, limits the interest of the system for conductive materials deposition. However, the microwave system remains interesting for the deposition of isolating materials for conditions of reactive sputtering. For the deposition of conductive titanium compounds, we transformed our reactor by replacing the microwave system by classical RF loops. A scheme of the reactor is represented on Fig. 10. The cathode is circular with a diameter of 10 cm. The device was used for the sputtering and deposition of titanium based thin films. We studied thin films obtained with Ti-X-Y (X = Al, Si, V, Zr, Nb, et Ca; Y = C, O, et N) systems and particularly with the $(TiC_{0,5})_x(NbC_{0,5})_y$ system. The targets have been provided by the Moscow

Steel and Alloys Institute (Moscow Technical University) with which we established collaboration. We report on Fig. 11 the results for the ionization fraction that we obtained with the RF loop device described above. We observe that the ionization fraction varies from a few percent without RF to about 15 % with RF at a pressure of 40 mTorr.

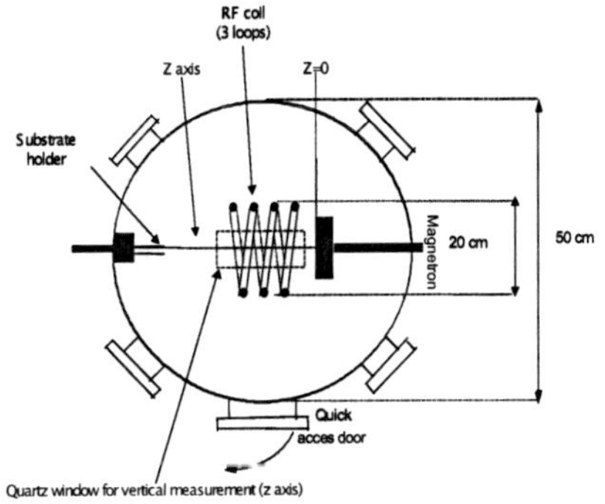

Figure 10. Experimental device for IPVD sputtering with RF loop system.

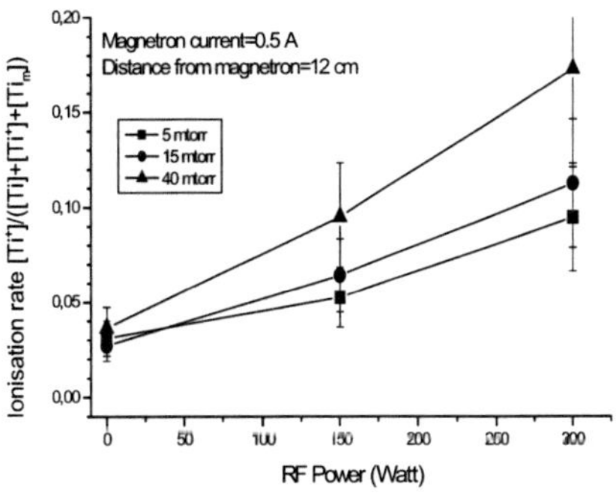

Figure 11. Variation of the ionization rate versus the RF power in front of the substrate for various pressure conditions.

The fact that the ionization fraction increases with the pressure is clearly connected to the residence time of the sputtered vapor within the plasma created by the RF excitation.

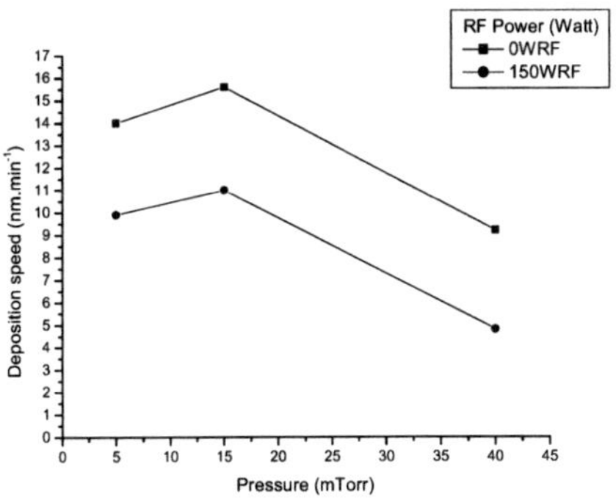

Figure 12. Deposition rate obtained with and without RF assistance.

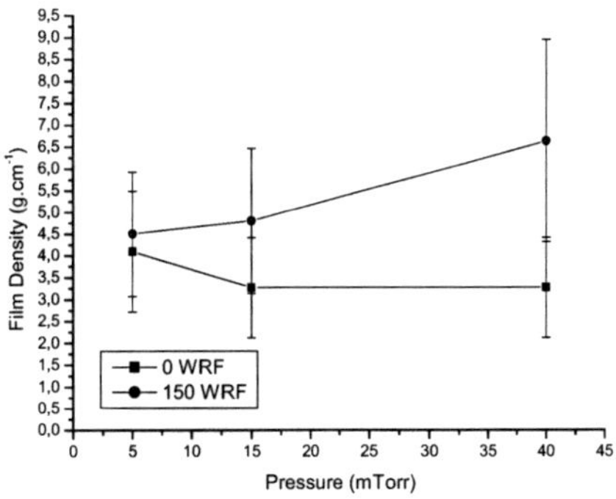

Figure 13. Thin film density obtained with and without RF assistance.

The present device has not already been optimized: the loop system was modified in order to limit the coupling with the magnetron cathode. This prevents the modification of the magnetic field geometry which leads to a strong decrease of the DC magnetron voltage and, therefore, of the sputtering yield. The results obtained for the plasma phase are in accordance with those obtained for deposited films. We present on Figs 12 and 13, respectively, the results as a function of the pressure for the deposition rate and the film density with 150 W and without RF power.

4. CONCLUSIONS

We reported results obtained with two designs of IPVD reactors, a new one using a microwave additional excitation and the other with a classical RF loop system. We determined the ionization fraction by using Optical Absorption Spectroscopy. Results obtained for deposited thin films proved the efficiency of the two devices. New concepts of high voltage pulsed magnetron discharges can also be considered [14,15].

ACKNOWLEDGMENTS

The microwave assisted IPVD study was supported by French Ministry of Industry with the DIVA project. The study of deposition of Ti-based biomaterials was conducted within NATO and INTAS projects.

REFERENCES

1. S.M.Rossnagel and J.Hopwood, J. Vac. Sc. Technol. B. 12 (1994) 449.
2. R.A. Powell, S.M.Rossnagel, PVD for microelectronics: sputter deposition applied to semiconductor manufacturing, Ac. Press, San Diego (1999).
3. M. Yamashita, Y. Setsuhara, S. Miyake, M. Kumagai, T. Shoji and J. Musil, Jpn. J. Appl. Phys. 38 (1999) 4291.
4. S. M. Rossnagel, J. Vac. Sci. Technol. B 16 (1998) 2585.
5. J. Lu and M. J. Kushner, J. Vac. Sci. Technol. A 19 (2001) 2652.
6. C.Nouvellon, S.Konstantinidis, J.P.Dauchot, M.Wautelet, P.Y.Jouan, A.Ricard and M.Hecq, J. Appl. Phys. 92 (2002) 32.
7. A. Ricard, C.Nouvellon, S.Konstantinidis, J.P.Dauchot., M.Wautelet and M. Hecq, J. Vac. Sci. Technol. A 20 (2002) 1488.
8. J.Q.Zhang., Y.Setsuhara, T.Arityasu and S.Miyake, J. Vac. Sci. & Technol. A 14 (1996) 2163.
9. P.Vasina, L.de Poucques, C.Boisse-Laporte, O.Leroy, D.Pagnon, L.Teulé-Gay, M.Touzeau, ISPC, Italy, June 2003.
10. C.Boisse-Laporte, O.Leroy, L.de Poucques, B.Agius, J.Bretagne, M.C.Hugon, L.Teulé-Gay and M.Touzeau, Surf. Coat. Technol. (in press).
11. O.Leroy, L.de Poucques, C.Boisse-Laporte, M.Ganciu, L.Teulé-Gay and Touzeau M., J. Vac. Sci. Technol. A (in press).
12. A. Konstantinidis, A.Ricard, M.Ganciu, J.P.Dauchot and Hecq, Proceedings of CIP2003 14th Int. Coll. Plasma Processes, Antibes, p.95 (June-July 2003), SFV Editor, France.
13. De Poucques L., Boisse-Laporte C., Bretagne J., Hugon M.C., Imbert J.C., Leroy O., Teulé-Gay L., Touzeau M., Vasina P., Proceedings of CIP2003 14th Int. Coll. Plasma Processes, Antibes, p.104 (June-July 2003), SFV Editor, France.
14. J.Musil, J.Lestina, J.Vlcek and T.Tolg, J. Vac. Sci. Technol. A 19 (2001) 420.
15. A.P.Ehiasarian, R.New, W.D.Munz, L.Hultman, U.Helmersson, V.Kouznetsov, Vacuum 65 (2002) 147.

Chapter 12
NUCLEATION OF DIAMOND FROM VAPOR PHASE AND SYNTHESIS OF NANOSTRUCTURED DIAMOND FILMS

B.V. Spitsyn
Institute of Physical Chemistry RAS, 31 Leninsky Pr., 119991 Moscow, Russia

A short review on diamond heterogeneous nucleation as initial stage of activated CVD of diamond was given. Special emphasis was addressed to the most important process parameters like substrate nature, its temperature, geometry, and to content and pressure of the vapor phase, as well. Substrate biasing was indicated as one of the most influencing parameters, and several independent explanation of the effect were discussed. Some important fields of application, critically dependent on knowledge and skill around nucleation, have been indicated. Several approaches for and results of nanostructured diamond film formation have been discussed.

Keywords: diamond nucleation, substrate, geometry, biasing, nucleation rate, nanostructured diamond.

1. INTRODUCTION

The CVD diamond nucleation has been intensively studied but still is a not fully understood phenomenon. A joint consideration of experimental and theoretical data directly concerned with the problem of diamond nucleation from activated vapor phase is, in our opinion, vital for understanding and control of initial stages of activated CVD of diamond (ACVD). Progress in this field will reduce the share of pure empirical research with the aim of synthesis of polycrystalline and more specifically nanostructured diamond films (DF) growth on non-diamond materials and will be important for science and practice. This movement to new knowledge and technologies still remains one of the basic tasks of the diamond community.

2. FEATURES OF THE DIAMOND NUCLEATION BY ACVD

In comparison with supersatureted vapor condensation on a not identical to its structure solid surface the diamond nucleation as the initial stage of the ACVD-process has some remarkable features.

A. A. Voevodin et al. (eds.),
Nanostructured Thin Films and Nanodispersion Strengthened Coatings, 123–136.

In the volume of a vapor phase and on its boundary, on a non diamond substrate homogeneous and on the surface of a substrate with centers of diamond growth heterogeneous chemical reactions take place [1] very far from equilibrium conditions. The channels of these reactions, as a rule, are not mutually independent and their interaction determines a non-stationary or stationary course of the nucleation process.

For a comprehensive consideration of the nucleation process (taking place in most cases on carbide-forming substrates) it is necessary to take into account the following peculiarities:

- chemical content, crystallographic and phase structure, and geometry of a carbide-forming surface during nucleation are changing,
- there are solid state chemical reactions between formed diamond phase and the varied surface of a substrate,
- both diamond sub-nuclei and the parts of a surface not covered with diamond, as well, are in a field of simultaneous action of two thermodynamic driving forces: super-saturation towards any kind of solid carbon from side of gaseous carbon-containing species and under-saturation also to any kind of solid carbon in relation to atomic hydrogen.

The super-saturation or chemical potential difference between activated vapor phase and diamond phase needs special comment. The thermodynamic driving force, or super-saturation of condensed carbon formation by ACVD consists of two members. In the case of hot filament (HF) diamond CVD-process with stationary activated vapor phase CH_3-H-H_2 composition, we shall write down $\Delta\mu = RT \ln P_{CH3}/P^0_{CH3} - RT \ln P_H/P^0_H$, where P_{CH3} and P_H are working partial pressures of methyl radical and atomic hydrogen, and P^0_{CH3} and P^0_H are their equilibrium values.

However, together with the thermodynamic driving forces by diamond nucleation the kinetics factors should play a detrimental role. Therefore now it is difficult to make realistic numerical estimations of the probability of spontaneous formation of critical diamond nuclei.

3. INITIAL FINDINGS

Heterogeneous diamond nucleation [2, 3] was first observed under nearly the same conditions of the chemical transport reaction (CTR) used for regrowth of first epitaxial diamond films [4]. Initially as non-diamond substrates metallic copper and gold were used, followed by silicon, tungsten and molybdenum. The crystallization medium in CTR contains C_2H_2, H and H_2, in concentration ~1, ~10, and ~89 mol. %, respectively. It is remarkable, that in first publications [2, 3, 5] many peculiarities can be found, which were rediscovered much later. By comparison of the nucleation activity on single and polycrystalline substrates of non-carbide and carbide-forming nature could be observed [5], that the activity:

- situated in the 10^3 to 10^8 cm^{-2} h^{-1} range is influenced by deposition conditions, nature of substrate and surface finishing prior the deposition, e.g. polishing, etching, or annealing;
- is promoted by scratches and grain boundaries;
- is enhanced up to three orders of magnitude on carbide-forming materials compared with non-carbide forming ones;
- is several times lower for single crystal substrates, than for polycrystalline material of the same chemical composition;
- is reduced with deposition time and in presence of larger regrown diamond microcrystals.

Therefore, the above-mentioned data may be illustrated in the following graph (Fig. 1).

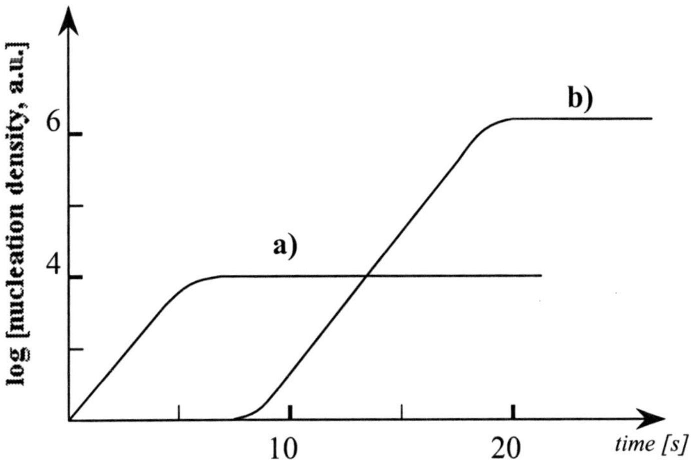

Figure 1. Approximate temporary dependence for nucleation density at initial stages of the diamond CVD on: a) non-carbide forming substrates (like Cu, Au); and b) carbide forming substrates (Si, Mo, W).

4. SUBSTRATE NATURE

Fig. 1 shows, that under comparable conditions the substrate with low affinity to carbon (Cu, Au) can initiate diamond nuclei formation with lesser time delay, then a substrate with remarkable carbon solubility and subsequent carbide formation. This takes place in latter case due to the absence of competition for carbon feeding between a nearly inert to carbon substrate and a diamond nuclei emerging on it surface.

Another important role of the substrate nature is not only bulk affinity of substrate to carbon and bulk diffusivity of carbon in the substrate [7], but surface diffusivity of carbon atoms and probably carbon-containing clusters.

As was presented in [7], with an increase of surface diffusivity, the incubation period is reduced and the nucleation rate increased. For example, at a substrate temperature of 860 ^{0}C, the surface diffusivity of carbon on Si, Mo, and Fe is calculated to be 1.21x10^{-13}, 7.07x10^{-11} and 1.32x10^{-6} (cm^{2}s^{-1}), respectively. Therefore nucleation on Fe starts nearly immediately after contact with the activated hydrocarbon-hydrogen vapor phase, while the begin of nucleation on Mo and Si occurs at a much later time.

The fact of critical nucleus formation was demonstrated [8] by AFM technique: no diamond islands with a size of less then 30 nm were observed on (100) Si after 10 minutes regrowth. Of course this observation [8] does not exclude, that a much smaller critical size of diamond nuclei is possible under other specific activated diamond (CVD) conditions.

The important rolc of the influence of the substrate was shown by S.-T. Lee and co-authors [9]. Substitution of surface oxide on silicon by Si-H, Si-I, and Si-Br compounds promotes the diamond nucleation, what may be beneficial for diamond-on-silicon heteroepitaxy and for nanostructured diamond film initiation, as well. In case of a brominated silicon surface a density of diamond nuclei up to 10^{10} cm^{-2} is possible.

A good early review on nucleation and selective growth of diamond films has been presented in [6].

Today it is generally accepted, that nucleation and following diamond film formation on carbide-forming substrates proceeds through 6 stages (Fig. 2).

Stages 1 to 5 can explain, just qualitatively, the time dependence of the nucleation density on the surface of carbide formed substrates (Fig. 1, b).

5. CARBON PRECURSOR CONCENTRATION AND TOTAL PRESSURE OF THE VAPOR PHASE

It seems clear, that when increasing the carbon precursor concentration and keeping the vapor phase activation at about the same level, the concentration of active carbon precursors, C_nH_m, should increase too.

In the same time the stationary atomic hydrogen concentration goes down in presence of other radicals, like atomic hydrogen. Therefore acting supersaturation (see Sect. 2), provides a much heavier nucleation. The experiment corresponds qualitative quite well with this expectation [10], but only in a restricted range of the total pressure. As was published recently [11], the nucleation density under mW reactor conditions may be very low near 1 Torr pressure, and then, after passing through a maximum, goes down again. Though the experiments were performed in [11] with bias application, at low pressure more energetic ions can be produced, with a more pronounced sputtering of the substrate surface, and, therefore, the destroying of diamond nuclei (Sect. 8) must happen.

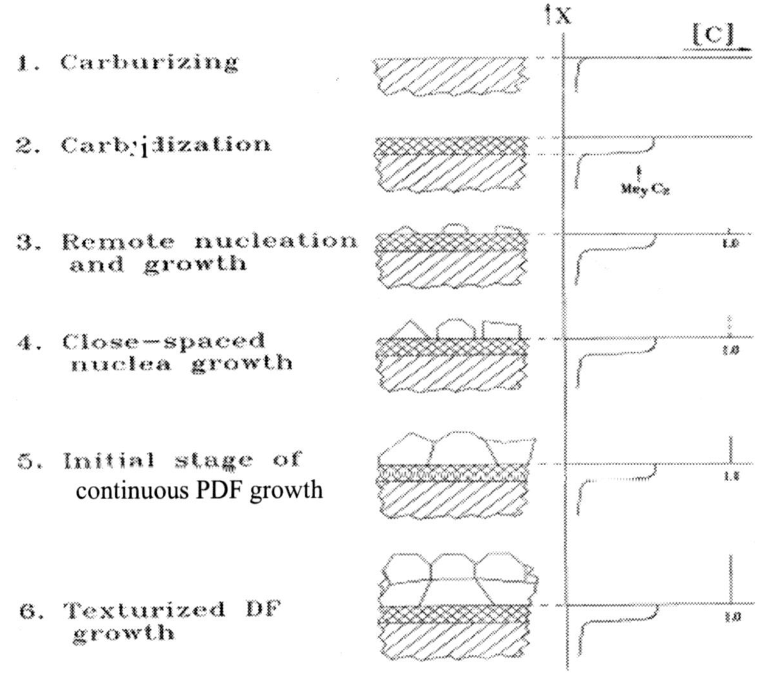

Figure 2. Stages of nucleation and polycrystalline diamond film growth by diamond CVD.

The total pressure of the vapor phase at comparable parameters of the CVD reactor can, on one side, enhance nearly linearly the C_nH_m yield. On the other side, the H concentration increases ~ $P^{1/2}$. Therefore diamond nucleation should be sensitive to total pressure of the vapor phase, too, and can grow vs. the pressure.

6. SUBSTRATE TEMPERATURE

The kinetics of new atom additions to diamond crystallization centers at its critical and sub-critical stage plays an important role, demonstrated typically by plotting nucleation density vs. substrate temperature. As it is well known, the growth rate of macroscopic diamond films drops after some maximum temperature (around 900 °C) [12]. A similar tendency shows nucleation density. It finds an explanation in the prevalence of destruction of pre-diamond carbon clusters under the action of atomic hydrogen at high temperatures. Due to the nanoscopic size of diamond sub-nuclei and nuclei populations, the decrease of nucleation rate starts in a lower substrate temperature range (~700 °C) [11, 13], than is typical for the growth rate of thick diamond films (around 900 °C) [13].

7. SURFACE GEOMETRY

The role of surface geometry in diamond nucleation probability includes several factors. First of all the sharp tips on the carbide-forming substrate surface should pass the carburizing/carbidization stage more early, and will then be preferentially ready for the diamond nuclei formation (see Sect.3, Fig.2). Secondly, due to the very high surface energy of diamond, 'wetting' by diamond nuclei on convex parts of the surfaces is much more pronounced, according to the Wentzel-Derjaguin formula [14, 15]:

$$\cos \Theta_r = r \cos \Theta_a \qquad (2),$$

where Θ_r and Θ_a are wetting angles for the rough and flat surfaces, respectively; $r = S_r / S_a$ is ratio of the real and appearing (projection to the real) surfaces. Fig.3 demonstrates such opposite influence on wetting contact angle.

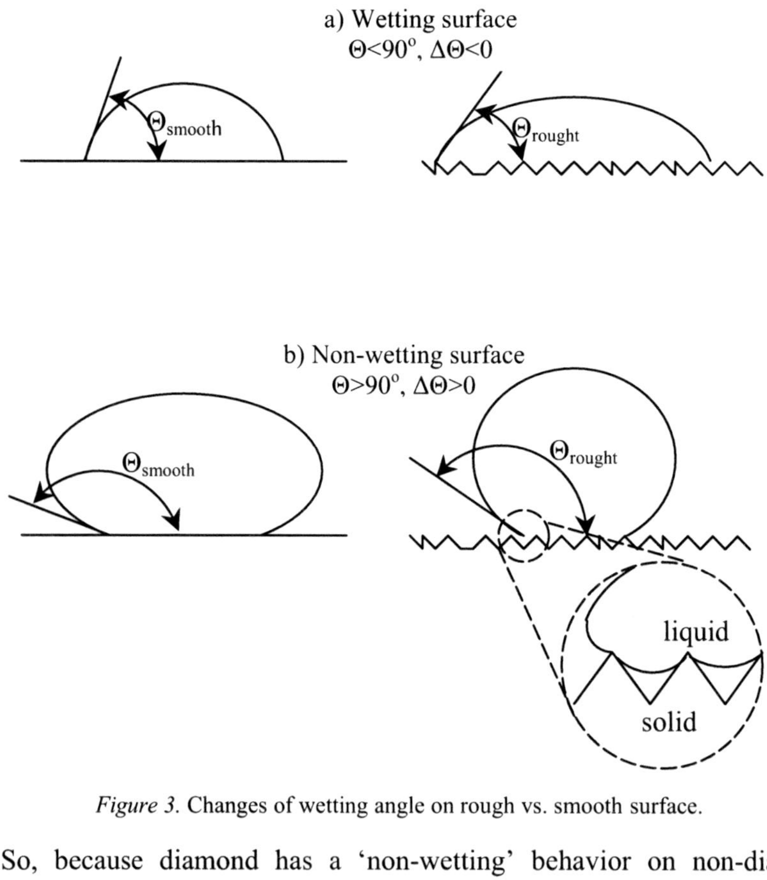

Figure 3. Changes of wetting angle on rough vs. smooth surface.

So, because diamond has a 'non-wetting' behavior on non-diamond surface, it should preferentially adhere to sharp convex surface features (edges, etc.), like demonstrated in [16] experimentally. The similar observation was done by Givargizov et al. [17]. They found in a HF CVD reactor preferential nucleation and growth of diamond nano-particles on edges, and particularly, on tips of the needle-like silicon crystals.

8. ELECTRICAL BIAS INFLUENCE ON THE DIAMOND NUCLEATION

The phenomenon was discovered by S.Yugo et al. [18]. The authors used a tantalum electrode located 4 cm apart from the substrate in a common mW reactor for the application of a dc-electrical field at the initial stage of the diamond deposition. The nucleation enhancement started only at a methane concentration in hydrogen of more than 5 %, and a bias, U_b is equal to –70 V relative to the smooth Si substrate. The most pronounced effect was achieved by 5 minutes action of the seeding conditions, with 40 % of methane in the incoming vapor phase and a bias, U_b = - 100 V. This procedure provides a great promotion of the nucleation up to 10^{11} cm^{-2}. Then usual growing conditions (methane concentration in hydrogen ~1%, no bias) provide a submicrometer continuous diamond film regrowth on Si substrates. Authors [18] explain the influence of negative biasing of the substrate by extra activation of precursors of diamond (or mixed sp^3/ sp^2 clusters) by impingement of positive ions, arriving at the surface with some optimal energy. As well, the effect just partially may originate from the enhancement of surface diffusion (Sect. 4).

As was demonstrated in [19], biasing of the substrate to + (75-85) V may increase the diamond film (DF) growth rate 2-3 times relative to non-biasing conditions. Such additional gain in the growth rate at the initial stage of diamond nuclei formation may be helpful very much for their surviving and passing through their 'critical' size towards stable diamond nuclei. Charging of the surface can cut activation energy of the activated complex conversion up to 20 kcal/mole, if the complex contains polar chemical bonds [20]. The effect should be independent on sign of excessive charge of the surface.

Now, Si substrate biasing at special concentration/temperature conditions is very popular for the initiation of preferential DF/Si orientation as one of inroads to the diamond heteroepitaxy on single crystalline Si substrates [21, 22].

9. NANOSTRUCTURED DIAMOND FILMS

It is obvious, that one of the most effective ways for polycrystalline diamond film deposition with small and, in particular, nanosized diamond crystallites, is the use of regularities of diamond nucleation on nondiamond substrates.

9.1. Role of Super-Saturation

As was explained in previous sections the value of carbon super-saturation of gas phase in relation to nucleating and growing diamond phase undoubtedly is the factor most strongly influencing frequency of formation of diamond nuclei on a surface. Even on a single crystalline diamond substrate, by it the regrowth from pure methane gas at ~1000 ^{0}C and pressure 0,2 Torr, actual super-saturation is so great, that already after a short period of epitaxial growth (up to thickness of 15 A), at the expense of multiple twinning and secondary nucleation, the growth of polycrystalline nanostructured diamond film begins [23].

Though the isothermal decomposition of methane on a diamond surface is impractical for nanocrystalline diamond film synthesis, owing to low rate of their growth, nevertheless because of relative simplicity of process proceeding by brutto- reaction: CH_4 (г) = C (diamond) + $2H_2$ (г),
it is possible to estimate working super-saturation at deposition of diamond, using expression for chemical potential: $\mu = \mu^0 + RT \ln P$.
In the diamond CVD case, interesting for us, the difference of chemical potentials or working super-saturation will be equal to

$$\Delta \mu = RT \ln[P(CH_4)/P^0(CH_4)],$$

where $P(CH_4)$ is given partial CH_4 pressure, and $P^0(CH_4)$ its equilibrium value.

For known conditions of CH_4 decomposition on a diamond substrate, the temperature 1200K and P (CH_4) =0.2 Torr; P^0 (CH_4) =10^{-4} Torr, it is easy to receive a rating of the numerical value of the working super-saturation, equal to 7.54 kJ/mol.

Considerably more complex is the task of super-saturation rating by activated CVD of diamond, as was mentioned earlier (Sect. 2): working super-saturation is the algebraic sum of two thermodynamic driving forces. The first one works on the side of all kinds of hydrocarbon particles at their decomposition towards to any kind of solid carbon.

The second term has an opposite sign and represents under-saturation, owing to the etching action of atomic hydrogen, thermodynamically permitted, also related to all kinds of solid carbon, including in their number the diamond.

At all not knowing absolute size of expression:

$$\Delta\mu=\sum RT\ln[P_i(C_nH_m)/P_i^0(C_nH_m)] - RT\ln[P(H)/P^0(H)],$$

it is possible to use it as the qualitative index at a choice of conditions for intensive creation of centers of diamond phase growth and, hence, for reception of nanocrystaline diamond films.

As already mentioned in a previous section, S.Yugo et al. [18, 22] for the first time have shown a super-saturation frequency of diamond of order (10^{11}-10^{12}) cm^{-2}, that allows to achieve diamond crystallite sizes at an initial stage of its growth at a level of 10 nm. The authors [18, 22] used a two-step process:

At its first stage, which is carried out in the mode of a standard mW-discharge, a methane rich (40 % CH_4 in hydrogen) initial vapor phase is used and low, ~ 600 ^{0}C, silicon substrate temperature, and more then -100 V substrate biasing., as well.

Under conditions of activation by mW-plasma, both in gas phase and on the silicon surface with high probability not only nuclei of diamond, but also nondiamond clusters of carbon formed. After the specified stage for the deposition of a thin composite diamond - carbon film it was changed to the usual mode of growth of a diamond film from the mW-discharge (~1 % CH_4 in H_2, temperature ~ 850^{0}C). Obviously, in this deposition mode all nanoislands of nondiamond carbon were actively etched by atomic hydrogen, and the further growth led to the formation of a nanostructured diamond films, which are not containing clusters of nondiamond carbon in the structure, as could be proofed by Raman-spectroscopy.

A similar effect like the increase of the partial pressure of the carbon-containing component in the gas phase can be achieved by lowering the total gas pressure. It is a fact that for the majority of carbon-containing molecules and radicals (with the exception of only C_2H_2 and C_2H) their decomposition takes place with increase of number of particles, for example:

$$CH_3\ (g) = C\ (diamond) + 3/2H_2\ (g).$$

Therefore, as experiment could show too, the average crystallite size in the diamond film obtained at reduced total pressure with all other experimental conditions identical can be reduced. This opens one more option to synthesize nanostructured diamond films.

9.2. Crystallization Temperature

Both, the growth rate of a diamond film and the nucleation frequency of centers of diamond growth have an extreme dependence on substrate temperature [7, 13]. However, the nucleation frequency usually passes through a maximum at a temperature ~200 degree lower, than the temperature of the maximal growth rate of the diamond film, which is in the 1200-1300 K range [12].

The most pronounced faceted morphology of a diamond film is shown near to extreme temperature, though the visible size of cubic and octahedral faces can exceed the size of coherent scattering 1-2 decimal orders of magnitude, determined by XRD [12].

At a lower crystallization temperature of the diamond film, than the temperature of a maximum, the secondary nucleation (at ~950^{0}C) is somewhat like in the appearance to raspberry, as it demonstrates the pattern in scanning electron microscope (Fig. 4, b).

Apparently, the secondary nucleation to a greater degree increases with the further decrease of crystallization temperature to 875 ^{0}C (Fig. 4, a).

The SEM resolution is not high enough already to observe the separate crystallite morphology. Their aggregation gives an external morphology, which in the literature is called “cauliflower”. It is important to note, that with a decrease of the crystallization temperature from 1400 to 1100 K the coherent scattering range of diamond crystallites reduced monotonously from 190 up to 30 nm (Fig. 5 and Fig. 6), respectively.

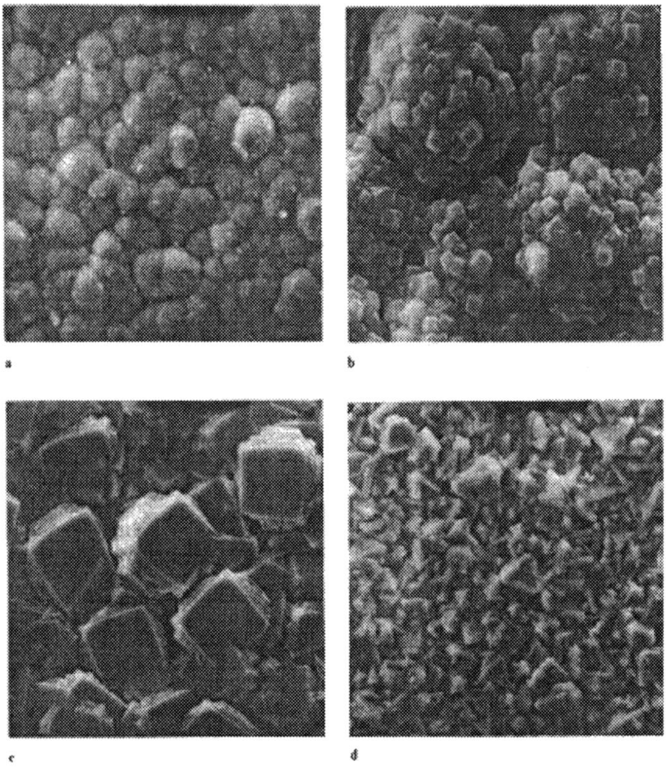

Figure 4. SEM-revealed surface morphology of polycrystalline diamond films grown at temperature of the Si substrate of: a – 1100 K, b – 1200 K, c – 1300 K, d – 1450 K.

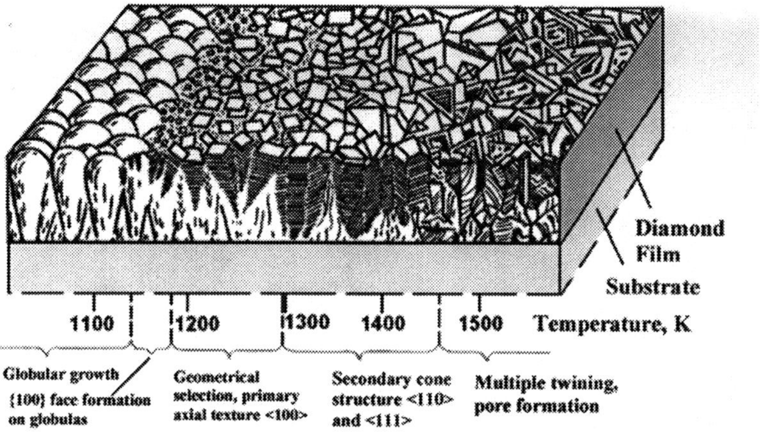

Figure 5. Evolution of the internal structure with increasing thickness at different crystallization temperatures.

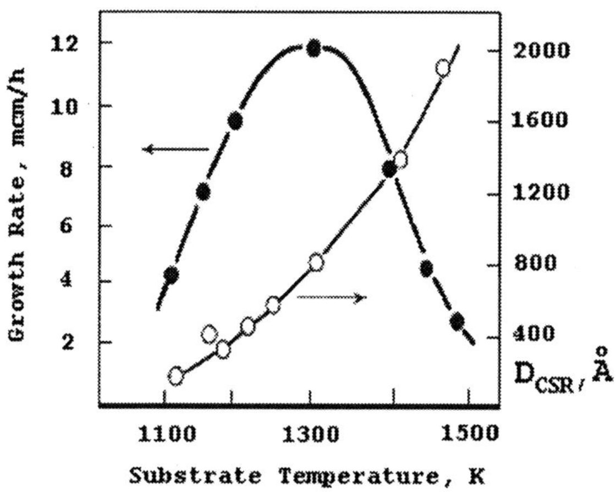

Figure 6. Crystallization-temperature dependence of the range of X-ray coherent scattering (D_{CSR}) expressed in A on the right-hand scale and the growth rate of polycrystalline diamond film on the left-hand scale.

Simultaneously and also monotonously the number of the dangling carbon bonds in the diamond film is increased, which was determined by the ESR method. However, their concentration is approximately 2 decimal orders of magnitude lower, which could be, if all atoms at grain boundaries had not coupled electronic orbital. Obviously, on one hand, significant share sp^3 - orbital of carbon atoms are mutually overlapping in the area of grain boundaries even at not optimal mutual angular orientation sp^3 - orbital and greater, than equilibrium in diamond (1.54 A), distance between carbon atoms.

At the same time bonded in a diamond film the hydrogen can compensates the broken bonds with C-H bond formation, which are found out by FT-IR -spectrometry. This process also is responsible for reduction of the broken bond density diamond film.

9.3. Role of the Precursor

A new nucleation mechanism with basic precursor C_2 molecule was discovered by Gruen [25] by using fullerene–argon or methane–argon mixtures with very low content of hydrogen. It was revealed an original path for new carbon atom addition through insertion of the C_2 molecules in C-C and C-H bonds. Because of much lower kinetic barriers of the reactions, the nucleation density grow up 6 decimal orders of magnitude, resulting in nano-diamond film formation with crystallite size of 3 to 15 nm [25].

9.4. Intentional Substrate Seeding by Diamond Nanoparticles

This method, under conditions for achieving a uniform and dense (close to a monolayer) seeding of the substrate surface by nanodiamond particles allows, basically, to achieve the crystallites in a diamond film at a level of a nanoparticle of diamond, i.e. about 4-5 nm. The method yet has not received appropriate application because pronounced and enough stable diamond nanoparticles are not inherent in nanodiamond powders, fabricated by detonation of explosive substances [26].

The first experiments in this direction [27] have shown, that individual particles of detonation diamond are twinned, which has been revealed at early stages of their CVD growth through observation in the high resolution TEM.

10. CONCLUSIONS

Basic results and current issues of metastable diamond nucleation from activated vapor phase have been discussed shortly.

It was revealed that diamond nucleation has special features compared with more simple cases of condensed phase nucleation.

The control of nucleation is crucial for obtaining the desired grain size, orientation, transparency, adhesion, etc., which is required for an optimal use of the CVD diamond in most applications. For example, smooth pinhole-free polycrystalline films for optics or lithographic masks can be formed only at high nucleation density and uniform crystal size on appropriate substrates. CVD diamond semiconductor devices need low defect density, single crystal heteroepitaxial diamond and selective area nucleation. Wear and cutting tool applications require uniform and dense nucleation over large and/or irregularly shaped substrates like working parts of the tool.

Progress in the understanding and control of diamond nucleation on flat and 3D substrate of non-diamond materials will initiate many kinds of other CVD diamond applications in novel and high technologies like diamond heteroepitaxy, cold electron emitters, nanotechnology, etc.

Therefore the solution of the diamond nucleation problem demands increased efforts for a better understanding and will open the way to effective and wide practical applications.

ACKNOWLEDGEMENTS

Author appreciate for help in the manuscript preparation to Roman G.Samarin, and Alexey B.Spitsyn.

REFERENCES

1. Bouilov L.L., A.E.Alexenko, A.A.Botev, and B.V.Spitsyn, Sovjet Physical Chemistry (1986) 302.
2. B.V.Derjaguin B.V., B.V.Spitsyn, L.L.Bouilov, A.A.Klochkov, A.E.Gorodetskii, and A.V.Smol'yaninov, Sovjet Phys. – Doklady 21 (1976) 676.
3. Derjaguin B.V., B.V.Spitsyn, L.L.Bouilov, G.V.Alexandrov, G.G.Alexandrov, V.P.Repko, A.E.Gorodetskii, and Z.E.Sheshenina, Dokl. Akad. Nauk SSSR 244 (1979) 388.
4. Derjaguin B.V., B.V.Spitsyn, A.E.Alexenko, et al., Dokl. Akad. Nauk SSSR 213 (1973) 1094.
5. Spitsyn B.V., L.L.Bouilov, and B.V.Derjaguin, J.Cryst. Growth 52 (1981) 219.
6. Popovici G., M.A.Prelas, Phys. Stat. Sol. (a) 132 (1992) 233.
7. Liu H. and D.S.Dandy, Electrochem. Soc. Proc., vol. 95-4, 112.
8. Jiang X. et al., in : The American Physical Society, vol. 50, No 12, 15 Sept., 1994 - II, 8402.
9. Lee C.H., Z.D.Lin, N.G.Shang, L.S.Liao, I.Bello, and S.-T.Lee, Phys Rev. B 62 (2000) 1734.
10. Kim J.W. et al., in: Application of Diamond Films and Related Materials, Proc.1-st Int. Conf. – ADC'91, Auburn, Alabama, U.S.A., Aug.17-22, 1991, ed. by Y.Tseng et al., 399.
11. Shu X.-S., Q.-C.Wu, R.-Q.Liang, Plasma Sci. Technol. (Hefei, China) 2, (2000) 475.
12. Spitsyn B.V., in: Handbook of Crystal Growth, D.T.J.Hurle, ed., Vol. 3A, Ch. 10, 401-456 (1994), Elsevier, Amsterdam.
13. Aksenov I.I. et al., in: Eurodiamond '96, A Workshop on CVD Diamond and its Application, Torino, 17-20 January,1996, ed. by C.Manfredotti et al., 17.
14. Wentzel R.N., Ind. Eng. Chem., 28 (1936) 988.
15. Deryaguin B.V., Dokl. Akad. Nauk SSSR 51 (1946) 257.
16. Denning P.A. and D.A.Stevenson, Appl.Phys. Lett. 59 (1991) 1562.
17. Stepanova A.N., L.L.Aksenova, A.V.Kuznetsov, E.V.Rakova, and E.I.Givargizov, Mater. Letts. 22 (1995) 285.
18. Yugo S., T.Kanai, T.Kimura, and K.Muto, J.Appl. Phys. 58 (1991) 1036.
19. Spitsyn B.V., L.L.Bouilov, A.E.Alexenko, Diam. Relat.. Mater. 8 (1999) 1371.

20. Kazansky V.B. and N.D.Ttshuvylkin, Dokl. Akad. Nauk SSSR 223 (1975) 910.
21. Bozeman S.P., B.R.Stoner, and J.T.Glass, Nucleation and epitaxy, in Handbook of Industrial Diamonds and Diamond Films, M.A.Prelas et al., eds., Marcell Dekker , Inc. , N. Y., (1998) 901.
22. Yugo S., Mechanism of diamond nucleation and heteroepitaxial growth by chemical vapor deposition, in: Proc. 5-th Int. Symp. " Diamond Films and Films of Related Materials", Kharkov Scientific Assembly, April 22-27, 2002, Kharkov, Ukraine, Ed. by V.I.Lapshin, M.A.Prelas, V.M.Shulaev, and B.V.Spitsyn, 16-19.
23. Tesner P.A., A.E.Gorodetsky, E.V.Denisevitch, A.P.Zakharov, T.V.Tekunova, Dokl. USSR, 222 (1975) 1384.
24. Spitsyn A.B., I.V.Galoushko, A.F.Belyanin, in: Proc. 4-th Int. Symp. On Diamond Films and Films of Relative materials, eds. A.F.Belayanin et al., Kharkov, Ukraine (1999).
25. Gruen D.M., Ann. Rev. Mater. Sci. 29 (1999) 211.
26. Dolmatov V. Y., Detonation Synthesis Ultradispersed Diamonds: Properties and Applications, Russian Chemical Reviews 70 (2001) 607.
27. Butler J.E., Growth and application of nanodiamond films – MEMS, NEMS, and bejoind, in14-th Eur. Conf. on Diamond, Strassburg, Austria, Sept.7-12,2003. Abstract Booklet, 1.1.

Chapter 13
NANOSTRUCTURED AND NANOLAYERED THIN FILMS DEPOSITED BY ION-PLASMA METHODS

V.V. Sleptsov, A. M. Baranov
"MATI" - Russian State Technological University by K.E. Tsiolkovsky, Orshanskaya str. 3, Moscow 121552, Russia

Theoretical discussion of surface influence on the properties of nanostructured and nanolayered films is provided. Correlations between mechanical, thermal, chemical and electrical contributions to the nanocomposite properties are shown. Results of investigations of quantum-dimensional multilayer carbon structures are presented. Perspectives of applications of multilayer carbon structures are discussed.

Keywords: nanostructure, nanocomposite, multilayer, carbon, mechanical properties, optical properties, electro-physical properties.

1. INTRODUCTION

A fundamental interest to nanostructured materials and devices is based on the reduction of the size of particles or structures to the level which is comparable with material characteristic parameters, such as, electron wavelength, free path of an electron, size of a magnetic domain, etc [1]. As a result, physical principles of operation of these electronic devices are changed.

An applied interest to nanomaterials is related to a possibility of an additional property variation as well as to a capability of nanotechnology in the creation of novel functional devices.

Property modification of solid state materials by nanoparticle synthesis or preparation of multilayer structures with nanosize layers is the main activity of researchers in nanotechnology [2-8].

As the size of a particle is progressively reduced, the particle volume properties become altered by its surface characteristics. The reduction of geometrical dimensions involves the reduction of the volume to surface area ratio, such that the properties of the substance (resistance -ρ, index of refraction - n, energy gap - E_g, hardness - H, plasticity, friction coefficient, etc.) become dependent on the particle size. It is quite natural that the surfaces properties dominate over the bulk properties when particle size decreases below 10-100 nm. This allows a wide range change of the

A. A. Voevodin et al. (eds.),
Nanostructured Thin Films and Nanodispersion Strengthened Coatings, 137–146.

properties of nanoparticle and nanolayerd structures by a variation of the individual particle size or layer thickness. In result, we can obtain a combination of specific properties (electrical, mechanical, and optical) in nanocomposite materials which can not be realized in bulk materials.

2. THEORETICAL DISCUSSION OF NANO-COMPOSITE MATERIALS

There is a difficulty in establishing a well defined interface between the surface and the bulk volume of material. Therefore, surface properties of a solid body begin at the interface of two media and spread into material bulk to a depth of about 100 nm [9-14]. Therefore, the surface shell can be considered as a special class of self-organizing materials, which properties are determined through a combination of properties coming from a three-dimensional bulk volume, a two-dimensional surface area, and, in some cases, an one-dimensional component.

Using this approach, one can consider and define a whole class of new self-organizing structures, which have fractural dimensionality from 0.5 to 3. For these structures, the surface properties dominate substantially over the bulk volume properties [15-21].

A question naturally arises, how hardness, mechanical strength, and electro-physical properties of a solid nanostructured body depend on the energy of its crystal lattice, which is strongly influenced by the surface energy. Theoretical and practical importance of this question attracts attention of many researchers.

Surface energy (σ_{hkl}) is the most important property of the surface. Specific surface energy σ_{hkl}, which is determined by the Gibbs-Curie [9] principle and the Wolf law for an equilibrium solid state, is

$$\sigma_{hkl} = \varepsilon_{hkl} - T(d\sigma_{hkl}/dT)p \quad (1)$$

where ε_{hkl}- is a specific total surface energy, T – is temperature, p –is a pressure.

It is important to show that variation in (σ) is well correlated with a variation of mechanical characteristics, e.g. hardness (Fig. 1).

Ormont [9] showed that σ_{hkl} depends on the bonding energy Ω in the following manner:

$$\sigma_{hkl} = M\varepsilon_{(hkl)} (\Omega / a^2) \text{ [erg/cm}^2\text{]} \quad (2)$$

where $M\varepsilon_{(hkl)}$ is a structure constant of the surface energy, which depends on structure type and crystallographic plane orientation (hkl); and d is an identity period (for example, a crystalline period, period of multilayer structure, etc.).

Equations (1) and (2) suggest that the bonding energy, which determines mechanical properties of the material, determines also its specific surface energy (σ_{hkl}). This conclusion is confirmed by experimental results (Fig. 1). This can be seen from Fig. 1, that one can change H by several orders of magnitude by varying σ. This is particularly pronounced for diamond, where σ_{100} =9304 erg/cm^2; H_{100} = 10,000 kg/mm^2; σ_{111} = 5378 erg/cm^2; H_{111} =5000 kg/mm^2 [22].

Using general equations of the first and second laws of thermodynamics applied to the surface, one can conclude that mechanical characteristics (micro-hardness H and wear resistance) increase with the increase of the internal energy of a solid body. This conclusion agrees with traditional approaches to calculate a solid body hardness [22]:

$$dU^F = TdS + \sigma\, dF + \varphi dq + \Sigma \mu_i dn_i \qquad (3)$$

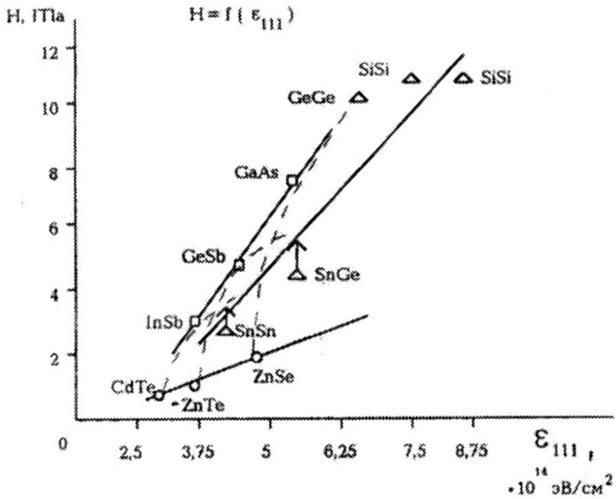

Figure 1. Relations between micro-hardness (H) and specific surface energy of semiconductors (E_{111}) [6].

where U^F is internal energy, S is entropy, σ is specific surface energy (surface tension), F is surface area, φ is potential, q is charge, μ_i is chemical potential, and n_i is molar fraction of a substance.

In a two-dimensional system the internal energy can also increase with the increase of the entropy. This fact is confirmed by experimental observation of phase transformations on solid surfaces and an increased thermal capacity of the surface For example, NaCl bulk thermal capacity is Cv=0.34 cal/(mol*K), while surface capacity is Cp= 0.6 cal/(mol*K).

Taking into consideration that

$$S = \int_0^T \frac{C}{T} dt \quad , \qquad (4)$$

where C - thermal capacity, we can see that the surface entropy is directly related to the surface capacity and, therefore, can be also much higher than that of the bulk volume. From eqs (3) and (4) we obtain that

$$dU^F = CdT + \sigma\, dF + \varphi dq + \Sigma \mu_i dn_i \qquad (5)$$

Thus, both thermal surface capacity and entropy are higher than volume capacity and entropy. The smaller is the grain size the large is the surface energy contribution to the hardness. At the same time the hardness is strongly influenced by the thermal capacity term in eq. (5). Fig. 2 demonstrate the surface thermal capacity effect by comparing indentation hardness at different loads (different surface penetration depths) for two nanostructured materials with different thermal surface capacity.

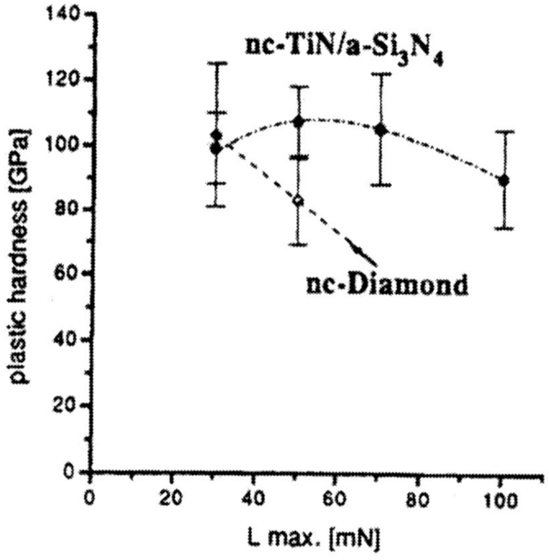

Figure 2. Hardness variation with indentation load for 3.5 μm thick nc-TiN/a-Si_3N_4 coating and single phase 1.5 μm thick nanocrystalline diamond coating.

The eqs (3)–(5) indicate a possibility of reversibility of internal energy in the process of deformations. The mechanical energy can be transformed into electrical, thermal, chemical, etc. These equations can be used as a theoretical base to predict mechanical characteristics of nanocomposites and predict self-organization of nanostructures.

3. EXPERIMENTAL

The experimental confirmation of conclusions in the previous section could be phase transitions on solid surfaces under mechanical deformation, hardness variation with gas absorption on surfaces, or elastic recovery of some surfaces after an applied mechanical load is removed. For example, Fig. 3 shows a variation of the indentation depth with the load in nanoindentation tests of an a-C:H coating. This figure demonstrates a complete recovery of the coating surface after indenter removal, which is an important factor for wear resistance improvement.

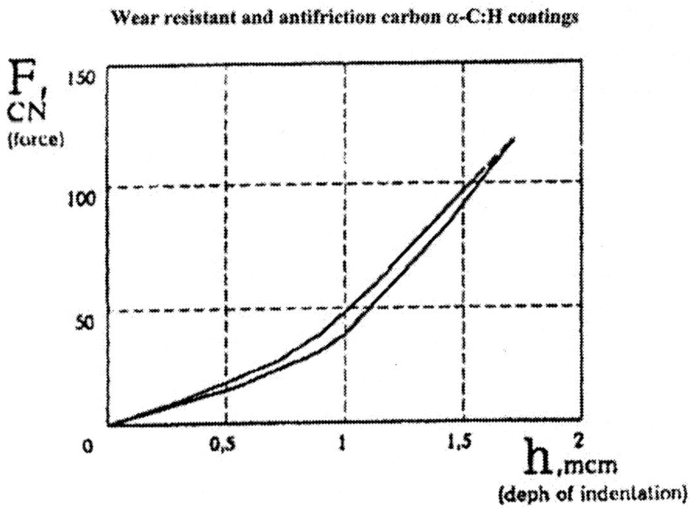

Figure 3. Indentation diagram F-h.

The effect of self-organization can most clearly seen in the properties of quantum size multilayer structures. Fig. 4 shows transmission electron microscope (TEM) image of a multilayer carbon structure with a period of 3.78 nm [23]. Such multilayer structures can be used for X-ray and neutron reflections (Figs 5 and 6) [24-26]. These multilayer structures are not stable thermodynamically, but can be produced in a metastable form with a very slow transformation rate, which provides their kinetic stability. They are stable under different external effects (in particular, X-ray and neutron radiation as well as thermal heating). This confirms that the self-organization may provide usefull stability to multilayer systems with a help of kinetic factors.

Figs 7 and 9 show optical and electrical properties of multilayer carbon structures with different thicknesses [27, 28]. An energy band diagram of multilayer carbon structures produced on a p-type Si is shown in Fig. 8.

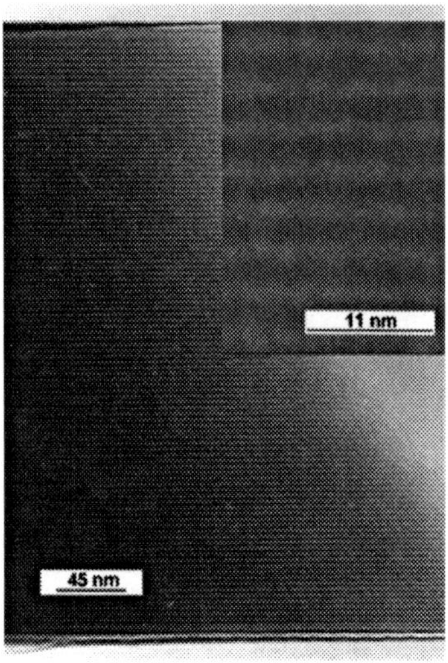

Figure 4. TEM cross section image of a C/C layer stack (PL-307) with d= 3.78 nm and N=100 in overview image and at higher magnification of the overview image (insert).

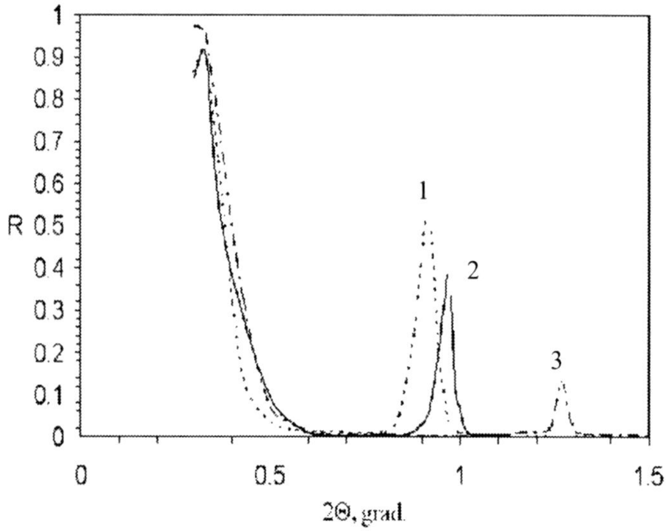

Figure 5. X-ray diffraction results $R_{(Cu\ K\alpha)}(\Theta)$ for C/C multilayers with different period of thicknesses d. 1) d=108 Å; 2) d=96 Å; 3) d=72 Å.

From Fig. 7, optical properties of multilayer carbon structures change with the layer thickness. It is important to note that the properties of the multilayer structure and the single layer of the same thickness are strongly different. This can be used to control the properties of nanostructured materials as well as to utilize the effect of self-organization. Similar effects are also observed in respect to electrical properties. A volt-ampere characteristic of the multilayer structure (Fig. 9) has a characteristic behavior indicating the presence of energy quantum effects.

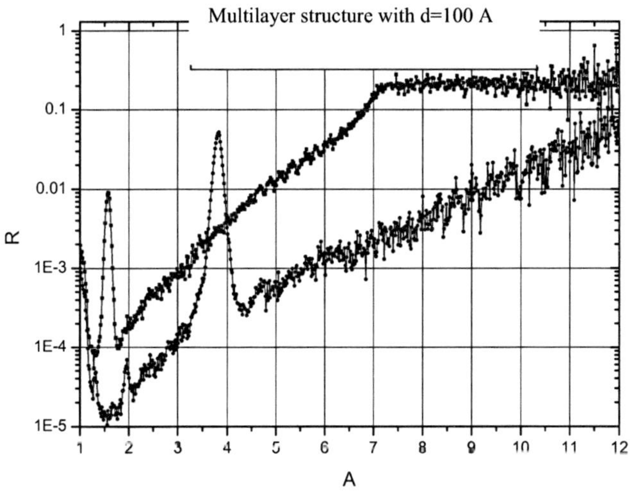

Figure 6. Spectral dependencies of neutron reflectivity.

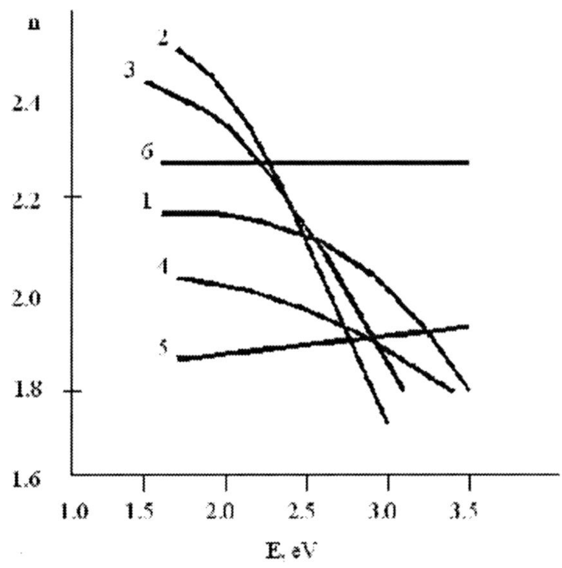

Figure 7. Reflectance (n) for α-C:H films obtained in Ar atmosphere d= (1)- 200 nm, (2) - 75 nm, (3) - 50 nm, (4) - 20nm, (5) - 10 nm, (6) - multilayer structure (19 layers, the layer thickness is 5 nm).

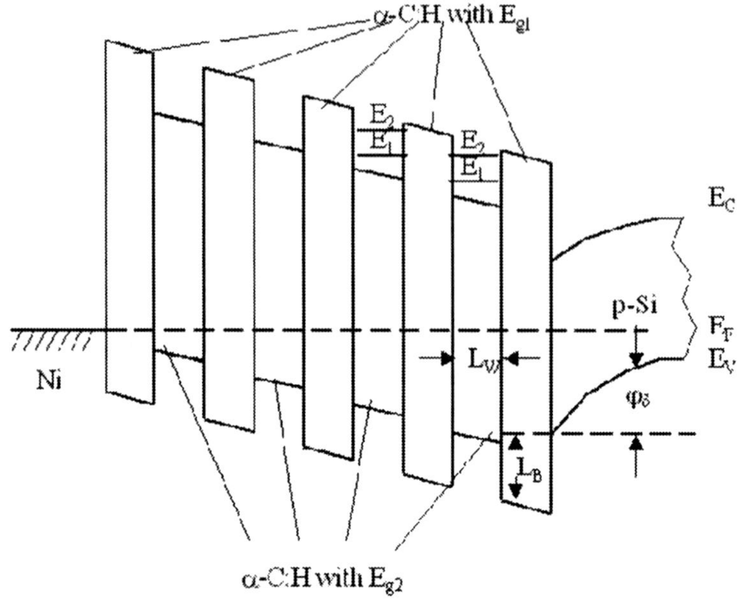

Figure 8. Multilayer structure.

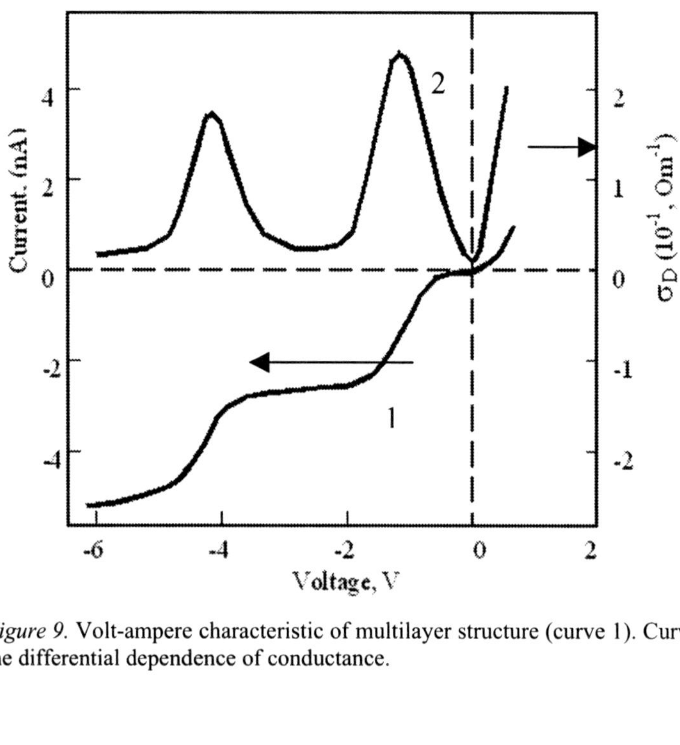

Figure 9. Volt-ampere characteristic of multilayer structure (curve 1). Curve 2 is the differential dependence of conductance.

Multilayer structures with nanometer layer thickness had already found extensive applications in various electronic and optical devices. Their chacteristic properties are ensured by alternating two or more layers of different materials. Examples of such multilayer structures are semiconductor superlattices and interference structures used in X-ray optics. These structures can be called "classical", owing to the fact that a change in the properties at the interface between different layers is due to a compositional change. Alternatively, multilayer structures can be produced in terms of carbon films deposited under different conditions. Their existence is due to the unique nature of the carbon atom which can form bonds with different hybridizations (sp^1, sp^2, sp^3), giving rise to carbon layers with a wide range of optical, electrical, and mechanical properties. Multilayer structures can, hence, be produced without varying elemental composition from one layer to another. This type of multilayer structures can be termed "non-classical". They exhibit small surface roughness. In addition, mutual diffusion between neighbouring layers, leading to spreading of the interface, is prevented by the lack of compositional gradients. Some features of the optical and electrical behavior of multilayer carbon structures can be explained in terms of quantum size effects.

The multilayer carbon structures attract today a great interest in view of a very large number of possible applications in various fields. In the electronic field, some interest is focussed on special optical and optoelectronic applications. Infrared photodiodes and light-emitting diodes based on multilayer carbon technology are currently considered for optical communications in very-large-scale integrated circuits to create infrared detectors and to develop optical information processing systems.

4. CONCLUSIONS

Theoretical discussion of surface influence on the properties of nano-composite films established a general correlation between mechanical, thermal, chemical and electrical properties of nanocomposites. All these properties were related to the surface effects, which are enhanced for nanostructured materials. Suggested correlations can be used to predict both mechanical properties and self-organization properties of these structures.

Investigations of quantum-dimensional multilayer carbon structures indicated a good kinetic stability of such composites under various types of irradiation. These structures have also very interesting quantum energy effects for both electrical and optical characteristics. One benefit is expected from the application of nanocomposite structures for hardness improvement and wear protection. Multilayer carbon nanostructures have also a strong potential for applications in x-ray optics and optoelectronics.

REFERENCES

1. Gusev A.I., Rampel A.A., Nanocrystalline Materials, Moscow, Physical-Mathematical Literature, 2000.
2. Workshop Report: SRC/NASA Ames Workshop on Emerging Opportunities and Issues in Nanotubes and Nanoelectronics; Workshop: November 12-13, 1998; (SRC/NASA), D. Herr, M. Meyappan and V. Zhirnov.
3. National Nanotechnology Initiative: Leading to the Next Industrial Revolution, Internal Government Report, Supplement to the President's FY 2001 Budget, February 2000.
4. Nanostructure Science and Technology, NSTC Report; Kluwer Academic Publ., October 1999.
5. Nanotechnology: Shaping the World Atom by Atom, NSTC Report.
6. Nanotechnology Research Directions: IWGN Workshop Report, NSTC Report; Kluwer Academic Publ., February 2000.
7. Russian Research and Development Activities on Nanoparticles and Nanostructured Materials, St. Petersburg, Russia, Proceedings 1997.
8. R&D Status and Trends in Nanoparticles, Nanostructured Materials, and Nanodevices in the United States, Arlington, VA, Proceedings 1997.
9. Ormont B.F., Introduction in Physical Chemistry and Crystalline Chemistry, High School, Moscow (1973) 389.
10. Petrov Yu.I. Clusters and Small Particles, Moscow, Science, 1986.
11. Abraham F.F. CRC Crit. Revs Solid State and Mater. Sci., May (1981) 169.
12. Wunderlich W., Ishida Y., Maurer R. Scripta Metall. Mater. 24 (1990) 403.
13. Thomas G.J., Siegel R.W., Eastmen J.A. Scripta Metall. Mater. 24 (1990) 201.
14. Mutschele T., Kirchheim R. Scripta Metall. 21 (1987) 135.
15. Sleptsov V.V., Elinson V.M., Baranov A.M., Tereshin S.A. Wide Band Gap Electronic Materials, NATO ASI Series 1 (1995) 257.
16. Polyakov V.I., Perov P.I., Ermakov M.G., Ermakova O.N., Sleptsov V.V., Elinson V.M. Thin Solid Films 212 (1992) 226.
17. Sleptsov V.V., Kuzin A.A., Elinson V.M., Baranov A.M., In book "Physics and Technology of Diamond Materials", Polaron Publishers, Moscow (1994) 80.
18. Sleptsov V.V., Kyzin A.A., Baranov A.M., Elinson V.M., Diamond and Related Materials 1 (1992) 570.
19. Meng F., Schlup J.R., Fan L.T., Chem Mater. 9 (1997) 2459.
20. Smirnov B.M., Physics of Fractal Clusters, Moscow, Science, 1991; Achievements of Physical Science, 6 (1991) 141; 163 (1993) 51.
21. Hou J.G., Wu Z.Q. Phys. Rev. B 42 (1990) 3271.
22. Ivanova V.C., Strength and Plasticity of Composite Materials, Materials Science and Heat Treatment, 1975.
23. Baranov A.M., R.Dietsch, T.Holz, M.Menzel, D.Weißbach, R.Scholz, V.Melov, J.Schreiber, High resolution carbon/carbon multilayers, Proc. SPIE 4782 (2002) 160.
24. Kondrashov P.E., Smirnov I.S., Novoselova E.G., Baranov A.M., Applied Physics Letters 69 (1996) 305.
25. P.E. Kondrashov, I.S. Smirnov, E.G. Novoselova, V.V. Sleptsov, Proc. SPIE 2517 (1995) 125.
26. Arkadiev V., Baranov A.M., Erko A., Kondrashov P.E., Langhoff N., Novoselova E.G., Smirnov I.S., Veldkamp M., Packe I., Proc. SPIE, Denver 3773 (1999) 122.
27. Polyakov V. I., Perov P.I., Rukovishnikov A.I., Khomich A.V., Rossukanyi N.M., Baranov A.M., Advances in Microcrystalline and Namocrystalline Semiconductors-1996, ed. R.W. Collins, P.A. Alivisatos, I. Shimizi, T. Shimada, J.C. Vial, MRS Symposium, Pittsburg, Pennsylvania, USA 452 (1997) 845.
28. Sleptsov V.V., Kuzin A.A., Ivanovsky G.F., Elinson V.M., Gerasimovich S.S., Baranov A.M., Kondrashov P.E., J. Non-Crystalline Solids 136 (1991) 53.

Chapter 14
PLASMA- AND BEAM-ASSISTED DEPOSITION METHODS

S. Grigoriev, A. Metel
STANKIN University, Vadkovsky Lane 1, Moscow 127055, Russia

Upgraded technology of combined substrate treatment (pre-nitriding plus hard ceramic film deposition) based on cathodic arc evaporation and ion acceleration with a bias voltage resulted in a substantial improvement of performance and increase in the useful lifetime of treated products. Meantime, as an alternative to the arc-based technology, two groups of new plasma- and beam-assisted deposition methods were developed. The first group allows deposition on dielectrics and is based on a modernization of the fast neutral molecular beam sources, which transforms the sources to universal devices able to produce high-power broad electron beams, metal vapor to be deposited on the substrates and beams of fast molecules. The second group may be applied only to conductive substrates. The substrates are immersed in the plasma of glow discharge with electrostatic confinement of electrons and ions from the plasma sputter quite homogeneously the water-cooled targets made of titanium sheets, which cover all the walls of the vacuum chamber. Losses of the target material and droplets are totally excluded.

Keywords: thin film, deposition, ion-beam, plasma, fast molecules, adhesion, microhardness, density, internal stress, wear reduction.

1. INTRODUCTION

One of the well-accepted industrial methods of hard ceramic film deposition is based on cathodic arc evaporation in vacuum of titanium, chromium, molybdenum, niobium, aluminum, other metals, their alloys and on the synthesis of their nitrides, carbides, carbonitrides and other hard materials on the surface of conductive substrates immersed in plasma under a negative bias potential. The bias potential is used to accelerate ions from the reactive gas plasma and to bombard the substrate surface during the deposition. The particles give their momentum to the surface atoms, and the increase of the surface atom mobility results in a substantial increase of film density, hardness and compressive internal stress in comparison with films deposited without ion bombardment.

In order to ensure an adequate adhesion of the films, the substrates are pre-treated in the arc metal plasma without reactive gas: etched and heated through metal ion bombardment with an energy of ~ 1 keV.

A. A. Voevodin et al. (eds.),
Nanostructured Thin Films and Nanodispersion Strengthened Coatings, 147–154.

The arc method is remarkable for a high film deposition rate of up to dozens of μm/h and is widely used to increase wear-resistance and lifetime of cutting tools. But the field of its applications is limited due to metal droplets being always emitted from the arc cathode spots simultaneously with highly ionized metal vapor. The method is not applicable in those cases when low surface roughness and low friction coefficient are required such as: microelectronics, friction pairs in machinery, forming tools, some cutting tools, for instance, broaches and others.

2. IMPROVEMENTS OF ARC TECHNOLOGY

In recent years, the arc method was sufficiently upgraded in the Physical and Technological Research Center (PTRC) of STANKIN University and a new technology of combined substrate treatment [1,2] (pre-nitriding plus hard ceramic film deposition) was developed. A great part of the work has been fulfilled within the framework of the NATO SfP-972420 project. The contribution of NATO partners from Leuven Catholic University and Istanbul Technical University [3] as well as the sponsorship by NATO has played an important role and allowed the development and production of new systems on the base of the most sophisticated control units, instruments and power supply units produced in NATO countries.

The Fig. 1 showes one of a new fully computerized arc PVD systems equipped with a movable screen, which prevents droplets from an arc-evaporated target to reach substrates during pre-treatment [3]. Metal plasma near the target is separated from the substrates by a screen, which permits only electrons to enter the working zone of vacuum chamber. Metal plasma is used as a high-current emitter of electrons, which ionizes reactive gas in the working zone on its way to a positively biased additional anode located at the opposite wall of the chamber. Substrates immersed in the gas plasma are heated and etched with gas ions instead of metal ions, which allows the bias voltage to be decreased from ~ 1 kV down to several hundred volts. During this treatment, deposition of metal droplets on the substrate surface is totally excluded, thus decreasing the number of total droplets in the films.

Pre-treatment with gas ions allows not only to etch surface with, for example, argon or krypton ions, but also a chemical-heat pre-treatment, for example, pre-nitriding of substrats when nitrogen or its mixture with argon or helium is used. Fig. 2 presents a fracture of a 5-μm-thick TiN-film deposited on ~ 20-μm-thick pre-nitrided layer of a HSS substrate. Tests showed that performance of the comparably thin TiN film with microhardness of ~ 2500 $HV_{0.050}$ deposited on a pre-nitrided surface layer with microhardness of ~ 1300 $HV_{0.050}$ is substantially improved in comparison with the same TiN film deposited without pre-nitriding on a HSS substrate with microhardness of ~ 800 $HV_{0.050}$.

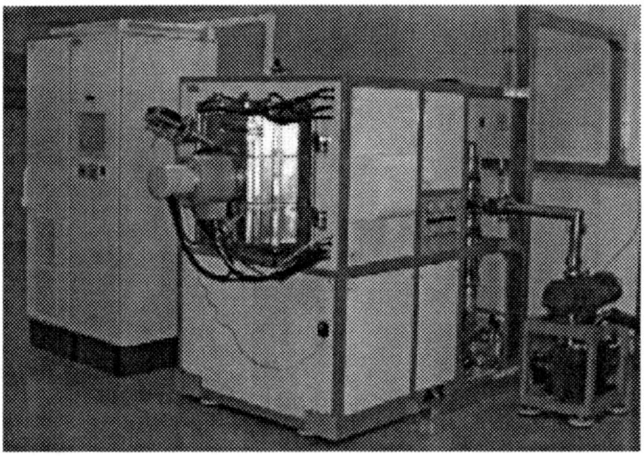

Figure 1. Picture of fully computerized arc PVD system for combined treatment (pre-nitriding plus hard ceramic film deposition) equipped with three arc evaporators and a movable screen inside the vacuum chamber.

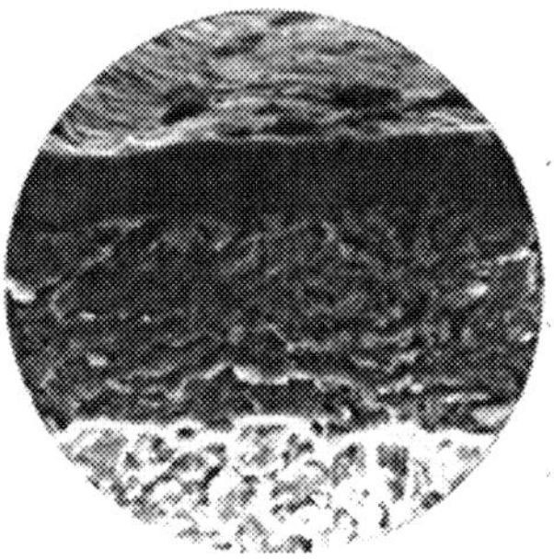

Figure 2. Fracture of 5-μm-thick TiN-film deposited on ~ 20-μm-thick pre-nitrided surface layer of HSS substrate.

The new arc PVD systems additionally equipped with sources of fast neutral molecules [4] facilitate homogeneous treatment of substrates with complex geometrical shapes and allow not only film deposition on negatively biased conductive substrates but also beam-assisted deposition on dielectrics. Fig. 3 presents a system equipped with two arc evaporators and two sources of crossed beams, the first being installed on the door and the second on the top of the working vacuum chamber.

Pre-cleaning and heating of the surface with homogeneous broad beams instead of ions accelerated with a bias voltage improves adhesion of the deposited films [5,6], eliminates excessive sputtering of sharp edges and blunting of tool cutting edges. Fig. 4 elucidates: acceleration of ions 1 in the sheath between homogeneous plasma 2 and negatively biased complex-shaped substrate 3. Due to variation of the sheath width and the electrical field at the substrate surface, the distribution of ion current density, j_i, is highly heterogeneous, and at sharp edges it may be many times higher than

Figure 3. Picture of arc PVD system for combined treatment (pre-nitriding plus hard ceramic film deposition) equipped with two arc evaporators, one movable screen and two beam sources on the door and on the top of the chamber.

in the slots. As to the homogeneous flow density distribution in a broad beam of fast neutral molecules, it cannot be disturbed with any fields and for this reason it guarantees homogeneity of substrate etching. The treatment with crossed beams in addition ensures an adequate treatment of the substrate slots.

When a substrate is positioned in simultaneously superimposed flows of metal plasma from an arc-evaporated target and of fast molecules from a beam source then beam-assisted deposition on a dielectric substrate is also available. Fig. 5 presents a micrograph of a 5.8-μm-thick TiN-film deposited on a flat Al_2O_3 substrate.

Before the film deposition the sample surface, with an initial microhardness of 1800 $HV_{0.050}$, was exposed to a 10-minute-long bombardment with 900-eV fast argon molecules at emission current 2 A of the beam source. During activating of the surface with argon molecules the sample temperature increased to 160 °C. After the ion-producing gas was changed to nitrogen the sample surface was treated for 3 minutes with 300-eV nitrogen molecules at the emission current of 0.4 A. An arc evaporator was then switched on and beam-assisted deposition started at an arc current of 100 A and nitrogen pressure of $4x10^{-3}$ mbar. During deposition the sample temperature was kept constant and equal to 160 °C. The adhesion of the TiN film, deposited at comparably low temperature, was quite good and its microhardness was 2500 $HV_{0.050}$.

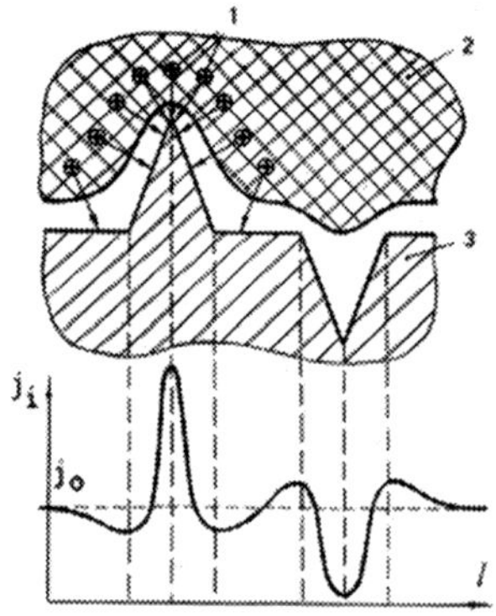

Figure 4. Ion current density distribution on substrate surface with sharp edges and slots.

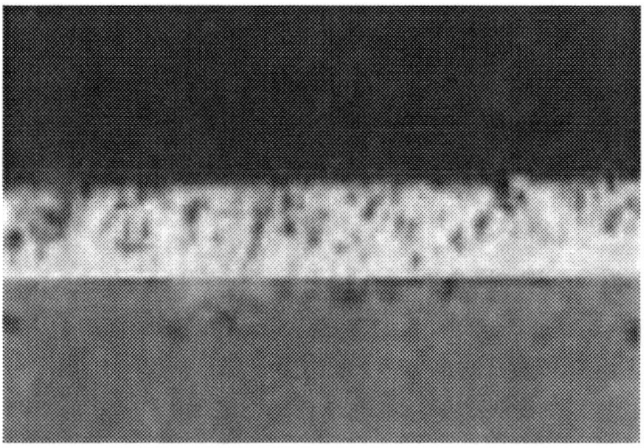

Figure 5. Micrograph of 5.8-μm-thick TiN-film deposited on an Al_2O_3 substrate with assistance of 300-eV nitrogen molecule beam.

3. BEAM-ASSISTED DEPOSITION WITH UNIVERSAL SOUCES OF BEAMS AND METAL VAPOR

All the above improvements of arc technology as well as plasma filtering with magnetic duct cannot totally eliminate droplets in deposited films. It is also very difficult to carry out beam-assisted deposition on substrates positioned on a planetary rotating system in any standard working vacuum chamber. The main requirement is a total and simultaneous superimposition on the substrate surface of metal plasma flow and of a fast molecule beam. The requirement may be adequately fulfilled, for instance, using groups of sources, each group comprising two sources of metal plasma with an angle between their axis of about 45° – 60°, and a beam source inserted between them. But in this case a special configuration of the working chamber is needed. For this reason some new approaches were proposed and new plasma- and beam-assisted deposition methods were developed. The new methods may be divided into two groups.

The first group is based on a modernization of the fast neutral molecule beam sources [4] by means of negatively biased target immersion in their plasma emitters [7]. As a result the sources turn into universal devices able to produce high-power broad beams of ~ 10-keV secondary electrons, coupled with flow of metal vapor to the substrates and beams of fast molecules. One such device is able to heat, if needed, massive substrates with electrons, to clean and activate their surface with energetic molecules and to deposit films containing material of the target, with the deposition being assisted by a high-current beam of low-energy molecules. All the particles follow the same path from the emission grid of the universal source to the substrate surface, and for this reason the requirement of their total superimposition is automatically fulfilled.

Fig. 6 presents an experimental beam-assisted deposition system with an A-series [7] universal beam source installed in the center of a 0.1-m^3 vacuum chamber. The design of the A-series sources allows simultaneous treatment of a great number of substrates positioned around the source. Tests showed that, in contrast to the magnetron method, the whole 600 cm^2 surface of cylindrical titanium target (6 cm diameter x 33 cm long), co-axial with the accelerating grid (20 cm diameter x 35 cm long) was sputtered homogeneously. At 3 kV bias voltage of the target and 2A current in its circuit the TiN film deposition rate on approximately 4000 cm^2 area of cylindrical substrate surface surrounding the source and 10 cm distant from the grid was homogeneous amounting to 2 μm/h.

Immersion in a plasma emitter of a negatively biased target is possible in all three series of beam sources described in [7]. Tests of a circular cross-section beam source are currently under way. This system has a 22-cm-diam flat accelerating grid and a 16-cm-diam disc-shaped titanium target immersed in its plasma emitter. Threefold decreases of the target surface down to 200 cm^2 in comparison with the source in Fig. 6, allowed the same power supply unit to increase the current density of target sputtering ions up to 30 mA/cm^2 and to increase deposition rate of TiN films up to 5 μm/h.

In contrast to the above arc method, the films can be deposited without bias voltage on various conductive and dielectric materials without any droplets. Independent regulation of the metal vapor flow density as well as the energy and flow of molecules assisting deposition allowed optimization of film density, hardness, internal stress and even stoichiometry. For instance, increase of the molecule beam energy from 200 eV up to 500 eV resulted in the change of the film color from the gold color of TiN to the dark color of Ti_2N. Sputtering of the accelerating grid outer surface is possible, if needed, which allows small additions of the grid material. If the grid is made, for instance, of copper, with an addition of 2 – 10 % of Cu, a nanocrystal TiN film may be produced.

Figure 6. Photograph of the experimental system for film deposition without droplets on dielectrics equipped with an A-series universal beam source.

4. FILM DEPOSITION WITHOUT DROPLETS AND LOSSES OF TARGET MATERIAL

The second group of new technologies is based on the glow discharge with electrostatic confinement of electrons [8] and may be applied only to conductive substrates. The substrates are immersed in the discharge plasma and ions from the plasma sputter quite homogeneously the water-cooled targets, which cover all the walls of the vacuum chamber. The most distinctive feature of these technologies is total elimination of the target material losses. In all previous cases, only a small part of the target material, arc-evaporated, magnetron-sputtered or sputtered in plasma emitters of the above universal beam sources, was deposited on the substrates and the rest was deposited on the walls of the working vacuum chamber. Because of a useless deposition of expensive target materials, the chamber walls should be periodically cleaned. In the experimental system presented in the Fig. 7 there is no need to clean the chamber because all the sputtered metal atoms missing the substrates are being deposited back to the target surface.

Figure 7. Picture of experimental system for film deposition on conductive substrates without droplets and losses of target material.

The experimental system has one cylindrical 50-cm-diam and 50-cm-long target and two disc-shaped 50-cm-diam targets mounted at the doors of the chamber. All the targets are made of 2-mm-thick titanium sheets and connected to the negative pole of a high-current DC arc-handling power supply unit. At an argon-nitrogen mixture gas pressure of $\sim 10^{-3}$ mbar, and a voltage of $\sim$ 700 – 800 V between the targets and a disc-shaped grounded anode positioned over the surface of the cylindrical target, the glow discharge current amounts to 20 A. The targets are being homogeneously sputtered with argon ions and, for this reason, there is no need to rotate the substrates in vacuum. In the deposited films there are no droplets.

The system allows two ways to negatively bias the substrates. To assist deposition with ions, a DC voltage of $\sim$ 200 V may be applied to a substrate holder positioned in the center of the chamber. In this case, the substrates are homogeneously coated with conventional TiN films. Before the deposition the substrate surface is etched with argon ions at 800 – 1000 V bias voltage, and at this time the surface of the targets is reactivated with ~ 300 eV ions after a slight oxidation caused by opening of the chamber door. In order to accelerate the reactivation a small addition of hydrogen is effective.

The system is additionally equipped with a high-voltage pulse generator. Applying to the substrate a series of negative 40-μs-wide pulses of up to 40 kV consequently with frequency of 25 Hz instead of the DC bias voltage, it is possible to deposit high quality Ti_2N films with a hardness of 5000 $HV_{0.04}$ and an interface width of up to several microns [9].

5. CONCLUSIONS

Tests of cutting tools and machine parts coated with the upgraded technology based on cathodic arc evaporation discussed in this paper exhibited a substantial improvement in their performance. Despite the impossibility of total elimination of the droplets in the films, this new, high deposition rate arc technology plays an important role in surface modification and still is able to keep its position in the world market.

In addition the new technologies of beam-assisted film deposition on dielectrics and of deposition on conductive substrates without droplets and losses of target material of new high quality films with a wide interface and high hardness may, in the nearest future, become serious competitors of the traditional arc technology.

REFERENCES

1. L. P. Sablev, A. A. Andreev, S. N. Grigoriev, A. S. Metel, US Patent No 5,503,725 issued on Apr. 2, 1996.
2. L. P. Sablev, A. A. Andreev, S. N. Grigoriev, A. S. Metel, European Patent No 0 583 478 issued on 14.10.1998.
3. S. Grigoriev, A. Metel, A. F. Cakir, M. Urgen, J.-P. Celis, M. De Bonte. Thin film deposition through complex ion-plasma treatment, Proc. of 14th Int. Conf. on Ion-Surface Interactions, Zvenigorod, 1999. Moscow, 2 (1999) 218.
4. A. Metel, S. Grigoriev, US Patent No 6,285,025 issued on Sep. 4, 2001.
5. M. Urgen, A. F. Cakir, U. Kahraman, O. L. Erilmaz, A. Metel, Effect of neutral molecule source sputtering on the adhesion of TiN coatings, Proc of 5th International Conference on Advances in Surface Engineering, 1999, Sao-Paulo, Brasil
6. A. F. Cakir, A. Metel, M. Urgen and S. Grigoriev, Galvanotechnik, 91 (2000) 768.
7. S. Grigoriev, Yu. Melnik, A. Metel, Surf. Coat. Technol., 156 (2002) 44.
8. A. Metel, Surf. Coat. Technol., 156 (2002) 38.
9. C. Ruset, E. Grigore, Surf. Coat. Technol., 156 (2002) 159.

Chapter 15
MULTICOMPONENT NANOSTRUCTURED THIN FILMS. DEPOSITION, CHARACTERIZATION, TESTING AND APPLICATION

D.V. Shtansky
Moscow State Institute of Steel and Alloys, Moscow, Russia

The current topics related to the deposition, characterization, testing and application of tribological coatings are reviewed and refer to our recent results on nanostructured (Ti,Cr)-(Al,Si,Zr,Ca)-(C,B,N,O) thin films. The PVD targets based on the systems TiN+TiB_2, TiB_2+TiC, Ti_5Si_3+Ti, Ti_5Si_3+TiN, TiB_2+Ti_5Si_3+Si, TiB_2+Si, Ti_5Si_3+TiC, TiAl+TiC, TiB+Ti_9Cr_4B+Cr_2Ti, CrB_2, TiB_2+TiAl+Ti_2AlN, TiC+Ti_3SiC_2+$TiSi_2$, $TiC_{0.5}$+ZrO_2, $TiC_{0.5}$+CaO, Ti_5Si_3+ZrO_2 were manufactured by means of self-propagating high-temperature synthesis (SHS). The BN and WSe_x films were deposited using a sintered hexagonal BN and a cold pressed WSe_2 target, respectively. Multicomponent films were deposited by DC magnetron sputtering of composite targets either in an atmosphere of argon or reactively in a gaseous mixture of argon and nitrogen. The WSe_2 films were deposited by PLD and ion-assisted PLD techniques. The microstructure and chemical composition of the films were studied by means of Auger electron spectroscopy (AES), energy-dispersive X-ray spectroscopy (EDXS), X-ray diffraction (XRD), scanning and transmission electron microscopy (SEM and TEM), high-resolution (HR) TEM, X-ray photoelectron spectroscopy (XPS), and electron energy-loss spectroscopy (EELS). The films were also characterized in terms of their hardness, elastic modulus, elastic recovery, adhesion, surface topography, friction and wear. Particular attention was paid to the analysis of the structure of the films in the nanometric scale and the evaluation of the properties of the individual nanostructures. Three groups of films are considered: 1) nanostructured hard, tribological coatings for civilian engineering Ti-B-N, Ti-Cr-B-N, Ti-Si-N, Ti-Al-B-N, Ti-Si-C-N, Ti-Si-B-N, BN and Cr-B-N; 2) double-layer self-lubricating coatings for civilian air force and space industry WSe_x/TiC, WSe_x/TiCN, WSe_x/TiSiN and 3) biocompatible coatings for medicine Ti-Ca-C-N-O, Ti-Zr-C-N-O, and Ti-Zr-Si-O-N.

Keywords: magnetron sputtering, nanostructure, thin film, surface fracture, structure, phase composition, surface topography, orientation relationships, physical properties, mechanical properties.

1. INTRODUCTION

Nanostructured thin films are of interest because of their remarkable characteristics. Despite a certain progress in the field of nanostructured thin

A. A. Voevodin et al. (eds.),
Nanostructured Thin Films and Nanodispersion Strengthened Coatings, 155–165.

films, many problems focusing on the fundamental understanding of their behavior in nanoscale and the measurement of film properties still remain unsolved, which so far has limited the applicability of nanostructured thin films.

2. FUNDAMENTALS OF NANOSTRUCTURED THIN FILMS

2.1. Nanoscale

Multicomponent thin films are usually characterized by a very small grain size down to several nm as estimated from either the dark field (DF) TEM image or XRD spectrum. Even at this rather small scale, the structure of the $TiAl_{0.2}B_{0.7}N_{0.7}$ film [1] was unique compared to other films investigated. The grain size was estimated from the DF TEM image to be approximately 3-8 Å. This is an extremely small size, which has hitherto been reported, being only one to three unit cells dimension, assuming the presence of TiN, TiB_2 or BN phases. Strictly speaking, the determination of grain size using DF image is not a perfect method, because a DF image of amorphous material also consists of a speckle of white spots, 0.2-1.5 nm in size, against a dark background. The selected area electron diffraction (SAED) pattern and HRTEM image provided further evidence of the crystalline nature of the film. Corresponding HRTEM images indicated an amorphous-like contrast with periodic lattice fringes of a few Å dimension.

The sp^2-bonded BN film is an example of the structure consisting of crystalline clusters of less than 1 nm in size. The sp^2-bonded BN usually grew in a layered fashion in such a way that a sub-layer normal made a relatively small angle with the film growth direction. The thickness of the individual sub-layer was of about 0.5-5 nm [2]. Each sub-layer consisted of parallel lamellae in both the hexagonal and rhombohedral configurations [3]. Although the local symmetry between the neighboring basal planes was described well by either the hexagonal of the rhombohedral configuration, the plane stacking was random. The crystallites were quite small with spacing as little as two to three lattice fringes being only ¾-1 unit cell in dimension.

2.2. Structure of Grain Boundaries

A significant amount of the atoms in nanostructured materials locates at the grain boundaries, raising the interest in this problem. The atomic structure of a grain boundary, the volume fraction of which is extremely large in nanocrystalline materials (NM), is also of great importance in the physical properties of material. Gleiter and co-workers [4,5] suggested determining the random grain-boundary structure in NM. A theoretical concept for designing nanocrystalline superhard material is also based on the assumption that a thin amorphous layer around the nanocrystallites prevents the formation and multiplication of dislocations [6]. Therefore, it is often speculated that the nanocrystals would be surrounded completely by a thin amorphous layer. In contrast, evidence for the ordered structure of grain boundaries was obtained in the Ti-Al-B-N and Ti-Si-C-N films [1,7]. Extra half planes and well-developed growth ledges were frequently seen on either side of the grain boundary. No intergranular precipitation of glassy phase was observed. The amorphous phase always formed as individual grains of several nm, rather than as a thin intergranular amorphous layer of uniform thickness. It should be noted that no special features of interface structures that could be attributed to the small scale were seen in these studies.

2.3. Identification of Phase Composition

In the case of multicomponent nanostructured thin films, the structure cannot be unambiguously determined by using only the diffraction technique. These films often show very broad diffraction peaks with low intensity, which is usually attributed to amorphous material. The composition of grains may vary from one grain to another [1,7] making the phase analysis even more complicated. A combination of various microanalytical techniques is needed for qualitative identification of phase composition [1]. In a number of studies, the electron energy-loss spectroscopy (EELS) [1, 8], high resolution TEM [1] and X-ray absorption fine structure analysis (EXAFS) [9] were successfully used to reveal the phase composition. It has been suggested that ternary phase diagrams provides a good basis from which the phase composition of any particular film composition can be estimated [10,11]. Nevertheless, it should be noted that equilibrium phase diagrams often do not describe the thermodynamics of the highly metastable states of the films, because many additional factors, such as high interdiffusion rates, high quenching rates and energetic ion bombardment, may affect the phase composition.

2.4. Correlation between the Surface Topography and the Film Structure

The film surface is covered with fully developed three-dimensional (3D) hills which usually originate from isolated film islands on the substrate surface. The correlation between the surface topography and the film structure can be summarized as follows.

a) The three-dimensional hills on the film surface can be treated as tops of the columns. Note that the grain size can be smaller than the column width [12].
b) In the case of a columnless course-grained structure, the surface islands can be treated as individual grains on the film surface.
c) The grain size and column width can be smaller than the 3D hills observed on the top of the film. Such films are characterized by a columnar structure with a wide range of systematically varying column size [13].
d) The SEM cross-sectioned fracture of the films often revealed a structureless film morphology, typical for amorphous structures. These films are usually characterized by a very small grain size that is much smaller than the size of hills on the film surface. This is the case when the surface roughness is not related to the size of the crystallites. It was shown [12, 14] that such thin films may possess latent columnar structure with grain boundary grooves consisting of voids and low-density regions that originate from isolated film islands on the substrate surface.

2.5. Role of Orientation Relationships

Recent results reveal the great role of crystallography during the films growth [2]. For instance, it is well known that c-BN films typically grow with a layered structure consisting of an amorphous layer at the film/substrate interface, followed by preferentially oriented sp^2-bonded BN, and a final layer of c-BN. It has been shown that the c-BN phase nucleates on the oriented graphitic BN layer in a semicoherent manner and obeys specific orientation relationships with the hexagonal and rhombohedral phases [3]. Part of the total driving free energy for nucleation must be spent on the creation of a new interface between the c-BN crystallite and its structural precursor. It is well known that the specific orientation relationship yields a low interface energy. Thus the formation of the intermediate sp^2-bonded BN layer provides the favorable boundary conditions for c-BN nucleation. The growth of highly twinned c-BN crystallites with twinning occurring about a single type of {111} planes is understood in terms of good atomic matching between the {111} close-packed plane of c-BN parallel to either the $(01\bar{1}2)_{r-BN}$ or $(01\bar{1}\bar{1})_{r-BN}$ plane of rhombohedral BN most parallel to the grain boundary, yielding a low interface energy. These crystallographic features for the c-BN nucleation and growth are consistent with the stress-induced solid-state $sp^2 \rightarrow sp^3$ transformation.

The atomic structure of an interface depends on the orientation relationship between adjacent grains and the boundary inclination. These conditions are also valid for nanostructured materials. The grain boundaries between adjacent c-BN grains was shown to consist of twin boundaries when two grains were oriented close to the $[110]_{c-BN}$ zone axis and the boundary plane was parallel to the $\{111\}_{c-BN}$ close packed planes of adjacent grains [3]. However, a thin layer, 1-2 nm, of sp^2-bonded BN formed between the c-BN grains when the boundary plane is inclined a few degrees from the $\{111\}_{c-BN}$ planes of adjacent grains. In the last case the sp^2-bonded BN precipitates are thought to reduce the grain boundary energy.

2.6. Mechanisms of Deformation

Recent fracture investigations have revealed that there are two types of deformation of nanostructured thin films, i.e., homogeneous deformation and localized inhomogeneous deformation with the formation of shear bands [12, 15-18]. The H^3/E^2 ratio (H – hardness, E – Young's modulus) which describes the resistance to plastic deformation was proposed to be ranking parameter for the prediction of shear banding under the localized deformation [12]. Both deformation mechanisms involve sliding of individual columns or columnar structural elements (CSEs) consisting of many crystallites along the direction of applied load. In the case of a weak chemical bond between neighboring grains, individual CSEs can be rejected outside as a result of relaxation of elastic stress upon unloading. The inhomogeneous deformation mechanism is associated with cooperative motion of CSEs due to the strong binding energy between the grains. Note that multicomponent nanostructured films deposited by magnetron sputtering have either a pronounced or a latent columnar structure that manifests itself under deformation [14].

2.7. Physical-Mechanical Properties

An important advantage of multicomponent nanostructured films is that one can fabricate superhard materials with identical hardness but different elastic modulus. This means that films with identical hardness can differ in their elastic strain to failure (H/E) and their resistance to plastic deformation. The elastic properties of these films can also be significantly different. A combination of high hardness and elastic recovery characterizes multicomponent nanostructured films as new unique hard, elastic material. It should be noted that different materials can possess similar mechanical properties, for instance Ti-Zr-C-N-O and Ti-B-N films [14]. Thus, superhard coatings can be produced with different combinations of elastic and plastic characteristics, which provide a wide choice of coating properties for various applications.

2.8. Film Growth Mechanism

It is well known that there are two basic types of film growth on atomically flat surfaces, which are termed "island growth" (Volmer-Weber mechanism) and "layer growth". The Later is often subdivided into two modes: two-dimensional layer growth (Frank-van der Merve) and layer-plus-island growth (Stranski-Krastanov). In the last case the growth begins as a thin layer but develops islands and particles as growth proceeds. The substrate surface structure is also important for film nucleation and morphology. The Ti-Si-N film growth on the atomically flat surface of single crystal silicon (001) was shown to proceed by a Stranski-Krastanov growth mode [19]. In contrast, the HRTEM analysis of the $LiNb_{1-x}Ta_xO_3$ film/Al_2O_3 substrate interface showed that the surface of (0001) sapphire substrate revealed well-defined polishing defects, ranging from 1-2 nm in height with smooth terraces of 25 nm in width [20]. In the case of columnar growth mode, the terrace width was shown to be a limiting factor controlling the lateral crystallite size. Nucleation was also observed to occur on terraces between ledges where either the amorphous or crystalline phase was formed. This type of film growth cannot be identified with any of the basic types of growth.

3. DESIGN OF TRIBOLOGICAL COATINGS

3.1. Hard Tribological Coatings

Nanostructured coatings based on carbides, borides and nitrides of transition metals are of interest because of their high hardness, thermal stability, oxidation resistance, wear- and corrosion resistance. An exciting and novel area of research deals with the synthesis of multicomponent nano-structured thin films by sputtering composite targets. Self-propagating high-temperature synthesis (SHS) was shown to be an effective method, which provides a highly dense, uniform structure that exhibits the required mechanical, thermal, and electrical properties needed for PVD target materials [21]. Over the last few years, several new types of hard coatings have been successfully developed and studied, for example Ti-B-N, Ti-Si-C-N, Ti-Al-C-N [7, 22], Ti-Si-B-N [23], Ti-Al-B-N [1], Ti-Si-N [19], Ti-Cr-B-N, Cr-B-N [24]. The structure of multicomponent coatings was shown to consist of mixtures of various phases (Ti,Al)N+TiB_2+h-BN (Ti-Al-B-N films); TiN+$TiSi_x$ (Ti-Si-N); TiN+TiB_2 and TiN+TiB_2+BN (Ti-B-N); TiN_{1-x}+CrB_2 (Ti-Cr-B-N); some of the phase can be in an amorphous state.

The films were characterized in terms of their hardness, elastic modulus, elastic recovery, friction coefficient and wear rate. The best Ti-Si-N films showed a hardness of 30-35 GPa, Young's modulus in the range of 220-250 GPa and percentage of elastic recovery of 72-75%. Similar to the previous results, the hardness of Ti-Si-N films increased with Si concentration, reaching the maximum value of 35 GPa at 4.8 at%, and then monotonically decreased [19]. The hardness of Ti-B-N and Ti-Cr-B-N coatings was in the range of 31-34 and 40-47 GPa, respectively [24]. The high hardness of the Ti-Si-B-N films was shown to be closely related to both the stoichiometry of the $Ti(B,N)_x$ phase and the residual compressive stress in the films. The maximum hardness of 70 GPa was measured for strictly stoichiometric $Ti(B,N)_2$ film [23]. The value of hardness was reduced by 10-15% of the initial value after aging at room temperature for 6 months.

Fig. 1 shows the wear rates and friction coefficients of various coatings. Wear tests were performed in ambient air at 25°C and 40-60% humidity. The tribological properties of the coatings were evaluated using a conventional ball-on-disc tribometer (CSM Instruments, Switzerland) under a normal load of 5 N with 3-mm diameter WC+6%Co ball as a counterpart material. The sliding speed was maintained at 10 cm/s. The wear rate of the Ti-B-N films was as low as $(3.4\text{-}4.6)\cdot 10^{-7}$ $mm^3N^{-1}m^{-1}$. Maximum critical load L_c=50 N was achieved for Ti-B-N coatings. No correlation between H/E or H^3/E^2 ratios and wear resistance were found; however the composition and properties of wear products as well as the film adhesion seems to be essential factors for wear resistance [22].

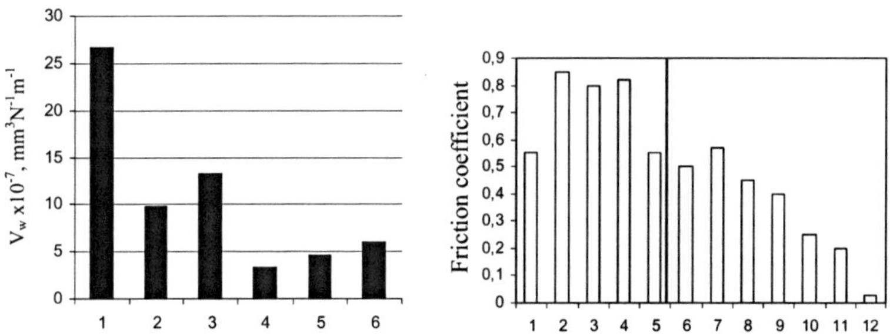

a) *b)*

Figure 1. Wear rate a) and friction coefficient b) of various coatings.
a) 1 – TiN; 2 – Ti-C-N; 3 – Ti-Si-N; 4 – Ti-B-N; 5 – Ti-B-N; 6 – Ti-Cr-B-N.
b) 1 – TiN; 2 – TiC; 3 – TiB_2; 4 – Si_3N_4; 5 – hard alloy TT8K6; 6 – Ti-Si-N; 7 – Ti-B-N; 8 – Ti-Cr-B-N; 9 – Cr-B-N; 10 – Ti-Zr-C-N-O; 11 – Ti-Ca-C-N-O; 12 - WSe_2/TiCN, WSe_2/TiSiN.

The friction coefficient of multicomponent coatings was lower than for the TiN, TiC, TiB_2 and Si_3N_4 coatings. The Ti-B-N, Ti-Cr-B-N, Ti-Si-B-N and Ti-Si-C-N films exhibited excellent abrasive wear behavior compared with TiN and TiCN films. The Ti-Si-B-N, Ti-Si-C-N and Ti-Al-C-N films were more resistant against high-temperature oxidation than TiN [22]. The combination of low friction properties with high hardness and wear resistance makes multicomponent films promising candidates for protective coatings to be used in various tribological applications.

3.2. Double-Layer Self-Lubricating Coatings

Hard coatings based on carbides, nitrides and borides of transition metals provide superior mechanical and tribological properties. However, these coatings show a high coefficient of friction (0.85 TiB_2, 0.8-0.9 TiC, 0.7-0.9 TiSiN) and require a lubricant to reduce friction. The friction of hard coatings can be improved by the deposition of a thin lubrication layer on top of the coatings. Recently, WSe_x coatings were obtained by pulsed laser deposition (PLD) and ion-assisted PLD [25]. The hard TiC, TiCN and TiSiN under layers were fabricated by DC magnetron sputtering of composite targets $TiC_{0.5}$, Ti_5Si_3+Ti and Ti_5Si_3+TiN, respectively, in an atmosphere of argon or reactively, in a gaseous mixture of argon and nitrogen. The adhesion between hard film and substrate was improved by low energy ion bombardment of the substrate at the initial stage of deposition using an additional Ar ion source operated at a fixed energy of 1.5 keV. The structure of WSe_x films consisted of various nanocrystalline mixtures of WSe_2 and W_3O phases in an amorphous matrix WSe_x (Fig. 2). Depending on deposition conditions, the WSe_2 crystallites were oriented with their basal planes either parallel (//WSe_2) or perpendicular ($\perp WSe_2$) to the substrate surface. A correlation between an amount of W_3O phase and *c*-axis-oriented WSe_2 phase was outlined. The incorporation of oxygen atoms between the basal planes was shown to cause lattice distortion and resulted in anisotropy

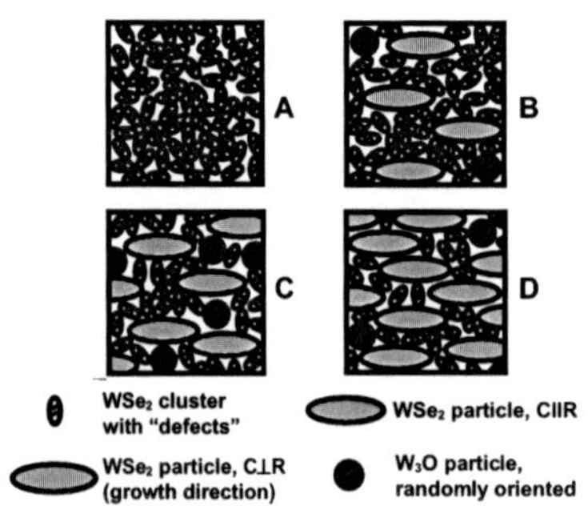

Figure 2. Various structure of WSe_x coatings.

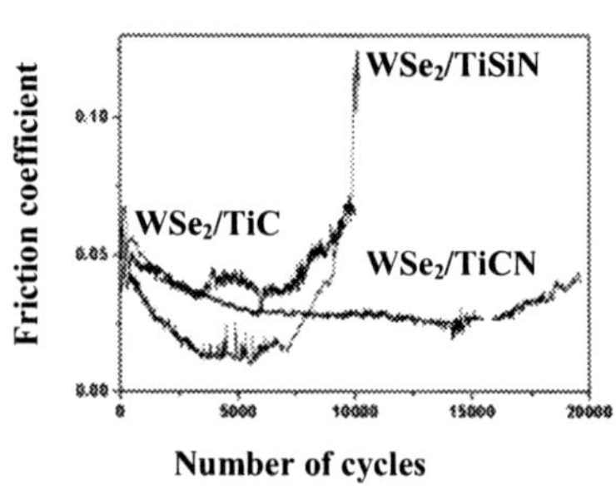

Figure 3. Friction coefficients of the double-layer coatings against 3 mm WC+6%Co ball.

of the lattice parameter. Improved endurance of WSe_x coatings was due to their composite structure. The presence of a hard underlayer, such as TiC, TiCN or TiSiN, was shown to be essential for extremely low friction coefficients down to 0.015-0.03 (Fig. 3). The WSe_x/TiC films showed low friction from the very start of the tests without an initial maximum. The superior performance of WSe_x/TiSiN coatings both in air and under water suggests that the double-layer coatings are suitable for use in terrestrial tribological applications.

3.3. Biocompatible Coatings for Load-Bearing Medical Applications

It is well known that implanted materials must form a stable interface with surrounding tissue. Titanium is a well-established material for orthopedic implants; however, it is a bioinert material with relatively poor resistance against wear. One approach to alleviate poor interface bonding is the use of HAP (hydroxyl apatite) coatings applied to the implant surface. Unfortunately, the high loads required for many orthopedic implants restrict the use of functionally graded materials with a top CaP (calcium phosphate) layer. Recently, it was shown that the combination of excellent mechanical properties with biocompatibility and non-toxicity makes Ti-Ca-C-N-O and Ti-Zr-C-N-O films promising candidates as tribological coatings to be used for various medical applications such as total joint replacements and dental implants [26].

A comparative investigation of multicomponent thin films based on the systems Ti-Ca-C-N-O, Ti-Zr-C-N-O, Ti-Si-Zr-N-O and Ti-Nb-C-N has recently been completed. SHS composite targets $TiC_{0.5}$+10%CaO, $TiC_{0.5}$+20%CaO, $TiC_{0.5}$+10%ZrO_2, $TiC_{0.5}$+20%ZrO_2, Ti_5Si_3+10ZrO_2, $TiC_{0.5}$+10%Nb_2C and $TiC_{0.5}$+10%Nb_2C were subjected to DC magnetron sputtering in an atmosphere of argon or in a gaseous mixture of argon and nitrogen. The films were characterized in terms of their structure, surface topography, hardness, elastic modulus, elastic recovery, zeta potential, friction coefficient, and wear resistance. The biocompatibility of the films was evaluated by both *in vitro* and *in vivo* experiments. *In vitro* studies involved the investigation of the proliferation of fibroblasts Rat-1 and epithelial cells IAR-2 at the tested films, morphometric analysis and actin cytoskeleton staining of the cells cultivated on the films. *In vivo* studies were fulfilled by subcutaneous implantation of teflon coated with the tested films in mice.

Depending on the deposition conditions, the films displayed either polycrystalline or nanocrystalline single-phase NaCl-structure. The films deposited under optimal conditions exhibited high hardness in the range of 30-37 GPa, significant reduced friction coefficient down to 0.1-0.2 (Figure

1b) both in air and physiological solution and low wear rates in comparison with conventional pure TiC and TiN films. The surface charge decreased systematically as the nitrogen content of the films was increased.

Statistically significant differences in the attachment, spreading and cell shape of cultured IAR-2 and Rat-1 cells of the Ti-Ca-C-N-O, Ti-Zr-C-N-O ($TiC_{0.5}$+10%ZrO_2 target), Ti-Si-Zr-N-O films and the uncoated substrata were not observed. IAR-2 epitheliocytes were discoid in shape. Rat-1 fibroblasts had elongated well spread form with broad leading lamella. The adhesion and proliferation of cells was good at all investigated films. However, in the case of the Ti-Zr-C-N-O film deposited using the $TiC_{0.5}$+20%ZrO_2 target the IAR-2 cells were polygonal or triangular in shape. Actin bundles were short and fine. The epitheliocytes and fibroblasts cultivated on the Ti-Nb-C-N films had different shape and disturbing actin cytoskeleton. The circumferential actin bundle in the lot of epitheliocytes was lost. Actin bundles in fibroblasts were irregular in shape and thin.

The Ti-Ca-C-N-O, Ti-Zr-Si-N-O and Ti-Zr-C-N-O ($TiC_{0.5}$+10%ZrO_2 target) films, demonstrated good cell/film interaction in-vitro, were deposited on a Teflon substrate and implanted subcutaneously into mice. Assessment of the population of cells covering the plate after 16 weeks of implantation revealed the high biocompatibility level of the tested films and the absence of inflammatory reactions. We observed just a single neutrophils (inflammatory response), giant multinucleated cells and polarized macrophages (cells unfavourable for biocompatibility) at the plates. At the interface there was a close contact between the capsule and film surface. Inflammatory reactions in capsules were not observed.

4. CONCLUSIONS

Based on recent experimental results, the fundamentals of nanostructured thin films are discussed. Various approaches to design and evolution of tribological coatings are considered. The composite SHS targets can be successfully used for deposition of hard, low friction and wear resistant coatings. The growth of multicomponent films using composite targets is of technological interest, because this method is suitable to produce nanostructures with different combinations of the deposited elements and phases. Particular applications vary from civilian engineering, air force and space industries to medicine.

ACKNOWLEDGMENTES

The work described above represents the research of many scientists who have worked on the deposition, characterization and testing of various films.

Their valuable contribution is greatly acknowledged. The author acknowledges the support from the International Science and Technology Center (ISTC), Award No. 1852 and from INTAS, Award No. 2001-2274.

REFERENCES

1. D.V.Shtansky, K. Kaneko, Y.Ikuhara and E.A.Levashov, Surf. Coat. Technol. 148 (2001) 204.
2. D.V.Shtansky, Y.Ikuhara, Y.Yamada-Takamura, T.Yoshida, Sci. Technol. Adv. Mater., 1/4 (2001) 219.
3. D.V.Shtansky, O.Tsuda, Y.Ikuhara, T.Yoshida, Acta Mater., 48 (2000) 3745.
4. P. Keblinski, D. Wolf, S.R. Phillpot, H. Gleiter, Scripta Mater., 41 (1999) 631.
5. S.R. Phillpot, D. Wolf, H. Gleiter, Scripta Metall. Mater., 33 (1995) 1245.
6. S. Veprek, S. Reiprich, Thin Solid Films, 268 (1995) 64.
7. D.V.Shtansky, E.A.Levashov, A.N.Sheveiko, .J.Moore, Metall. Mater. Trans. A, 30 (1999) 2439.
8. A. Gupper, A. Fernandez, C. Fernandez-Ramos, F. Hofer, C. Mitterer, P. Warbichler, Monatshefte für Chemie 133 (2002) 837.
9. T.P. Mollart, P.N. Gibson, M.A. Baker, J. Phys. D: Appl. Phys., 30 (1997) 1827.
10. W. Gissler, Surf. Coat. Technol., 68/69 (1994) 556.
11. M.A. Baker, T.P. Mollart, P.N. Gibson, W. Gissler, J. Vac. Sci. Technol. A, 15(2) (1997) 284.
12. D.V. Shtansky, S.A. Kulinich, E.A. Levashov, A.N. Sheveiko, F.V. Kiriuhancev, J.J. Moore, Thin Solid Films 420-421 (2002) 330.
13. R. Messier, A.P. Girl, R.A. Roy, J. Vac. Sci. Technol. A, 2 (1984) 500.
14. D.V. Shtansky, S.A. Kulinich, E.A. Levashov, J.J. Moore, Phys. Solid State 45(6) (2003) 1177.
15. R.A. Andrievskii, G.V. Kalinnikov, D.V. Shtansky, Phys. Solid State 42 (2000) 760.
16. R.A.Andrievski, G.V.Kalinnikov, D.V.Shtansky, Features of Fracture Surface and Grain Boundary Structure of Boride/Nitride Materials, Nanophase and Nanocomposite Materials III, Mat. Res. Soc. Symp. Proc., S. Komarneni, Y. Parker and H. Hahn (Eds.), 581 (2000) 583.
17. R.A. Andrievski, G.V. Kalinnikov, J. Jauberteau, J. Bates, J. Mater. Sci. 35 (2000) 2799.
18. R.A. Andrievski, G.V. Kalinnikov, Surf. Coat. Technol. 142-144 (2001) 573.
19. D.V. Shtansky, I.V. Lyasotsky, N.B. D'yakonova, F.V. Kiryukhantsev-Korneev, S.A. Kulinich, E.A. Levashov, J.J. Moore, Surf. Coat. Technol., 2003, (in press).
20. D.V.Shtansky, Y.Ikuhara, S.A.Kulinich, K.Terashima, and T.Yoshida, J. Mater. Res., 16 (2001) 2271.
21. E.A. Levashov, A.S. Rogachev, V.I. Yukhvid, I.P. Borovinskaya, Binom, Moscow, 1999, pp. 134 (in Russian).
22. D.V.Shtansky, E.A.Levashov, A.N.Sheveiko and J.J.Moore, J. Mater. Synth. Process., 1998, vol.6, No.1, pp. 61.
23. D.V.Shtansky, E.A.Levashov, A.N.Sheveiko and J.J.Moore, J. Mater. Synth. Process., 1999, vol. 7, No. 3, pp. 187.
24. D.V. Shtansky, F.V. Kiryukhantsev-Korneev, A.N. Sheveiko, M.I. Petrzhik, I.A. Bashkova, O.V. Malochkin, E.A. Levashov, I.V. Lyasotski, Phys. Solid State (2004) (to be published).
25. D.V. Shtansky, T.A. Lobova, V.Yu. Fominski, S.A. Kulinich, I.V. Lyasotsky, M.I. Petrzhik, E.A. Levashov, J.J. Moore, , Surf. and Coat. Technol, 2003 (in press).
26. D.V. Shtansky, E.A. Levashov, N.A. Glushankova, N.B. D'yakonova, S.A. Kulinich, M.I. Petrzhik, F.V. Kiryukhantsev-Korneev, F. Rossi, Surf. and Coat. Technol, 2003 (in press).

Chapter 16

STRESS DEVELOPMENT AND RELAXATION IN NANOSTRUCTURED FILMS DEPOSITED BY CATHODIC VACUUM ARC

M.P. Delplancke-Ogletree[1], O.R. Monteiro[2]

[1] *Industrial Chemistry, Université Libre de Bruxelle, Brussels, Belgium*

[2] *Lawrence Berkeley National Laboratory, University of California, Berkeley, USA*

The great interest in the use of diamond-like carbon (DLC) films as a coating material is justified by the superior wear resistance and hardness, chemical inertness, and very low friction coefficients of these coatings. Vacuum arc is well suited to prepare super-hard films with high sp^3/sp^2 ratios. However, the high level of internal stresses originating during growth prevents the deposition of thick films and limits their applications. To study *in-situ* the development of stress in the growing film, we developed a simple optical technique based on the deflection of cantilever. Our system can measure stresses as high as few tens of GPa with a precision of few kPa. The influence on stress of bias voltage and duty cycle, and of the pulse length was studied as well as the effect of highly charged ions. The results of these studies show the key role of the energy input per ion on the stress level. Strategies to deposit thick hard films with minimum stress are proposed.

Keywords: diamond-like carbon, vacuum arc, stress, hardness, nanostructure.

1. INTRODUCTION

Carbon is an unusual material in that it exhibits both metallic and nonmetallic characteristics. It exists in both crystalline and amorphous forms [1]. Hard amorphous carbon (a-C) coatings, also called DLC (diamond-like carbon) are mostly metastable amorphous materials that could include a nanocrystalline phase. These coatings are a random network of covalently bonded carbon in hybridized sp^2 (trigonal) and sp^3 (tetragonal) coordination. Depending on the deposition technique, some of the bonds are terminated by hydrogen. In this paper, the term DLC will be exclusively used to describe non-hydrogenated amorphous carbon with a significantly high content of tetragonal bonding.

The high hardness and wear resistance, chemical inertness, lack of magnetic response and low friction coefficient of DLC coatings can be tuned by choosing the deposition method and conditions. Important applications of these coatings are for examples overcoats for heads for magnetic storage

A. A. Voevodin et al. (eds.),
Nanostructured Thin Films and Nanodispersion Strengthened Coatings, 167–174.

devices and magnetic media [2,3] where they are used for protection against wear and corrosion, coatings on razor blades and on glass windows for laser barcode scanners.

Traditionally, filtered cathodic arc (FCA) [4,5], mass selected ion beam [6] and DC/RF sputtering [7] are used to prepare DLC coatings. These techniques are characterized by precursors with hyperthermal energies that control the sp^3 content and the stress in the films. Intrinsic stresses can be very high. For applications requiring thicker films, it is very important to reckon with the level of stresses.

In this paper, we will describe the preparation by filtered cathodic arc and characterization of DLC films. The development of intrinsic stresses will be examined and different means to reduce the stress level will be presented.

2. DEPOSITION PROCESS

The films used in this investigation were prepared using filtered cathodic arc with a simultaneous application of a pulsed bias voltage to the substrate. The process, named Metal Plasma Immersion Ion Implantation and Deposition (MePIIID) was described in detail elsewhere [8,9]. Fig. 1 gives a schematic description of the deposition set-up.

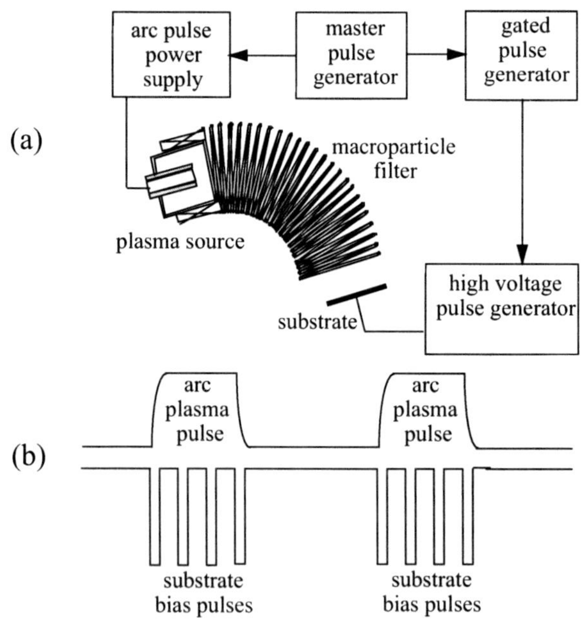

Figure 1. Schematic representation of MePIIID process: (a) source and electronics (b) arc current and bias voltage.

The plasma source was operated in a pulsed mode, with a pulse duration between 5 to 50 ms. During each pulse a train of negative bias voltage pulses was applied to the substrate to provide additional energy to depositing ions. The applied voltage varied from 0 to –2000 V and the duty cycle varied from 0 to 100% (DC bias). For films containing more than one element, we used two plasma sources.

Film properties were characterized by several techniques: transmission electron microscopy (TEM), Rutherford Backscattering (RBS), Auger Electron Spectroscopy (AES) and X-ray photoelectron Spectroscopy (XPS). Intrinsic stresses were calculated using Stoney's equation [10] after measuring the change in curvature radius of the coated substrates during the deposition and after it. A picture of the home-built device used to monitor *in*-situ the stress evolution is given in Fig. 2. A detailed description of the data interpretation can be found in reference [11]. Hardness and modulus were determined by nanoindentation using Hysitron Picoindenter and Triboindenter.

Figure 2. Optical device used to measure the deflection of a silicon cantilever.

3. HARDNESS AND STRESS

The properties of DLC films prepared by energetic deposition strongly depend on ion energy [12-14]. In general, hardness, sp^3 content, film density and stress achieve simultaneously a maximum value for a bias of –100 V. Ferrari et al. [15] challenged recently the direct correlation between sp^3 content and stress. R. Nieminen [16] has recently presented a modeling multiscale approach for computational studies of materials growth and processing. He showed that for carbon film grown by energetic particles, the major factor is a subplantation growth. It is accompanied by a strong scattering induced densification for incoming particle energy around 100 eV. Schultrich [17] showed that the structure of such amorphous carbon films is highly dependent on a competition between subplantation and relaxation. The densification is responsible of the high intrinsic stress level that can reach values higher than 10 GPa in the hardest films. Several approaches were proposed to reduce this stress in DLC films. Vacuum annealing can reduce the stress below 1GPa or better [18] without changing drastically the bonding characteristics. Repetitive growth - pulsed laser annealing cycles could be used for producing thicker films [19] but this approach increases the deposition time and the equipment investment. Another approach consisting in alternating fully developed layers of hard and soft DLC produces properties that are some average of the properties of individual layers.

To investigate yet another approach, we studied *in-situ* the development of stress in DLC films deposited in different conditions. Fig. 3 presents the average stress (a) and instantaneous stress development in a DLC film deposited in the best conditions for hardness (b) the graph giving access to the instantaneous stress in the same film.

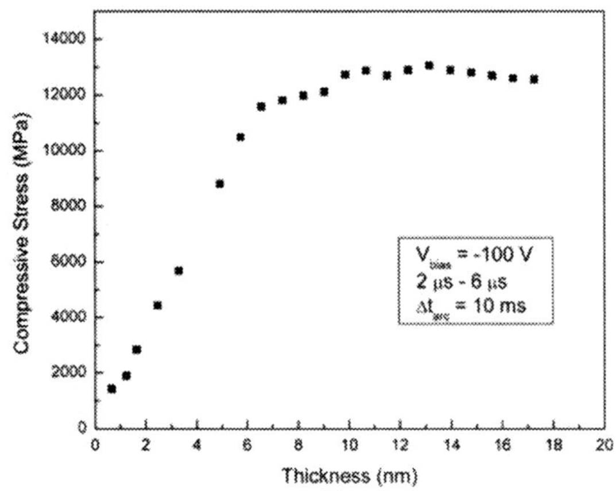

(a)

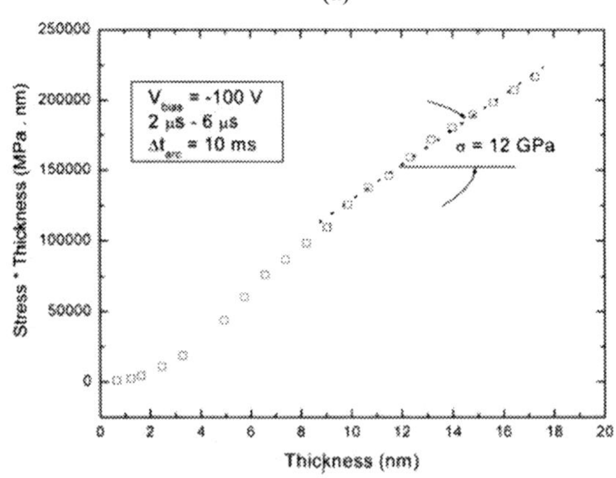

(b)

Figure 3. (a) Average stress evolution during deposition of monolithic DLC film (b) Stress*thickness curve for monolithic DLC film.

We observed a low instantaneous stress at early stage of growth that is probably due to a low density layer as observed by Siegal et al. [20]. But very quickly, a high stress value is reached, in agreement with the models cited previously.

The variation of the instantaneous stress and density with the substrate bias is presented in Fig. 4.

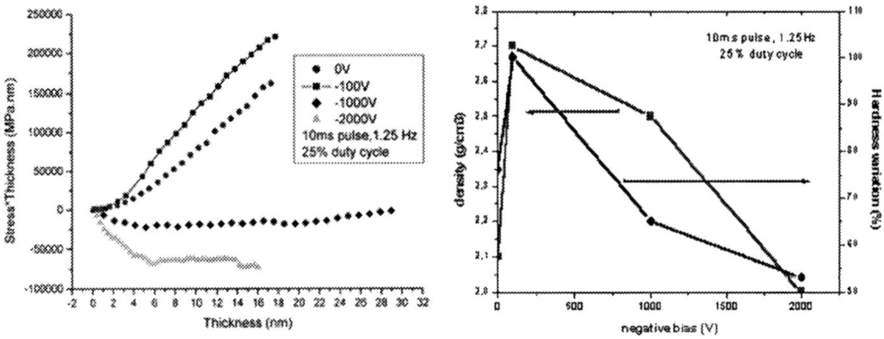

(a) (b)

Figure 4. Influence of substrate bias on (a) stress development (compressive stress is taken positive) (b) density and relative hardness.

Compressive stress is observed for no bias and –100 V bias. At –1000 V bias, the stress is around zero, while at –2000 V bias the stress is tensile. As expected the hardest film and the highest density are observed for the highest compressive stress. The relaxation due to the input of energy is dominating the densification due to the subplantation for high energy (~1000 – 2000 eV) ions and the sp^3/sp^2 ratio is reduced.

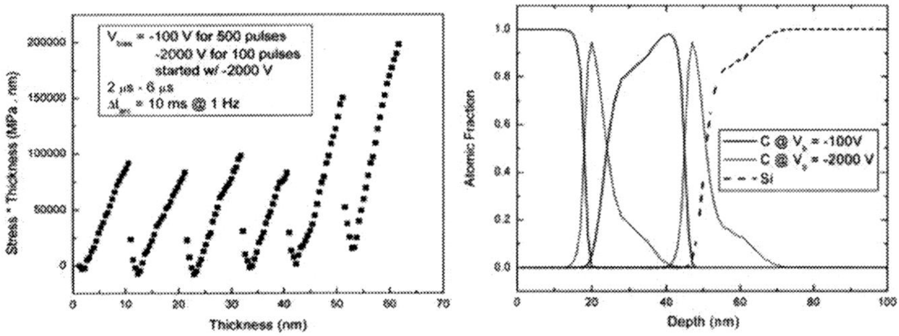

Figure 5. (a) Development of instantaneous stress (slope of the curve) during the deposition of a DLC layer. The sequence of biasing is the following: 100 plasma pulse at –2000V followed by 6 sequences (500 pulses at –100V + 100 pulses at –2000V), (b) T-Dyn simulation of the penetration depth of 2000 eV and 100 eV ions in a C-C multilayer on silicon.

This fact could be used to grow thick films by alternating fully developed hard layers of DLC (produced at –100 V bias for optimal hardness) with short sequence of high bias, -2000 V as suggested by Bilek and coworkers [21-22].

Fig. 5a illustrates the variation of instantaneous stress during the deposition of a hard DLC coating. The process described here is different from conventional IBAD as the ion bombardment is accomplished by accelerating depositing ions through the plasma sheath. The high energy ions do not contribute to the formation of a fully developed soft layer but penetrate through the entire hard layer, as illustrated in Fig. 5b, allowing local transformations. Local stress relaxation occurs primarily within the thermal spike volume according to the proposed mechanism [23]. This is accomplished without loosing the high hardness of the 100 eV layers. The succession of low and high bias could easily be implemented in an industrial set-up.

Another approach to produce superhard carbon films with low intrinsic stress calls for the incorporation of few percent of highly charged carbide forming ions in the DLC films. Incorporation of few percents of tungsten, for instance, reduces strongly the stress but only slightly the hardness as shown in Fig. 6a and b. The mean charge state of tungsten ions is higher than 3 while the charge state of carbon ions is around 1. The energy input per tungsten ion will be three times the energy input of a carbon ion for the same substrate bias. This higher energy input is responsible for the decrease in intrinsic stress. It is realized without any modification of the deposition conditions. A carbide forming element is chosen to prevent any major loss of hardness that could result from the formation of a separate phase with very different mechanical properties. The tungsten atoms are well dispersed in the film for concentrations up to 20 at.% and bonded to the carbon atoms.

Similarly to the soft-hard DLC multilayers, coatings alternating layers of DLC and DLC:Me (Me=metal) were investigated. DLC-DLC:Ti multilayers with 3, 5 and 7 layers were prepared and tested. The interfaces between the layers are diffuse due to a higher polarization (-2000 V) during few shots before returning to the classical –100 V bias. The titanium content of the DLC:Ti layers varied between 20 and 50 at. % Ti. At those concentration levels, the formation of TiC nanocrystallites was observed. The hardness is reduced by the incorporation of such titanium quantities. The 3 and 5 layer structures present a hardness of the order of 40 GPa but their mechanical performances in a pin-on-disk test are superior to a monolithic DLC layer. No delamination of the DLC-DLC:Ti layers is observed while the monolithic film starts to crack and peal. The 7 layer structure is softer. *In-situ* stress measurements showed that titanium incorporation contributes to stress relaxation in a similar way to what we observed with tungsten incorporation. The mean charge state of titanium ions is 2,1. The second factor to contribute to the stress relaxation in the DLC-DLC:Ti layers is the increased bias to produce the diffuse interface. The combine effect of the two factors is responsible of the poor performances of the 7 layer structure. The energy input is such that the sp^3 content is reduced in this film and the density is low.

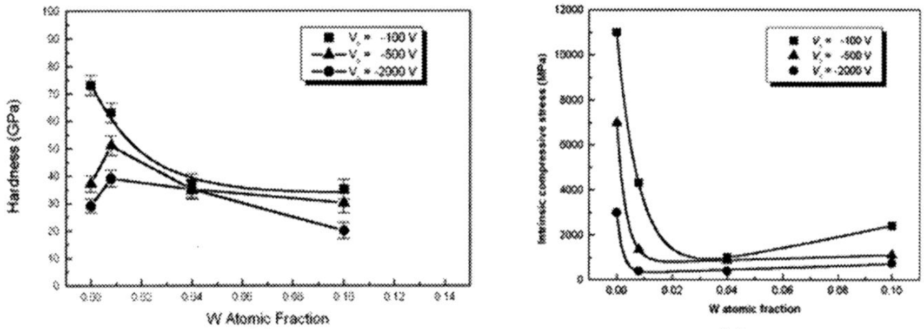

(a) (b)

Figure 6. (a) Hardness of DLC:W at several W content prepared using different bias voltages applied to the substrate, (b) Intrinsic stress of DLC:W at several W content prepared using different bias voltages applied to the substrate.

4. CONCLUSIONS

DLC is a very interesting material, as protective coating in applications where superior hardness is required due to its outstanding mechanical properties. They are directly correlated to the high sp^3 content of the films that can be obtained by energetic ion deposition. These deposition conditions are unfortunately also responsible for the development of high intrinsic stresses in these layers. These stresses are a major challenge for implementing thick DLC coatings in industrial applications.

Various techniques are proposed in the literature to reduce the stresses. For DLC films prepared by filtered cathodic arc, we propose different methods based on a local stress relaxation in the thermal spike volumes created by the bombardment with energetic depositing ions. In monolithic films, the incorporation of small amounts of alloying elements is efficient and rely on the higher mean charge state of metallic ions. Another solution if the incorporation of another element is prohibited, is to tune the substrate bias to alternate fully developed hard layers of DLC with short sequence during which the high energy ions penetrate the entire hard layer allowing local transformations. The depth of interaction can be tuned by adjusting the polarization. Multilayer structures which alternate chemically distinct phases is an acceptable solution. The three methods can easily be implemented in an industrial process and are cost effective.

ACKNOWLEDGMENTES

This work was supported in part by the US Department of Energy under contract number DE-AC03-76SF00098 and the FNRS. The authors would like to thank R.A. Mac Gill for his assistance, F. Ogletree, M. Bilek and A. Anders for helpful discussions.

REFERENCES

1. B. Bhushan, Diamond and Related Materials 8 (1999) 1985.
2. B. Bhushan, S.T. Patton, IEEE Trans. Magn. 34 (1998) 1883.
3. B.K. Gupta, B. Bhushan, Wear 190 (1995) 110.
4. P.J. Fallon, V.S. Veerasamy, C.A. Davis, J. Robertson, G.A.J. Amaratunga, W.I. Milne, Phys. Rev. B 48 (1993) 4777.
5. S. Anders, J. Diaz, J.W. Ager, R.Y. Lo, D.B. Bogy, Appl. Phys. Lett. 71 (1997) 3367.
6. Y. Lifshitz, G.D. Lempert, E. Grossman, Phys. Rev. Lett. 72 (1994) 2753.
7. J.J. Cuomo, D.L. Pappas, J. Bruley, J.P. Doyle, K.L. Seagner, J. Appl. Phys. 70 (1991) 1706.
8. A. Anders, Surface and Coatings Technology 93 (1997) 157.
9. I.G. Brown, Ann. Rev. Mater. Sci. 28 (1998) 243.
10. G.G. Stoney, Proc. Roy.Soc. London A 82 (1909) 172.
11. M.P. Delplancke-Ogletre, O.R. Monteiro, Diamond and Related Materials 12(12) (2003) 2119.
12. G.M. Pharr, D.L. Callahan, S.D. McAdams, T.Y. Tsui, S. Anders, A. Anders, J.W. Ager, III, I.G. Brown, C.S. Bhatia, S.R.P. Silva, J. Robertson, Appl.Phys. Lett 68 (1996) 779.
13. A. Anders, S; Anders, I.G. Brown, M.R. Dickinson, R.A. MacGill, J. Vac. Sci. Technol B 12 (1994) 815.
14. D.R. McKenzie, D. muller, B.A. Palithorpe, Z.H. Wang, E. Kravtchinskaia, D. Segal, P.B. Lukins, P.J. Martin, G. Amaratunga, P.H. Gaskell, A. Saeed, Diamond and Rtelated Materials 1 (1991) 51.
15. A.C. Ferrari, S.E. Rodil, J. robertson, W.I. Milne, Diamond and Related Materials 11 (2002) 994.
16. R. Nieminen, invited talk, ICMCTF 2003, San Diego.
17. B. Schultrich, communication, ICMCTF 2003, San Diego.
18. O.R. Monteiro, J.W. Ager, III, D.H. Lee, R. Yu Lo, K.C. Walter, M. Nastasi, J. Appl. Phys. 88 (2000) 2395.
19. S. Weissmantel, G. Reisse, D. Rost, communication, ICMCTF 2003, San Diego.
20. Siegal, M. P., Barbour, J. C., Provencio, P. N., Tallant, D. R., Friedmann, T. A., Appl. Phys. Lett. 73(6) (1998) 759.
21. Bilek, M. M. M., McKenzie, D. R. Tarrant, R. N., Lim, S. H. M. and McCulloch, D. G., Surface and Coatings Technology 156 (2002) 136.
22. Bilek, M. M. M, Tarrant, R. N., McKenzie, D. R., Lim, S. H. N. and McCulloch, D. G., Proceedings of the XXth International Symposium on Discharges and Electrical Insulation in Vacuum, 95-102 (2002), Ed. IEEE.
23. R.N. Tarran, C.S. Montross, D.R. McKenzie, Surface and Coatings Technology 136 (2001) 188.

Chapter 17
SYNTHESIS AND PROPERTIES OF TiB_2/TiN AND TiB_2/B_4C FILMS

R.A. Andrievski, G.V. Kalinnikov
Institute of Problems of Chemical Physics, Russian Academy of Sciences, Chernogolovka, Moscow Region, 142432 Russia

The TiB_2/TiN and TiB_2/B_4C films with a thickness of approximately 1-2 μm were prepared by magnetron non-reactive d.c./r.f. sputtering on Si and stainless steel substrates. Sintered and hot-pressed targets of the individual compounds (TiB_2, TiN and B_4C) and their mixtures (25/75, 50/50 and 75/25 in weight percents) were used in the film preparation. These films were deposited without and with additional external magnetic fields. The films were characterized by means of their chemical composition, structure, grain size and microhardness. The main result of the experiments is the increase of hardness by 15-65% as revealed both for crystalline films and amorphous ones. The AFM images also showed a more fine-grained morphology for the film deposited with external magnetic field. In the case of crystalline films, the decrease of grain size and especially of standard deviation of the grain size values were also fixed. Some samples were vacuum-annealed up to 1000°C.

Keywords: TiB_2/TiN films, TiB_2/B_4C films, magnetic field, microhardness, grain size, thermal annealing, roughness.

1. INTRODUCTION

Beginning in 1985-1990, that is from the works of Holleck, Knotek, Mitterer and their coworkers [1-3], there are extensive studies of superhard films within the system Ti-B-C-N (e.g. see some recent comprehensive reviews [4-10]). Hardness values more than 40-50 GPa have been observed by many investigators. However, many details of the hardness increase and stability (i.e., the more precise definition of optimal grain size, the role of the non-equilibrium phase formation and existence of residual stresses and so on) need further study.

In this work, we present the initial results of experiments designed to the study of quasi-binary systems TiB_2-TiN and TiB_2-B_4C. There are many results regarding to the first system but there are not so many information about the films based on B_4C. The additional external magnetic field (EMF) as a method for hardness increase [11] was also used. Thermal stability of prepared films was studied for temperatures up to 1000 °C.

A. A. Voevodin et al. (eds.),
Nanostructured Thin Films and Nanodispersion Strengthened Coatings, 175–182.

2. EXPERIMENTAL

The TiB_2, TiN, B_4C, Ti(B,N) and Ti(B,C) films with a thickness of approximately 1-2 μm were prepared by magnetron non-reactive r.f. sputtering on Si and stainless substrates [12, 13]. The substrate temperature was 150°C in 0.4 Pa argon pressure. Powders of the individual compounds and their mixtures (25/75, 50/50, 75/25 in weight percents) were hot-pressed and sintered to targets with ∅ 7 cm. The deposition time was 90 or 120 min depending on the film type and the desired film thickness. The film composition and structure were characterized by X-ray diffraction (XRD), Auger electron spectroscopy (AES), atomic force microscopy (AFM), transmission electron microscopy (TEM) and selected area diffraction SAED analyses. Hardness measurements were made by Vickers microhardness tester (P=0.2-0.5 N), using the Jonsson-Hogmark relationship [14]. Such procedure permits to avoide the influence of different film's thickness and hardness of substrate on film's hardness values. Thus, this procedure provides a possibility to compare the hardness of films with different thickness. Ten indents were made for each of the loads.

The distribution of grain size was defined using plan-view dark-field TEM examination (using JEOL-200CX) with calculation of 1,000-2,000 reflexes having a size more than 1 nm (operating under programs Image Pro Express 4.0 and Statistica).

A Digital Instruments NanoScope IIIa AFM was employed to study the surface topography and roughness parameters of some films.

The additional EMF (of about 400 Oe) was created by a coil with core which located directly behind the substrate. No bias potential was used during the deposition.

The data of composition and structure of the experimental films are listened in Table 1.

Table 1. Some characteristics of investigated films.

Target	Film approximate formula (from AES ahalysis [12,13])	Structure type
TiB_2	$Ti(B_{0.73}N_{0.2}O_{0.05}C_{0.02})_{1.56}$	AlB_2
$75TiB_2$+25TiN	$Ti(B_{0.56}N_{0.29}O_{0.05}C_{0.1})_{1.32}$	AlB_2
$50TiB_2$+50TiN	$Ti(B_{0.36}N_{0.42}O_{0.15}C_{0.07})_{1.57}$	NaCl
$25TiB_2$+75TiN	N/d	NaCl
TiN	$Ti(N_{0.6}O_{0.2}C_{0.2})_{1.58}$	NaCl
$75TiB_2+25B_4C$	N/d	AlB_2
$50TiB_2+50B_4C$	N/d	Amorphous
$25TiB_2+75B_4C$	N/d	Amorphous
B_4C	N/d	Amorphous

XRD and SAED studies revealed that all Ti(B,N) films after deposition were crystalline. For the case of TiB_2+B_4C targets just those coatings deposited from the $75TiB_2+25B_4C$ target were crystalline, all other coatings were amorphous.

As is evident from Table 1, our films contain some impurities and non-stoichiometric phases. Hexagonal TiB_2-based films are sub-stoichiometric compounds and cubic TiN-based films are superstoichiometric with a high content of non-thermal (structural) vacancies. Other experimental details can be found elsewhere [12, 13].

3. RESULTS AND DISCUSSION

Table 2 and Table 3 show the effect of an additional EMF on hardness and grain size of deposited films. As pointed out before, the hardness values are consistent with those obtained from experimental measurements and using the relationship of Jonnson-Hogmark. Unlike to the average grain size values with standard deviations, values of hardness are indicated in the whole value intervals because of difficulties to measure very small indents after loading in the interval of 0.1-0.5 N.

Table 2. Average grain size *L* (nm) and microhardness H_V (GPa) of Ti (B,N) films.

Target	Without EMF		With EMF	
	L	H_V	*L*	H_V
TiB_2	3.6±1.8	32-39	3.3±1.1	42-48
$75TiB_2$+25TiN	2.3±1.1	28-34	N/d	44-48
$50TiB_2$+50TiN	2.9±1.5	31-36	1.8±0.3	52-58
$25TiB_2$+75TiN	5.9±4.0	25-30	6.5±2.5	44-49
TiN	29.0±15.0	23-28	8.8±2.2	36-42

Table 3. Average grain size *L* (nm) and microhardness H_V (GPa) of Ti (B,C) films

Target	Without EMF		With EMF	
	L	H_V	*L*	H_V
TiB_2	3.6±1.8	32-39	3.3±1.1	42-48
$75TiB_2+25B_4C$	8.5±4.0	40-46	2.9±0.3	54-58
$50TiB_2+50B_4C$	Amorphous	44-50	Amorphous	60-68
$25TiB_2+75B_4C$	Amorphous	46-54	Amorphous	62-71
B_4C	Amorphous	42-50	Amorphous	48-56

It should be noted that our hardness values for Ti(B,N) films are in good agreement with literature results [15], proving that the above-mentioned procedure [14] usage is reasonable. The interval of the reported hardness values for B_4C films is wide (about 30-60 GPa; e.g. see [5, 16]). Our results are in this interval too.

It can also be seen from Table 2 and Table 3 that the significant hardness increase has been observed both for crystalline and amorphous films. In the case of crystalline films most of them show a decrease in mean grain size and in all cases the standard deviation was considerably decreasing. In Fig. 1

the grain size distributions are shown for Ti(B,N) films deposited from $50TiB_2$+50TiN target without EMF (Fig. 1a) and with EMF (Fig. 1b). It is clearly visible, that the grain size decreases with an EMF and also the grain size distribution becomes narrower.

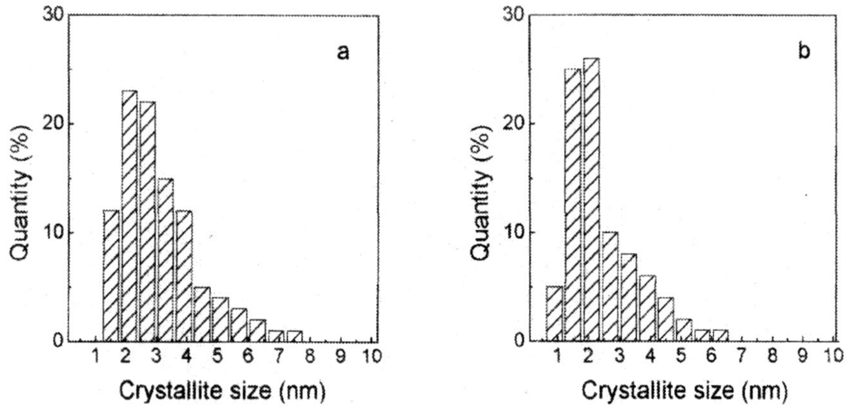

Figure 1. Distributions of grain size in crystalline Ti(B,N) films deposited without EMF (*a*) and with EMF (*b*).

The crystalline size reduction as a result of the EMF effect is also consistent with AFM studies. The surface topography characteristics calculated from AFM observations are shown for some films in Table 4.

Table 4. The roughness parameters determined from AFM analysis on a 1 μm×1 μm area for crystalline TiN films and amorphous Ti(B,C) films.

Target	EMF Presence	R_{rms} (nm) [1]	R_{max} (nm) [2]	R_a (nm) [3]
TiN	–	1.021	4.97	0.805
TiN	+	0.734	3.81	0.582
$25TiB_2+75B_4C$	–	0.175	0.75	0.138
$25TiB_2+75B_4C$	+	0.139	0.7	0.107

[1] R_{rms} is the root-mean-square roughness; [2] R_{max} is the maximum height; [3] R_a is the mean roughness.

As the data of Table 4 show, all roughness parameters decrease when EMF used during deposition. Fig. 2 shows the AFM images of TiN films deposited without EMF and with that. In these images it is possible to observe the typical columnar structure (the dome-rounded shape) as its often happens for TiN films [4] but deposition with EMF results in a column height/width decrease. While comparing Fig. 2a and Fig. 2b with the results in Table 2 one can conclude that there is agreement but a closer look at the data is necessary for better understanding.

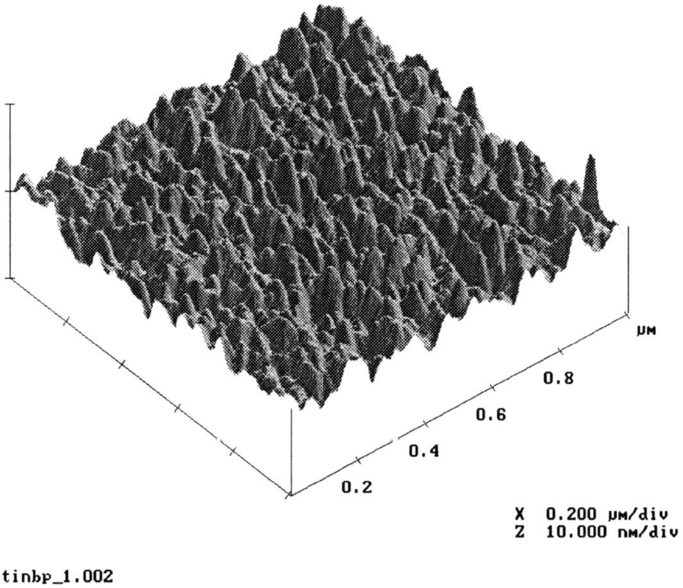

a)

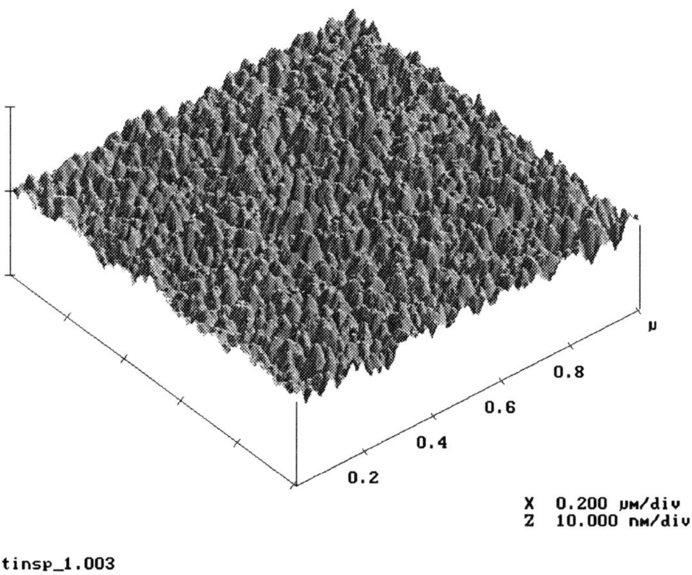

b)

Figure 2. Three-dimensional AFM 1 µm × 1 µm images of about 1 µm thick TiN films deposited without EMF (a) and with that (b).

The hardness increase in the case of deposition of crystalline TiB_2, TiN, Ti(B,N), ZrN and ZrB_2 films with EMF has been described in Refs [11, 15, 17]. This effect was attributed to the increased number of ions (ion/neutral flux ratio). It is anticipated that a high-enough ion flux results in the increase of the preferential nucleation site number and atomic surface mobility with a consequent formation of more dense nanostructure and reducing porosity. The change of preferred orientation was also fixed. In order to elucidate the EMF effective mechanism, it seems to be very interesting to study the initial film growth in the EMF presence. Initial growth and texture formation features during magnetron sputtering of TiN films were investigated by authors [18].

Table 5 shows the effect of annealing on the average grain size and microhardness of crystalline films Ti (B,N) prepared by d.c. regime without EMF [19].

Table 5. Change of average grain size L (nm) and microhardness H_V (GPa) for Ti (B,N) films after vacuum annealing at 700 ° and 1000 °C for 15 min.

Target	As prepared		T=700 °C		T=1000 °C	
	L	H_V	L	H_V	L	H_V
TiB_2	3-5	41-46	n/d	40-46	n/d	35-39
$75TiB_2$+25TiN	2.3±1.1	40-46	4.4±1.4	40-45	n/d	32-38
$50TiB_2$+50TiN	2.9±1.1	46-52	3.7±1.2	47-54	3.4±1.9	38-45
$25TiB_2$+75TiN	5.4±4.0	53-59	8.1±4.4	52-58	7.9±3.8	32-42
TiN	9.9±8.8	36-40	13.0±6.6	38-42	10.2±7.0	33-40

As compared with Table 2, the hardness values in Table 5 for as prepared films are higher which is connected with the sputtering power density difference in the case of r.f. regime (N=9 W/cm^2) and d.c. regime (N=40 W/cm^2) [12, 13, 20].

From Table 5 it is obvious that the grain size change is not pronounced but there is some decrease of hardness. This hardness evolution can be likely attributed to the residual compressive stresses relaxation. The availability of significant residual compressive stresses in boride and nitride films has been observed by many authors (e.g. see reviews [4, 5, 8] and other Refs [21, 22]). The effect of compressive stresses is believed to play an important role in the hardness values of deposited films.

The effect of annealing on grain size for films prepared in r.f. regime is near the similar as that for films prepared in d.c. regime. These results will be published later.

Recrystallization and grain growth of nanocrystalline $TiB_{0.55-1.25}N_{0.95-0.7}$ films have been recently studied by authors [8, 23]. They showed by differential scanning calorimetry in argon atmosphere using heating rates between 27-50 K/min, that these films start to recrystallize between 1030 and 1070 °C. The results are in agreement with those in Table 5.

The experiments of annealing other crystalline and amorphous films deposited with and without EMF are now in progress.

4. CONCLUSIONS

In addition to literature data [11, 15, 17], our results reveal that the EMF use tends to a hardness increase not only for crystalline films but also for amorphous ones. In both cases the roughness parameters decrease by the action of EMF. For crystalline films the decrease of grain size and especially of the standard deviation has been observed.

An annealing of crystalline films Ti(B,N) at temperature up to 1000°C results in slightly changes of grain size and some decrease of hardness, that can be likely connected with the residual compressive stresses relaxation.

ACKNOWLEDGEMENTS

This research is sponsored by NATO's Scientific Affairs Division in the framework of the Science for Peace Program (Project No 975529) and the Program of Russian Academy of Sciences "Fundamental Problems of the Physics and Chemistry of Nanosystems and Nanomaterials" (Project No 453) as well as by the Russian Program "Integration" (Project No BO-115).

Authors would like to thank D.L. D'yakonov (TEM), Dr. M. A. Kozodaev (AFM) and Prof. S. S. Ordan'yan (targets) for their active and friendly assistance.

REFERENCES

1. H. Holleck, Ch. Kuhl, H. Schultz, J. Vac. Sci.Technol. A 3 (1985) 234.
2. O. Knotek, R. Breidenbach, F. Jungblut, F. Loffler, Surf. Coat. Technol. 43/44 (1990) 107.
3. C. Mitterer, M. Rauter, P. Rodhammer, Surf. Coat. Technol. 41 (1990) 351.
4. R. A. Andrievski, J. Mater. Sci. 32 (1997) 4463.
5. S. Veprek, J. Vac. Sci. Technol. A 17 (1999) 2401.
6. J. Musil, Surf. Coat. Technol. 125 (2000) 322.
7. L. Hultman, Vacuum 57 (2000) 1.
8. C. Mitterer, P. H. Mayrhofer, J. Musil, Vacuum 71 (2003) 279.
9. S. Zhang, D. Sun, Y. Fu, H. Du, Surf. Coat. Technol. 167 (2003) 113.
10. R. A. Andrievski, J. Mater. Sci. 38 (2003) 1367.
11. P. Losbichler, C. Mitterer, Surf. Coat. Technol. 97 (1997) 567.
12. R. A. Andrievski, G. V. Kalinnikov, N. P. Kobelev, Y. M. Soifer, D.V. Shtansky, Phys. Solid State 39 (1997) 1661.
13. R. A. Andrievski, G. V. Kalinnikov, Surf. Coat. Technol. 142-144 (2001) 573.
14. B. Yonsson, S. Hogmark, Thin Solid Films 114 (1984) 257.

15. C. Mitterer, P. Losbichler, F. Hofer, P. Warbichler, P. N. Gibson, W. Gissler, Vacuum 50 (1998) 313.
16. O. R. Monteiro, M. -P. Delplancke-Ogletree, C. C. Klepper, J. Mater. Sci. 38 (2003) 3117.
17. E. Kelosoglu, C. Mitterer, M. K. Kazmanli, M. Urgen, Surf. Coat. Technol. 116-119 (1999) 133.
18. T. Q. Li, S. Noda, Y. Tsuji, T. Ohsawa, H. Komiyama, J. Vac. Sci. Technol. A 20 (2002) 583.
19. R. A. Andrievski, G. V. Kalinnikov, A. S. Oblezov, D. V. Shtansky, Doklady Phys. 47 (2002) 353.
20. R. A. Andrievski, G. V. Kalinnikov, A. F. Potafeev, M. A. Ponomarev, S. Y. Sharivker, Inorg. Mater. 31 (1995) 1395.
21. E. Kelesoglu, C. Mitterer, Surf. Coat. Technol. 98 (1998) 1483.
22. R. Wiedemann, V. Weihnacht, H. Oettel, Surf. Coat. Technol. 116-119 (1999) 302.
23. P. H. Mayrhofer, H. Willmann, C. Mitterer, Thin Solid Films 440 (2003) 174.

Chapter 18
CONTROL OF STRUCTURE AND PROPERTIES OF COATINGS DEPOSITED BY PULSED MAGNETRON SPUTTERING

R.D. Arnell[1], P.J. Kelly[2], J.W. Bradley[3]
[1] *University of Central Lancashire, Preston, UK*
[2] *University of Salford, Salford, UK*
[3] *UMIST, Manchester, UK*

The structure and properties of magnetron sputtered films depend on the homologous temperature; the ion/atom arrival ratio, and the energy of the bombarding ions. Pulsing the target voltage can strongly influence these parameters and can increase the energy flux delivered to the substrate. The energies of the bombarding ions, and, hence, coating structure and properties are influenced by both the reverse time/duty factor and the pulse frequency. The increasing flux of higher energy ions at increasing pulse frequency causes structure, surface topography and properties to improve up to a certain frequency, and then to deteriorate with further increase. With mid-frequency pulsed substrate bias, the substrate current does not saturate, but increases with increasing bias voltage, and this effect becomes more marked as pulse frequency is increased. Pulsing the substrate bias voltage can thus control the substrate ion current. This has implications for film growth, sputter cleaning, and substrate preheating processes.

Keywords: magnetron sputtering, pulsed sputtering, pulsed substrate bias, coating properties.

1. INTRODUCTION

This paper is not concerned, in particular, with nano-composite coatings; rather, it aims to demonstrate the potential generic benefits of using pulsed magnetron sputtering and pulsed bias sputtering to control the microstructure and properties of any sputtered coating.

In magnetron sputtering, coating properties such as film density and microstructure are largely determined by the rate at which energy is delivered to the growing film through simultaneous ion bombardment. The structure and properties of magnetron sputtered films depend on three parameters: the homologous temperature; the ratio of the fluxes of bombarding ions and depositing atoms; and the energy of the bombarding ions (1). The purpose of the work reported here was to investigate the effects

A. A. Voevodin et al. (eds.),
Nanostructured Thin Films and Nanodispersion Strengthened Coatings, 183–192.

on these parameters, and on the resulting coating properties, of applying pulsed power to the target and the substrate in a magnetron sputtering system

The flux and energy of the charged bombarding species depend on the nature of the plasma sheath adjacent to the substrate, which, in turn, depends on the plasma properties across the whole discharge. It is therefore, ultimately, the plasma properties that largely determine the coating quality, and these have therefore been studied in detail.

Pulsed power in the frequency range 100 to 350 kHz has been applied to a magnetron target and the resulting plasma parameters and coating properties have been investigated for both metallic and reactively sputtered ceramic coatings.

Time averaged Langmuir probe measurements have been used to investigate the effects of pulsing on electron temperatures, ion and electron densities and ion fluxes to the substrate. Time resolved mass-energy spectrometry at the substrate position has been used to study the energy distributions and concentrations of plasma ions, sputtered atoms, post-ionized sputtered species and gas atoms. Measurements have been made in both reactive and non-reactive environments, concurrently with film deposition.

The effects of applying mid-frequency (100–350 kHz) pulsed dc power at the substrate, in particular the effects of pulse bias and frequency on the substrate ion current, have also been studied. It has been shown that variation of the pulse frequency offers a new way of independently controlling the ion current; which has significant implications in relation to film growth, sputter cleaning, and substrate preheating processes.

The experimental procedure is briefly described in Section 2; effects of pulsing the target voltage are described in Section 3; and effects of applying pulsed bias to the substrate are described in Section 4.

2. EXPERIMENTAL PROCEDURES

The plasma analysis experiments were carried out using the equipment shown in Fig. 1. A 5 kW, dual output Advanced Energy Industries, Inc. Pinnacle™ Plus+ unit, was used to drive the magnetron. At 1 kW output power in power regulation mode. The cathode was an unbalanced rectangular magnetron fitted with a Titanium target. The target driving waveforms were recorded using a Tektronix TDS 3014 oscilloscope with a x100 voltage probe, Tek P 5100, and a TCP 202 current probe.

During deposition the chamber pressure was 0.2 Pa, the operating gas was Argon and, for reactively deposited coatings, the reactive gas flow was controlled by optical emission monitoring, using conditions known to produce stoichiometric films.

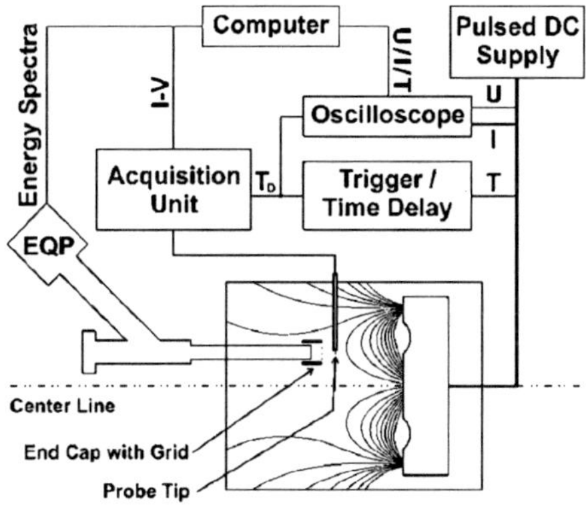

Figure 1. Schematic view of equipment for plasma analysis.

For the deposition of Titanium films, the substrate holder was connected to an Advanced Energy MDX power supply and the substrate voltage and current were recorded; for the reactive depositions, the substrate was floated electrically.

The Hiden EQP System is a high transmission ion energy analyser and quadrupole mass spectrometer. The spectrometer acquires mass spectra and energy spectra, providing detailed analysis of positive ions, negative ions and neutrals.

The Langmuir probe was connected to a Hiden data acquisition unit with a time-resolution of typically 0.3 μs.

All deposition experiments, including those with pulsed bias, were carried out in a Teer Coatings Ltd. UDP 450 rig which has been described elsewhere (2) Two 300mm×100mm vertically opposed unbalanced magnetrons were installed, although only one was driven, the second being installed only to provide a closed magnetic field between the magnetrons. The pulsed power system used was a dual channel Advanced Energy Pinnacle Plus unit, which can operate at frequencies up to 350 Hz. The normal mode of operation for these experiments was to have channel A, acting as the magnetron driver, with channel B acting as the bias supply. When the output from the Pinnacle Plus is pulsed, it operates in the asymmetric bipolar mode, in which, during the pulse-off period, the voltage reverses to approximately 10% of the nominal negative pulse-on voltage. For each channel, the unit displays the average voltage and current during the pulse-on periods only, so that the current displayed by channel B is an approximate measure of the ion current drawn at the substrate during the pulse-on period.

Substrate materials, including glass, tool steel and silicon wafers were used for various measurement and analytical techniques, including electron microscopy, microhardness testing, surface profilometry, thrust washer wear testing, and scratch adhesion testing. Continuous DC films were deposited for comparison purposes.

3. EFFECTS OF PULSING THE TARGET VOLTAGE

The effects of detailed applied discharge voltage waveforms on the intrinsic plasma parameters have been investigated as a function of pulse frequency and duty cycle, and the resulting film properties have then been directly related to these parameters.

Time averaged Langmuir probe measurements showed that electron temperatures, ion and electron densities and ion fluxes to the substrate all increase with increasing frequency (at 350 kHz, Te increased from 3 eV to 5 eV, Ni increased from $5x10^5$ to $1.2x10^6$, ion flux increased by 32%.

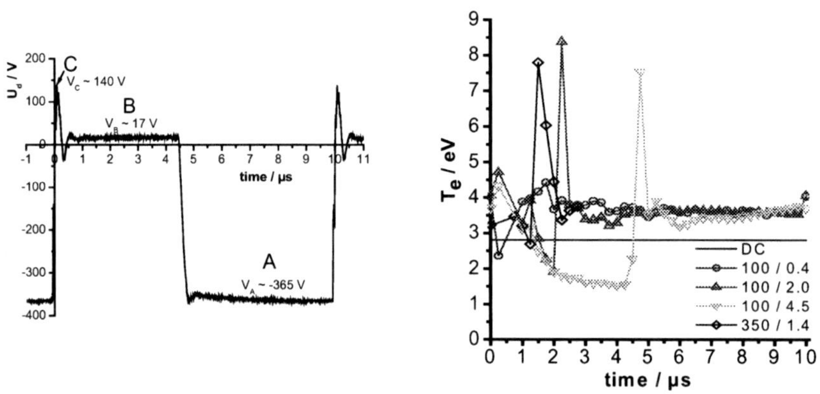

(a) *(b)*

Figure 2. Voltage waveform from pulsed power supply: (a) full cycle at 100kHz; (b) the temporal evolution of the electron temperature showing a burst of electrons at the 'on' initiations.

A typical voltage waveform, for pulsing at 100 kHz, is shown in Fig. 2(a), and the time resolved electron temperatures measured for a range of frequencies and reverse times, in Fig. 2(b). It can be seen that the waveform exhibits three peaks: region A, the pulse on period; region C, the large overshoot on reversing the voltage; and region B, showing additional peak(s) in the period 0.5 to 2.0 µs after reversal.

The plasma potential Vp reflects the voltage waveform, as illustrated in Fig. 3(a).

As the plasma potential always sits above the most positive surface, Fig 3 (a) implies that it can reach several hundred Volts +ve during the short-lived pulse-off overshoot period. The floating potential V_f will tend to follow changes in Vp at a floating substrate, but the exact relationship is not yet known.

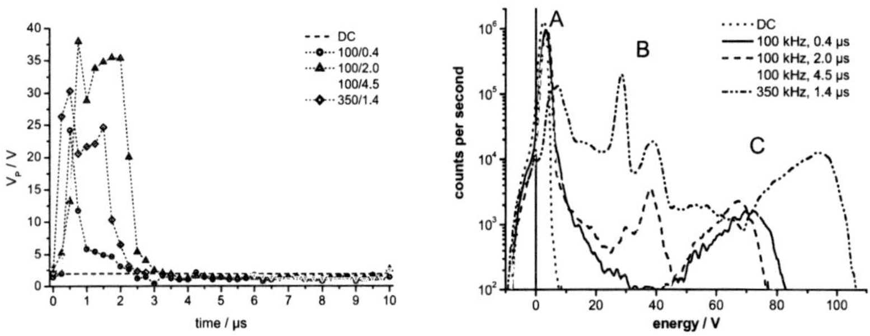

(a) *(b)*

Figure 3. (a) Plasma potential over the period 0 – 10 μs of the pulse cycle; (b) energy distribution of Ar^+ ions over the pulse cycle.

Time and energy resolved mass spectrometer studies have been performed and up to three distinct populations of ions with energies extending over a range of 150eV have been detected in the pulsed mode, as shown in Fig. 3(b). These energies reflected the target voltage waveform shown in Fig.1. The time-resolved measurements show that the high energy ions occurred only at the overshoot period at the beginning of pulse off (region C of Fig. 1), mid energy ions occurred only during the steady state pulse off period (region B of Fig.1), and low energy ions were created during pulse on (region A of Fig. 1). The variations with frequency and duty of the energy fluxes in the three regions are shown in Table 1. It is clear that the energies of the ions arriving at the substrate are influenced by both the reverse time/duty factor and the pulse frequency. In particular, at the highest frequency, 350 kHz, the majority of the ions are in the medium and high-energy bands, with energies of ~ 60eV and ~ 100 eV, respectively. It is shown below that these energy variations have very significant effects on coating structure and properties.

Table 1. Fluxes of ions in low (A), medium (B) and high (C) energy phases.

Frequency kHz	RT μs	Duty %	FA %	FB %	FC %
DC		100	100	0	0
100	0.4	96	95.2	0	4.8
100	2.0	80	92.9	2.7	4.4
100	4.5	55	35.8	59.7	4.6
350	1.4	50	28.7	47.3	24

To relate the properties of the films to the plasma characteristics during pulsing, both titanium metal and titanium dioxide films were investigated. Typically, the pulsed sputtering process has been used only for the deposition of insulating layers; the effects of pulsing had not previously been investigated for metallic materials

For the titanium films, no trend was seen in crystalline structure, hardness, or texture of the films as the pulse frequency was increased. More significant variations were observed in the surface roughness, argon content and adhesion of the films. Surface roughness was higher at low frequency <50 kHz (Ra ~ 0.15μm) and at frequencies >300 kHz (Ra ~ 0.12μm). In the 50-300 kHz range the surface roughness remained unchanged at ~ 0.06μm. Fig. 4 shows a comparison of the surface appearance of two of the films grown at 100 and 350 kHz, respectively. The surface of the 100 kHz film is very smooth in comparison to the heavily pitted surface of the 350 kHz film. The roughness of the 350 kHz film may be attributed to the increased argon content of the coating; this is undetectable in the DC sputtered film but increases exponentially with pulse frequency to a maximum of 3.5% at 350kHz.

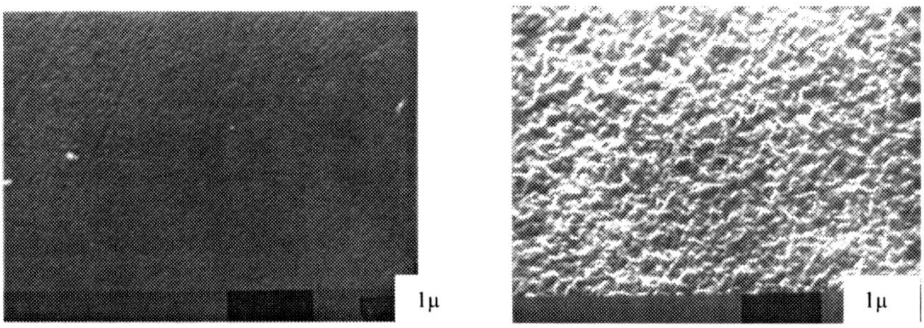

(a) *(b)*

Figure 4. Top surface SEM micrographs of Titanium deposited by pulsed magnetron sputtering: (a) 100kHz; (b) 350kHz.

The TiO_2 films showed no significant trend in hardness with frequency, giving values in the range 8-10 Gpa, and all the films were found to be X-ray amorphous. Surface roughness was similar to the glass substrates (<0.005 μm) for all frequencies except 350 kHz, which exhibited a fourfold increase in the Ra value (0.02 μm).

There was a general increase in wear resistance using pulsing in the range 50-250 kHz compared to the DC sputtered film. However, films deposited at frequencies >250 kHz showed very poor wear resistance. The optical properties of TiO_2 were also seen to vary with pulsing. Pulsing gave a higher refractive index than DC in all cases. The highest value of refractive index, 2.50, was obtained at a frequency of 50 kHz, with a general decrease in this value being observed with increasing frequency.

The observed variations in the properties of the Ti and TiO_2 coatings with pulse frequency can be related to the frequency dependent characteristics of the pulsed plasma that have been elucidated here.

The structure, surface topography and properties of both the metallic and insulating coatings were found to improve with pulse frequency up to a certain level, and then to deteriorate with further increase in frequency. This deterioration can be attributed to two effects; the increase in argon incorporation and the increase in coating intrinsic stress. Both these effects are due to the increased flux of higher energy ions at increasing pulse frequency, as shown in Fig. 3(b).

4. THE EFFECTS OF PULSING THE SUBSTRATE BIAS

In the initial batch of experiments, a 99.5% pure aluminium target was sputtered in a pure argon atmosphere. Standard operating conditions were: target current 4 A dc; sputtering pressure: 1.25 mTorr; and substrate-to-target separation: 110 mm. Under these conditions, the dc bias voltage was increased in stages from –100 to –300 V. At each stage, the current drawn at the substrate holder, as displayed by the power supply, was recorded. This process was then repeated while pulsing the bias voltage at frequencies of 100, 200, 300, and 350 kHz. In all cases a duty of 50% was used, i.e., the pulse-on time was equal to the pulse-off time. *I–V* waveforms at the substrate were stored for each pulse frequency at –100 and –300 V bias.

Following this, a fiber optic probe was installed in the rig to measure substrate temperatures and heating rates. The probe tip contains an interferometer device with the signal from this device being transferred to the control unit via a fiber optic cable, thus obviating adverse effects of high frequency interference. For each set of operating conditions, the probe temperature was logged at 1 s intervals, up to a maximum run time of 600 s or a maximum temperature of 300 °C, whichever occurred first.

The results for ion current are shown in Fig. 5.

It is known that, with dc bias the ion current saturates at ~ -100 V. However, when the bias is pulsed in the range 100-350 kHz, the current drawn at the substrate does not saturate. It is greater than the current drawn during dc biasing, it continues to increase with increasing bias voltage, and this effect becomes more marked as the pulse frequency is increased. This phenomenon is believed to be due to the initiation of a second discharge at the substrate, with a consequent change in plasma properties. The change in plasma properties has been confirmed by time-averaged Langmuir probe measurements showed that pulsing of the substrate bias typically increased electron temperature and ion saturation current by factors of 2–3 compared with the dc case. The I-V waveforms at the substrate showed a change from high voltage/low ion current during the pulse on period to high electron current/low voltage during the reverse voltage period, and the high electron current was reflected in rapid heating of the substrate.

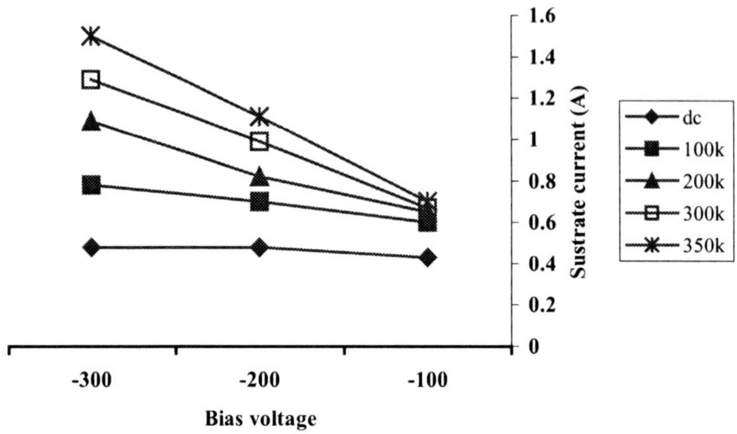

Figure. 5. Effect of bias voltage and pulse frequency on substrate ion current.

Pulsing the substrate bias voltage, therefore, offers a novel means of controlling the ion current and electron current drawn at the substrate and, hence, the ion to atom ratio and the heat input to the substrate. Clearly, this has significant implications in relation to film growth and morphology, sputter cleaning, and substrate pre-heating processes.

Following these results, experiments were carried out in which both target voltage and substrate voltage were pulsed. Three modes of operation were used: both channels in-phase (synchronous mode), 180° out-of-phase (dual cathode mode), and no with no specific phase relationship (asynchronous mode).

In addition to the general trends described above, additional trends were found. In synchronous mode the substrate ion current was ~ 20% higher than that in the dual bi-polar mode but the heating rate in dual bipolar mode, from 100–200 °C, was twice that for the synchronous mode. This is because, in the dual bipolar mode, the target pulse-on period coincides with the substrate pulse-off period. Thus, a large electron current can be drawn during the substrate pulse-off period. In contrast, in the synchronous mode, both pulse-off periods coincide and the flux of electrons to the substrate is reduced.

The final stage was to deposit TiO_2 and TiN films under conditions similar to those used for alumina. The films were deposited by dc reactive sputtering from a Ti target, and were, deposited at –100 V dc bias, –100 V at 100 kHz pulse frequency, and –300 V at 350 kHz pulse frequency. The reactive process was controlled using the optical emission monitoring technique, with conditions known to produce stoichiometric compositions.

Pulsed dc power at the substrate significantly influenced film structure, morphology and properties. The titania films had dense columnar structures which densified as bias voltage and frequency increased. XRD revealed that a film grown with –100 V dc bias had a weakly crystalline antase structure, a -100 V/100 kHz bias film, though predominantly anatase, had some second phase, and a -300 V/350 kHz film had a strongly crystalline single phase rutile structure. These experiments also showed that pulsed dc biasing can discharge insulating surfaces and can replace rf.

The effect on the structure of TiN films was also very marked. Fig. 6 shows the cross sections of two TiN coatings grown with respectively, dc and 350 kHz bias, with all other deposition parameters being identical.

The change from columnar to featureless cross section, as a result of the change in ion to atom ratio is very marked.

In changing from dc to pulsed operation, there were marked changes in morphology. The dc sputtered alumina films contained many growth defects and had Ra values of ~50 nm, whereas pulsed films at all frequencies were defect free and had Ra values of ~1nm, with profiles indistinguishable from those of the glass substrates. For titania the effect was smaller with Ra being ~ 3 nm for pulsed films and 10 nm for dc films. This difference in roughness, compared with alumina, reflects the fact that the titania films are crystalline whereas alumina films are amorphous.

Hardness was not affected by pulsing but critical load increased from ~14 to >40 N for both alumina and titania–as the pulse frequency changed from dc to 350 kHz.

Optical properties were found to be sensitive to pulse frequency for both alumina and titania, with refractive index increasing across the frequency range from 1.62 to 1.72 for alumina and from 2.44 to 2.52 for titania.

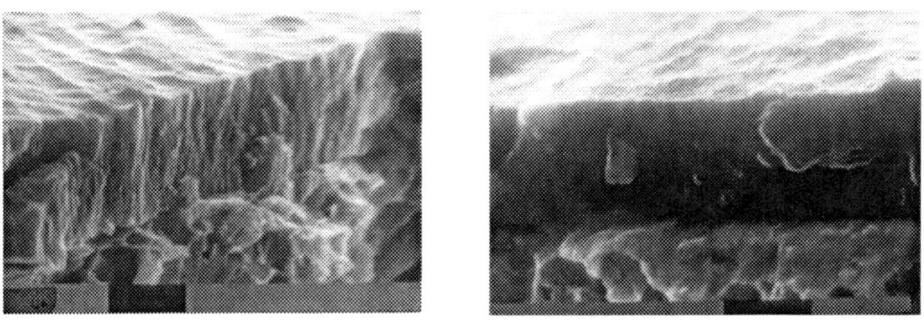

(a) (b)

Figure 6. Cross sections of TiN coatings deposited with (a) dc bias and (b) 350 kHz bias with all other deposition parameters being the same.

5. CONCLUSIONS

Pulsing of a magnetron target causes marked changes in the flux and energy of the bombarding ions at the substrate. These changes are reflected in very significant changes in the structure and properties of the coatings.

Many of these changes are frequency-dependent and some properties can be optimized by selecting an appropriate frequency.

During pulsed biasing of the substrate, the ion current does not saturate; it continues to rise with increases of both bias voltage and pulse frequency at the substrate. This has significant implications in relation to film growth and morphology, sputter cleaning, and substrate pre-heating processes.

REFERENCES

1. P.J. Kelly and R.D. Arnell, J. Vac. Sci. Technol. A17 (1999) 945.
2. P.J. Kelly, P.S. Henderson, R.D. Arnell, G.A. Roche and D. Carter, J. Vac. Sci. Technol. A18 (2000) 2890.
3. P.S. Henderson, P.J. Kelly, R.D. Arnell, H. Bäcker and J.W. Bradley, accepted for publication in Surf. Coat.Technol.

Chapter 19
ION IMPLANTED NANOLAYERS IN ALLOYS AND CERAMIC COATINGS FOR IMPROVED RESISTANCE TO HIGH-TEMPERATURE CORROSION

Z. Werner[1], W. Szymczyk[1] , J. Piekoszewski[1,2]
[1] *The Andrzej Soltan Institute for Nuclear Studies, 05-400 Świerk/Otwock, Poland*
[2] *Institute of Nuclear Chemistry and Technology, Dorodna 16, 03-145 Warsaw, Poland*

Ion implantation effects on resistance of alloys and ceramic coatings to the high-temperature corrosion have been reviewed. The most significant results on implantation of reactive elements (Y, La, Ce and other rare earth elements) into alloys and aluminum, boron, silicon, tantalum, and titanium into ceramic coatings have been cited. Ion implantation affects not only the oxide growth rate, but also seems to modify the growth mechanism and the oxide structure.

Keywords: ion implantation, high-temperature corrosion.

1. INTRODUCTION

1.1. Ion Implantation

Ion implantation consists in ballistic introducing foreign atoms into thin surface layers of solid substrates by using beams of accelerated ions. Accumulation of ions and radiation damage created by incident ions within the crystal lattice modifies material properties. Atomic concentration accumulated in the surface layer can reach several tens percent. Ion energies of up to few MeV are sometimes used, but most frequently the energy is within 10-300 keV range. One can note that projected range of 200 keV Ni ions in Fe is 67 nm, whereas the range straggling (i.e. width of the ion distribution) is 23 nm. Thus ion implantation can be considered as one of the methods belonging to the broad field of nanotechnology.

Although preliminary applications of ion implantation were limited to the field of semiconductor doping (dose range 10^{13}-10^{15} ions cm^{-2}), it was soon recognized that by extending the dose range to higher values (10^{17}-10^{18} cm^{-2}) one can also modify mechanical, optical, chemical, and other properties of materials [1-3].

A. A. Voevodin et al. (eds.),
Nanostructured Thin Films and Nanodispersion Strengthened Coatings, 193–202.

1.2. High-Temperature Corrosion

High-temperature corrosion is understood primarily as an oxidation process leading to formation of surface scales. This definition can be easily extended over other chemical reactions occurring in gaseous atmospheres, like carbon dioxide or sulfur dioxide. Such processes are often of great practical importance.

Metal corrosion, being a phenomenon related to the surface, was early identified as the field where application of ion implantation may be advantageous as a protection method. Although it was soon recognized that layers modified by ion implantation are usually too thin to be of practical value, this method is an excellent tool for studying effects of various additives on the corrosion resistance of materials. Further on, several examples of excellent work done on this subject will be presented.

1.3. Ceramic Coatings

The main purpose of a ceramic coating is usually to increase wear resistance of the substrate material. However, in many applications involving operation at elevated temperatures another equally important aim is to provide an oxidation barrier, as shown e.g. in Fig. 1 [4], where oxidation rate of a TiCrN coating on alloyed steel is compared with that of bare steel.

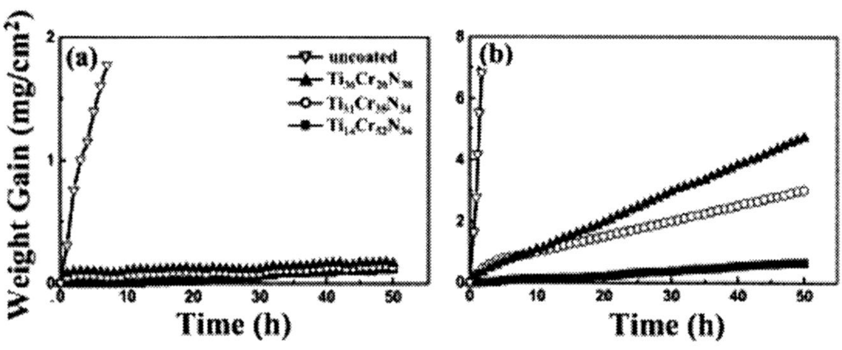

Figure 1. The oxidation kinetics of uncoated and TiCrN-coated specimens at (a) 700^0 C and (b) 900^0 C in air (after D.B. Lee et al. [4]).

A good example of great technological importance is light-alloy die casting industry where such coatings are extensively used as linings of very expensive high-pressure moulds. The moulds must withstand 10^5 or more cycles at a high temperature (700^0C), frequent temperature cycling, mechanical abrasion by a high-velocity stream of molten light alloy and chemical attack of molten aluminum [5,6]. Not all these conditions are easily simulated in the laboratory tests, especially when an accelerated test is desired. TiN and CrN coatings are traditionally used in this application and although they perform quite well, it is important to know whether their performance could not be further improved.

2. IMPLANTATION INTO STEELS AND METAL ALLOYS

Reactive elements (RE), such as yttrium, cerium, lanthanum and other rare-earths have a high affinity to oxygen. When implanted into the surface layer of a steel or a metal alloy they may improve oxidation behavior in three main ways [7]:

- **Selective oxidation of elements forming stable oxides of low diffusivity is promoted** e.g. Cr_2O_3 or Al_2O_3 on many alloys with Cr and Al content.
- **Growth rate of the Cr_2O_3 and NiO scale is reduced**. Typical example is shown in Fig.2. The benefits were maintained during extended exposures (e.g. 20 kh at 750^{0}C). Range of mass gain (1-2 mg/cm^2) corresponds to a scale thickness of several μm, whereas the thickness of implanted layer does not exceed 100 nm. The inhibition mechanism is effective in the temperature range 0.4-0.6 T_m, where T_m is the absolute melting temperature of the scale. The principal transport process controlling scale growth in this temperature range is believed to be cation movement via grain boundaries or other fast diffusion paths. Thus it is essential to establish the exact location and form of the reactive element within the scale.

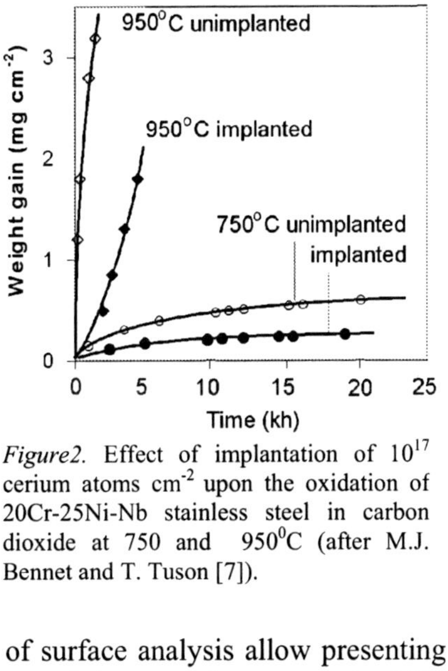

Figure2. Effect of implantation of 10^{17} cerium atoms cm^{-2} upon the oxidation of 20Cr-25Ni-Nb stainless steel in carbon dioxide at 750 and 950^{0}C (after M.J. Bennet and T. Tuson [7]).

Modern sophisticated techniques of surface analysis allow presenting the following picture. The scale of Cr-containing steel is composed of Cr_2O_3 layer, grains of $FeCr_2O_4$ spinel, and (in Y-implanted or Y-containing material) grains of Y_2O_3 oxide. At elevated temperatures these oxide grains act as a source of Y diffusing away from the oxide grains and segregating at spinel grain boundaries. This is shown in Fig. 3 presenting EDX results obtained in microanalysis of grain boundary regions in such scale by a scanning transmission electron microscope.

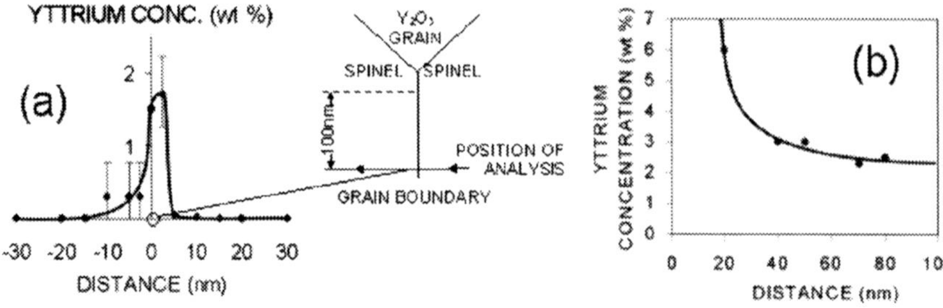

Figure 3. (a) EDX analysis using a field emission scanning TEM of the yttrium concentration profile across a spinel-spinel grain boundary at 100 nm from a Y_2O_3 grain. (b) Variation of the yttrium concentration along a spinel-spinel grain boundary with distance from Y_2O_3 grain (after M.J. Bennet and T. Tuson [7]).

• **Failure of Cr_2O_3 and Al_2O_3 scales by through-scale cracking and scale-alloy decohesion (spallation) is inhibited.** Scales growing at the metal surface tend to spall upon thermal cycling as a result of:

- excessive mechanical stresses in the oxide layer, leading to fracture;
- reduced interfacial adhesion between the scale and the substrate.

Implantation of reactive elements may potentially affect these properties in numerous ways. However, only reduction of stress has been proven to be consistent with experimental data. It results from the modified mechanism of scale development e.g. replacement of fast cation outward diffusion by slow anion inward diffusion because of segregation at grain boundaries.

It is generally believed that reduced spallation at a given temperature is caused by a slower scale growth in RE-implanted substrates and its tougher structure. These two effects combined together give a dramatic 4 orders of magnitude improvement in the time to initiate spall, observed in Ce- and Y-implanted 20/25/Nb SS, as shown in Fig. 4.

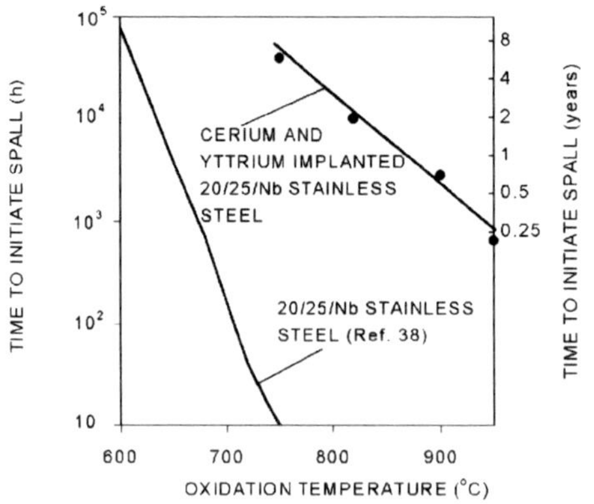

Figure 4. Influence of cerium and yttrium implantation upon the initiation on cooling to 25^0C of spallation of the oxide scale formed on the 20Cr-25Ni-Nb stainless steel (after M.J. Bennet and T. Tuson [7]).

Reactive elements are not the only additives that show beneficiary effects on oxidation resistance of metals. Dramatic improvements have been observed for noble metals like Pd and Nb, but also for less obvious candidates like Si and Al [8,9]. Stability of oxides plays a decisive role in the latter case. This is illustrated in Fig.5 showing the effect of implantation of various species upon oxidation rate of Ti-48Al-2Cr alloy oxidized in air at 800^{0}C.

3. IMPLANTATION INTO CERAMIC COATINGS

3.1. Aluminum into TiN

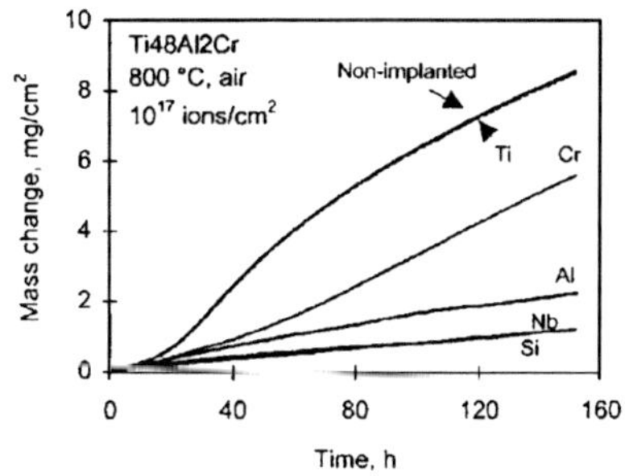

Figure 5. Mass gain vs. time of unimplanted and Ti-, Cr-, Al-, Nb-, and Si-implanted ($10^{17}cm^{-2}$) upon oxidation in air at 800^{0}C (after M.F. Stroosnijder [9]).

A pioneering work on improving high-temperature corrosion resistance of ceramic coatings by means of ion implantation was presented by A. Mitsuo et al [10]. They drew attention to aluminum, which has obvious merits in forming extremely stable oxide (heat of formation ΔG_0= -770 J/mol) of high hardness and has been used as a component of ceramic coatings for a number of years. Mitsuo's main results (Fig. 6) demonstrate that at a moderate oxidation temperature (700^{0}C) ion implantation reduces oxidation rate by a factor of 3. Even more remarkably, mass gain values correspond to oxidation of no less than 0.6 μm of coating thickness, which is much more than the projected range of the implanted ions. This means that inhibition of the oxidation rate is not associated with resistance of the implanted layer to oxidation but rather to composition and/or structure of the oxide layer formed in the implanted material. XPS spectrum of implanted and oxidized films revealed the presence of Al_2O_3 oxide states even if no aluminum oxides were found in XRD spectra.

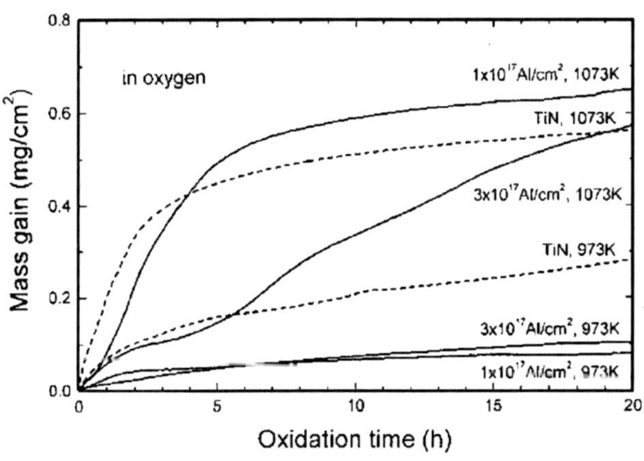

Figure 6. Oxidation curves of TiN films implanted with and without Al, oxidized at 973 K and 1073 K in dry oxygen (after A. Mitsuo et al. [10]).

Effects of Al implantation into

TiN coating were also studied by Wang and Chiu [11], who used MEVVA ion source operated at 45 kV (mean ion energy 76.5 keV) and ion dose $2x10^{17}cm^{-2}$. Studies of the oxidation kinetics at 873K revealed that the oxidation rate vs. time dependency was initially parabolic (comparable to that observed for unimplanted samples), but for longer times it changed to less steep, though still parabolic. Since no rutile signs were found in XRD spectra of implanted and oxidized samples (in contrast to the unimplanted ones), the authors concluded that ion implantation prohibited the formation of TiO_2.

In a recent paper [12] Mitsuo group further clarified the mechanism of oxidation inhibition by implanted Al ions. They used higher energies (100 keV) and higher doses ($4.5x10^{17}cm^{-2}$) than previously (50 keV, $1\text{-}3x10^{17}$ cm^{-2}). Using sophisticated analytical techniques like high-resolution EDX and selected area electron diffraction (SAED) they were able to demonstrate that the top layer of oxidized coating implanted with the highest aluminum dose contained fine-grain sized crystallites of Al_2O_3 with lattice distance matched with corundum (α-Al_2O_3). They concluded that in high-dose Al-implanted TiN coatings a fine-grain aluminum oxide layer was formed upon oxidation, which acted as a barrier against further oxygen diffusion.

3.2. Boron and Silicon into TiN

Aluminum is by no means the only candidate for oxidation inhibition in ceramic coatings. In this respect, it is interesting to refer to the results obtained by Shtansky et al [13] on oxidation resistance of reactively sputtered Ti-Si-B-N films containing about 30% of B and about 10% of Si. As shown in Fig. 7, oxidation resistance of such films is increased by a factor of 15 with respect to TiN coating at 550^0C in air.

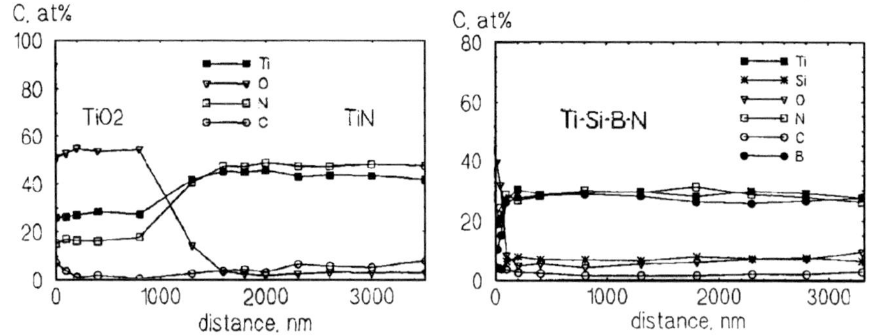

Figure 7. AES depth profiles for TiN and Ti-Si-B-N films after 2h exposure in air at 550^0C (after D.V. Shtansky et al. [13]).

Deposition of such coatings requires sputtering targets of complicated composition and a question arises whether a similar effect could not be obtained by ion implantation. The results of such approach [14] are presented in Fig. 8, where RBS spectra of TiN-coated H13 steel samples implanted with $10^{17}cm^{-2}$ and $2x10^{17}cm^{-2}$ of 80 keV boron and 200 keV silicon ions, and next oxidized in air at $630^{0}C$ for 1h are shown. The spectra of an unimplanted sample before and after oxidation are also shown for comparison. Since the energy scale of RBS spectra can be easily converted into the depth scale, one can see that the oxidized region depleted of Ti as a result of $TiN \rightarrow TiO_2$ transition is 6 times shallower in the case of implanted samples as compared to the unimplanted one.

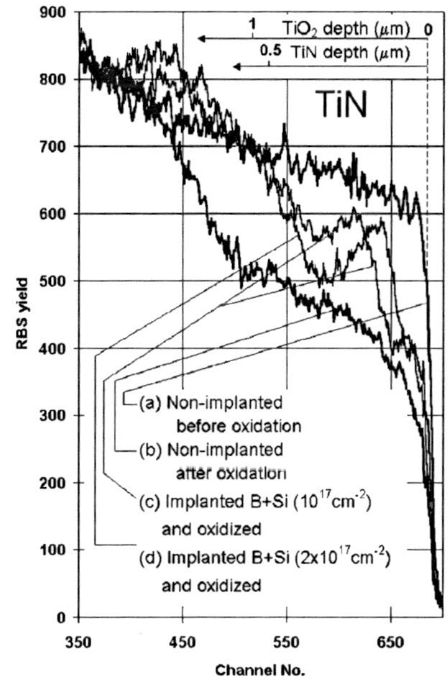

Figure 8. RBS spectra of TiN samples (non-implanted and implanted) before and after oxidation for 1h at $630^{0}C$ in technical air (after Z. Werner et al. [14])

Another conclusion which can be drawn from Fig. 8 is that oxidation has apparently stopped long before the maximum concentration in the implanted impurity profile was reached. Therefore, it is likely that the dose required to inhibit oxidation may be much lower than that used in the experiments.

It is difficult to manufacture conducting boron targets for MEVVA ion sources. On the other hand, MEVVA machines bear a potential for widespread applications of the ion implantation technique. Therefore, in view of encouraging results obtained for B+Si implantation, an attempt was made to check the effect of implantation of silicon alone upon oxidation

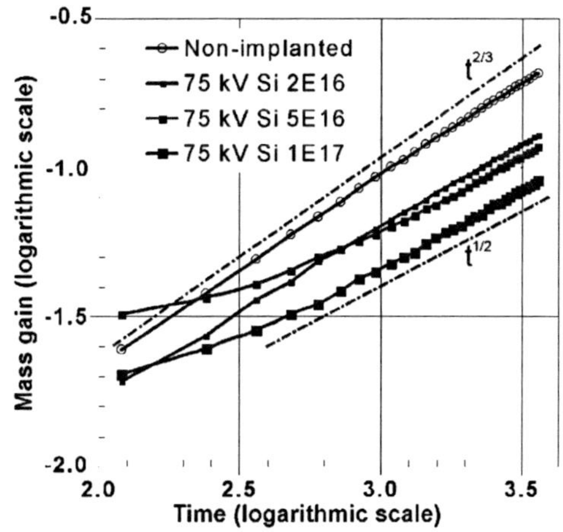

Figure 9. TG curves of TiN coated steel samples implanted with various doses of Si ions for 1h oxidation in technical air at $630^{0}C$, plotted in a log-log scale (after Z. Werner et al. [15]).

characteristics using MEVVA ion source [15]. The main results are shown in Fig. 9 as the isothermal thermo-gravimetric (TG) curves measured for one-hour oxidation time at 630^{0}C and plotted in log-log coordinates. The curves confirm more than 4-fold reduction of oxidation rate at early stages of the process.

Interesting conclusions can be deduced from an analysis of the kinetics of oxidation processes. It may be seen that for longer times the implanted samples follow very well a $t^{1/2}$ dependence, anticipated for a process, which is rate-limited by diffusion of reactants through a growing thickness of oxide layer. However, for an unimplanted sample, the curve follows very exactly a $t^{2/3}$ dependence (faster than the parabolic one indicating that a simple parabolic law model must be replaced by a more sophisticated one.

The oxide mass M apparently grows faster than the corresponding oxide thickness D. A hint how this may happen is given by inspecting a high-magnification SEM image of TiO_2 oxide layer grown on a TiN substrate (Fig. 10) [16]. One can see that because of the lattice mismatch, the oxide film does not grow like an ordered, dense layer, but forms a "cauliflower-like" structure with visible cracks and voids. Therefore, oxygen has free access to deeper layers and the thickness it has to diffuse to reach unoxidized Ti atoms is much smaller than that given by the mass increment.

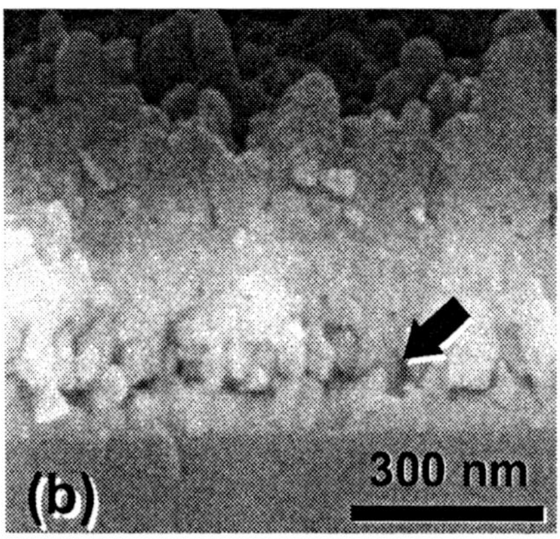

Figure 10. SEM image of TiN/AlN multi-layer annealed at 800^{0}C in air for 1h (after Do-Geun Kim et al. [16]).

Thus a tentative explanation of the observed kinetics is that M grows with "dense oxide thickness D" as $D^{4/3}$ or, in other words, that D grows with M as $M^{3/4}$.

A conclusion that can be drawn from the above discussion is that ion implantation not only inhibits oxidation, but also creates conditions for growing denser, more compact oxide layers. In such layers the normal diffusion mechanism of oxide growth is preserved.

3.3. Titanium and Tantalum into TiB_2

TiN and related protective coatings are by no means the only ones, which deserve oxidation resistance studies. Another interesting example is TiB_2 coating considered as a good candidate for protective coatings on metal interconnects employed in solid oxide fuel cells under development for future clean and efficient power generation. However, this material suffers from insufficient resistance to high-temperature oxidation. It has been speculated that addition of elements having higher valencies than Ti^{4+} may reduce concentration and mobility of oxygen vacancies, responsible for the TiO_2 oxide growth, to maintain electro-neutrality in the oxide. Hence, an attempt was undertaken to implant TiB_2 with higher (penta-) valence species of Ta and with Ti (to study self-ion damage effects) [17].

TiB_2 coatings about 500 nm thick were RF sputter-deposited onto SS 430 substrates from TiB_2 target and next implanted with Ta or Ti ions from MEVVA ion implanter at an extraction voltage of 65 kV. XRD analysis showed that the deposited films were amorphous. The most striking results are presented in Fig. 11, showing SIMS spectrum after ion implantation, vacuum annealing at 900^0C for 30 min and oxidation at 900^0C in air (a) and without vacuum annealing (b). In case (a) the coating has been retained with segregation of B at the coating-substrate interface whereas severe spallation resulting in a complete loss of coating occurs in case (b)

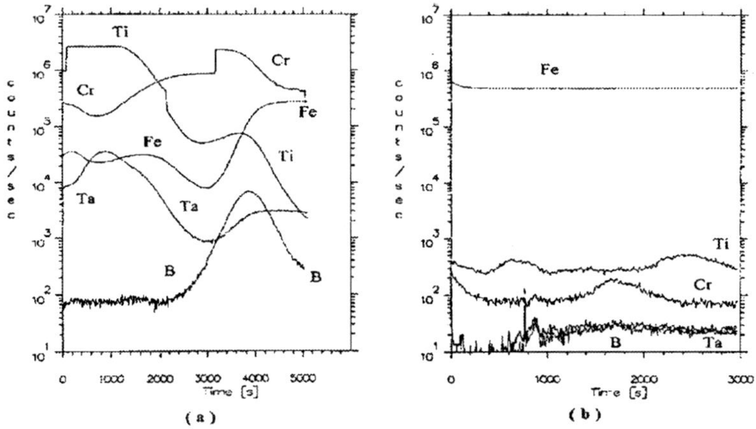

Figure 11. SIMS measurements on TiB_2 samples implanted with $1x10^{17}$ Ta ions cm^{-2} oxidized after vacuum annealing (a) and without vacuum annealing (b) (after D.K. Sood et al. [17]).

Thus vacuum annealing is necessary to obtain an increase of oxidation resistance by means of Ta implantation, although the vacuum-annealed films are still "XRD amorphous". The authors conclude that a fully crystallized TiB_2 film may be needed to achieve a further improvement in its oxidation resistance.

4. CONCLUSIONS

The use of ion implantation for enhancing oxidation resistance of steels, metal alloys and/or protective coatings is a relatively new research field, in which several promising results have been already obtained. Ion implantation affects not only the oxide growth rates but also seems to modify the growth mechanism and the oxide structure. Hence, implantation may influence the entire process of oxidation rather than be limited with its effects to a shallow depth corresponding to the implanted ion range.

REFERENCES

1. G. Dearnaley, Practical Applications of Ion Implantation, in: Ion Implantation Metallurgy, C.M. Preece and J.K. Hirvonen (editors), The Metallurgical Society of AIME, New York, 1980.
2. P.D. Townsend, P.J.Chandler and L. Zhang, Optical Effects of Ion Implantation, Cambridge University Press, Cambridge 1994.
3. E. McCaffery, Corrosion 57 (2001) 1011.
4. D.B. Lee, M.H. Kim, Y.C. Lee, S.C. Kwon, Surf. Coat. Technol. 141 (2001) 232.
5. S. Gulizia, M.Z. Jahedi, E. D. Doyle, Surf. Coat. Technol. 140 (2001) 200.
6. M. Nastasi, X.M. He, K.C. Walter, M. Hakovirta, M. Trkula, Surf. Coat. Technol. 136 (2001) 162.
7. M.J. Bennett, A. T. Tuson, Mat. Sci. Eng. A 116 (1989) 79.
8. M.F. Stroosnijder, Surf. Coat. Technol. 100-101 (1998) 196.
9. M.F. Stroosnijder, Surf. Engineering 13 (1997) 323.
10. A. Mitsuo, S. Uchida, N. Nihira, M. Iwaki, Surf. Coat. Technol. 103-104 (1998) 98.
11. Da-Yung Wang, Mingh-Chien Chiu, Surf. Coat. Technol. 156 (2002) 201.
12. T. Akhadejdamrong, A. Mitsuo, C. Iwamoto, T. Yamamoto, Y. Ikuhara, T. Aizawa, Mat. Trans. 43 (2002) 1291.
13. Shtansky D.V., Levashov E.A., Sheveiko A.N., Moore J.J., J Mat Synth & Proc 6 (1998) 61.
14. Z. Werner, J. Piekoszewski, W. Szymczyk, Resistance to high temperature oxidation in Si–implanted TiN coatings on steel, in Emerging Applications of Vacuum-Arc-Produced Plasma, Ions and Electron Beams, E.Oks and I. Brown (Eds) NATO Science Series, II. Mathematics, Physics and Chemistry, Vol.88, Kluver Academic Publishers, 2002, 191.
15. Z. Werner, J. Piekoszewski, R. Grötzchel, E. Richter, W. Szymczyk, Vacuum 70 (2003) 93.
16. Do-Geun Kim, private communication.
17. D.K. Sood, S. Mukherjee, G. Katselis, I.G. Brown, K.E.Prince, K.T. Short, P.J. Evans, Surf. Coat. Technol. 103-104 (1998) 304.

Chapter 20
QUANTUM DOTS IN Si-Ge STRUCTURES SYNTHESIZED BY GE ION IMPLANTATION INTO Si WAFERS

Yu.N. Parkhomenko[1] and N.N. Gerasimenko[2]
[1] *Moscow State Steel and Alloys Institute (Technological University), Russia*
[2] *Moscow State Institute for Electronics Engineering (Technological University), Russia*

Quantum dots (nanosize structures with quantum properties) have been obtained by ion bean synthesis using Ge ion implantation into a silicon wafer. Atomic force and scanning electron microscopies and Auger electron spectroscopy have shown that the germanium distribution in the implanted layer is inhomogeneous. In the germanium rich regions, the concentration is 10-12 at.% above the average concentration in the layer (i.e. beyond such regions), the structures being ordered in space and by sizes. The size quantizing effects have been studied using combination light scattering and luminescence spectroscopy.

Keywords: quantom dots, structure ordering, germanium, ion beam, ion implantation.

1. INTRODUCTION

Many recent publications and reviews [1-3] and a monograph [4] have dealt with the formation of quantum size structures in the Si-Ge system.

The most widely used method of synthesizing self-organized nanostructures is molecular beam epitaxy (MBE). This technique has doubtless advantages in the synthesis of nanosize structures for electronics, but it also has serious drawbacks, primarily, in that this technique is expensive, requires complex setup, and consumes much experimental time. It is hard to imagine commercial or any other large scale manufacturing of quantum size effect devices on the basis of MBE.

We have for the first time suggested an ion beam synthesis technique for the formation of ordered cluster arrays in the surface region of a wafer. These ordered clusters are germanium rich nanosize structures [5-6].

In this work, we present a study of ordered quantum size structures in the Si-Ge system formed using ion beam synthesis.

A. A. Voevodin et al. (eds.),
Nanostructured Thin Films and Nanodispersion Strengthened Coatings, 203–208.

2. SPECIMENS AND METHODS

The nanosize structures were synthesized by implanting $^{74}Ge^{+}$ ions at doses D = $5\cdot10^{16}$ and 10^{17} cm^{-2}, energy E = 50 keV and current density 5 μA/cm. The wafers were (111) p-type silicon, because Ge islands in the Ge-Si system are potential traps for holes. To avoid channeling, the ions were implanted at an angle of 7° to the normal of the wafer on a Balzers SCI 218 high-current ion accelerator. The post-implantation pulse anneal for 3 seconds was at 900 °C in a nitrogen atmosphere. The structure, composition and properties of the clusters were studied using a set of complementary techniques.

The composition of the nanosize structures was characterized by a high spatial resolution Auger electron spectroscopy (AES), and the distribution of these clusters in the surface region was examined using scanning electron microscopy (SEM), both on a PHI-680 Auger electron spectrometer (Physical Electronics, US). The primary electron acceleration voltage was 10 keV, the current was 10 nA, the primary beam diameter 20 nm, and the electron penetration depth within 5.0 nm.

Secondary ion mass spectroscopy (SIMS) on a Perkin Elmer PHI-6000 instrument was used to characterize the germanium concentration depth profile. The surface was etched within a 700 μm area with cesium ions at an accelerating voltage of 7 kV and a current of 150 nA.

The implanted layer depth was studied using a Tencor Instruments Alpha-Step 200 profilometer and SIMS.

The surface topography was studied under a NT-MDT Solver P47 atomic force microscope in cantilever resonance mode. This topography characterization technique provides the highest surface resolution.

Combination light scattering spectra were obtained on a Jobin-Ivon T 64000 spectrometer (France). The measurement setup included vertical incidence and a cooled GaAs detector. The wafer was irradiated with a 457.9 and 514.5 nm argon laser.

Photoluminescence was studied over a wide range of temperatures (11 to 300 K) by exciting the implanted quantum dot wafer with a He-Ne laser at λ = 632 nm, P = 20 mW and $\varnothing$ = 0.3 mm.

For the AES chemical composition study of the nanosize structures and surface relief characterization, the wafers were chemically etched in a 33% KOH solution for 25 s at 100 °C to strip the oxide cap from the silicon surface and remove the surface silicon layer down to the maximum germanium content layer.

3. RESULTS AND DISCUSSION

Both the directly implanted region and the implanted Ge containing layer were studied. The diagram also illustrates the implantation and anneal modes and the structural and optical techniques used for implanted layer characterization. An atomic force microscope was used to study the implanted Ge containing layer with nanosize structures.

Fig. 1 illustrates the spatial distribution of the nanosize structures and the configuration and size of an individual structure.

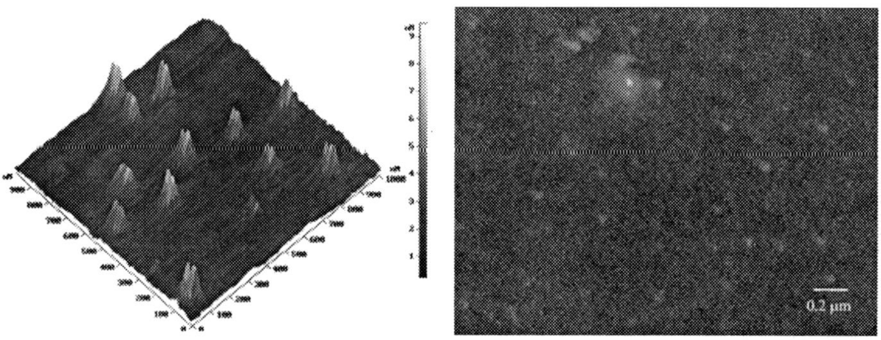

a b

Figure 1. AFM (a) and SEM (b) images of the specimen implanted with $D=10^{17}cm^{-2}$.

Data on the thickness and depth of the Si-Ge solid solution were obtained with SIMS. The germanium concentration peaks at a depth of 30 nm, the total layer thickness being 90 nm.

Atomic force microscopy (AFM) revealed a size ordered structure, the relief of which depends on the silicon and germanium etching rates (germanium etches slower). The structure consists of two types of pyramid shaped features with lateral sizes of 50±10 and 100±10 nm and heights of 5±2 and 10±2 nm, respectively.

Secondary electron images also indicated the presence of inhomogeneity in the germanium distribution with lateral sizes of 30-70 nm, which is in a good agreement with the AFM data.

The composition of the nanosize structures was characterized using the atomic force microscope that allows studies of surface topology and (during the reverse travel of the cantilever at a constant distance from the surface) surface capacitance. Fig. 2 shows that the surface relief and capacitance fluctuations are quite similar, suggesting that the distribution of the implanted Ge atoms is inhomogeneous over the implanted layer area.

A more authentic characterization of the Ge distribution over the implanted layer was obtained using high resolution Auger spectroscopy (Fig.

3). It can be seen from this figure that the Ge concentration in the quantum dots is 10-12 at.% higher than in the area between the dots.

The as-annealed Si-Ge solid solution has a completely crystalline structure without amorphous phase inclusions that were observed in the unannealed specimens. The crystalline structure is further confirmed by the absence of the combination light scattering peaks (Fig. 4) at 270 cm^{-1} (Ge-Ge), 370 cm^{-1} (Si-Ge) and 480 cm^{-1} (Si-Si) that correspond to the longitudinal optical phonons (LO) in the amorphous material [7].

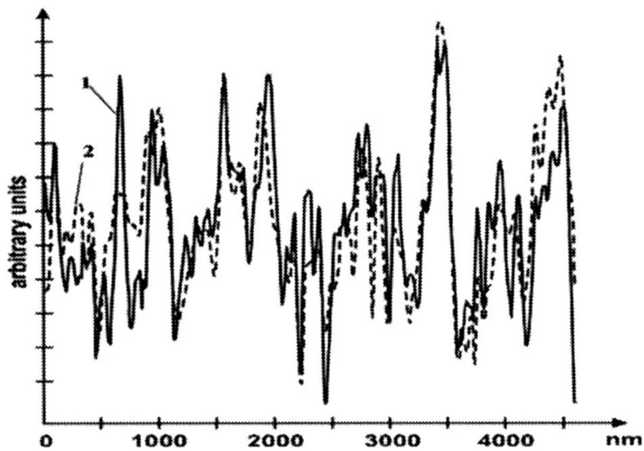

Figure 2. Comparison of implanted Si surface relief and capacitance fluctuations on specimen surface *p*-Si(111), Ge^+ ($D=6\cdot10^{13}$ cm^{-2}) 1 – Scanning capacitance mode; 2 – Atomic force mode.

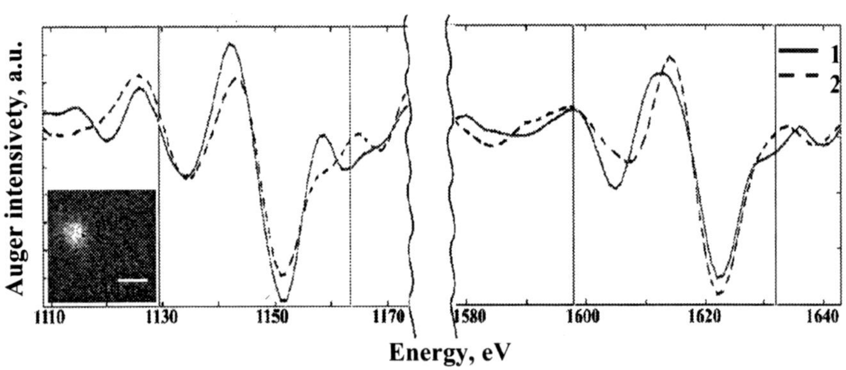

Figure 3. SEM image and Auger-spectrum for the specimen implanted with $D=10^{17}cm^{-2}$.

The lack of a Si-Si bond peak shift from 316 cm^{-1} suggests that the Si-Ge solid solution layer is not stressed [8] and hence the formation of the germanium rich areas is related to mechanical stress relaxation.

The presence of additional combination light scattering peaks in the regions of the Si-Si (495 cm^{-1}) and Ge-Ge (295 cm^{-1}) bonds that are typical of quantum dot oscillations, along with the peaks of the Si-Ge solid solutions (520 cm^{-1} for the Si-Si bonds, 420 cm^{-1} for the Si-Ge bonds and 300 cm^{-1} for the Ge-Ge bonds) suggests spatial restriction of the main optical phonon modes [9, 10]. The 5-4 cm^{-1} shift of the Ge-Ge bond peak in the low frequency spectral region is evidence for the size quantizing effect.

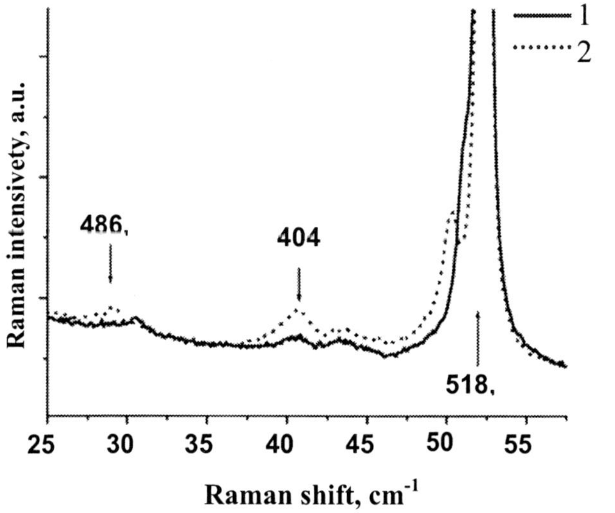

Figure 4. Comparison of Raman spectra for the specimens implanted with different doses.

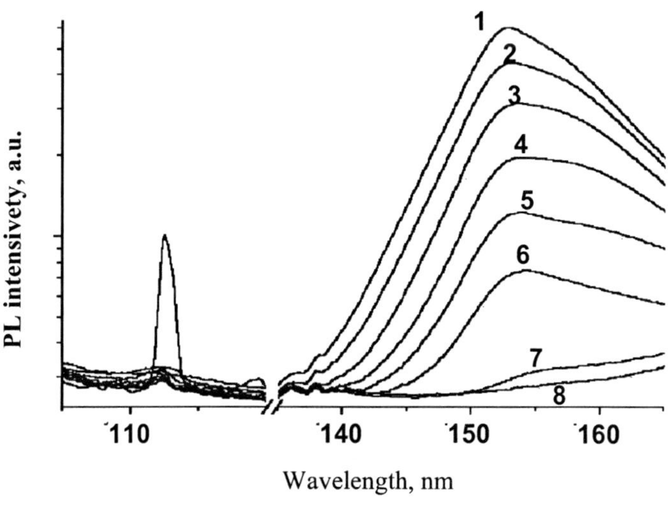

Figure 5. Different temperatures PL spectrums from QD's structures. 1-T=11K; 2-T=30K; 3-T=45K; 4-T=60K; 5-T=75K; 6-T=100K;7-T=150K; 8-T=300K.

The most convincing confirmation of the quantum size effect are the photoluminescence data presented in Fig. 5. Note the intense 1.54 μm luminescent line, which persists to room temperature. This line is usually attributed to the presence of quantum dots in the Ge-Si system. Stripping of the Ge containing layer removes this line from the photoluminescence spectrum.

By line intensity analysis, it can be clearly seen that the emitting region of this line is many times narrower than the excited layer that generates the intrinsic emission line. Note that the sizes of the germanium rich cluster (nanostructure) conform to the size quantizing effect condition only in one dimension (in height).

4. CONCLUSIONS

Quantum dots (nanosize structures with quantum properties) have been obtained for the first time using ion bean synthesis. Their formation mechanism agrees well with experimental data and literature reports. Our luminescence data suggests that these structures can be used in many fields of technology, e.g. in fiber optics.

REFERENCES

1. A.V. Dvurechenski and A.I. Yakimov, Izv. Vuzov. Electronics Engineering Materials 4 (1999) 4.
2. O.P. Pchelyakov, Yu.B. Bolkhovityanov, A.V. Dvurechenski, L.V. Sokolov, A.I. Nikiforov and B. Foigtlender, FTP 34 (2000) 1281.
3. K.S. Maksimov, N.N. Gerasimenko, M.N. Pavlyuchenko and I.V. Verner, Izv. Vuzov. Electronics 2 (2001) 3.
4. N.N. Gerasimenko, K.K. Jamanbalin and N.A. Medetov, Self-Organized Structures on the Surface and in the Bulk of Semiconductors, LEM, Almaty, 2002, 192.
5. N.N. Gerasimenko, Yu.V. Troitskii, A.A. Valyaev, Proc. of Int. Conf. on Ion Implantation Technology, 2000 Alpbah, Austria.
6. N.N. Gerasimenko, Yu.V. Troitskii, A.A. Valyaev et al., Perspektivnye Materialy 5 (2002) 26.
7. S. Kumar and H.J. Trodahi, J. Appl. Phys. 70 (1991) 3088.
8. T.M. Burbaev, FTP 35 (2001), No. 8.
9. M. I. Alonso and K. Winner, Phys. Rev. B. 39 (1989) 10056.
10. M.Ya. Ballakh, N.V. Vostokov et al., Izv. AN, Ser. Phys. 66 (2002) No. 2.

Chapter 21
CVD DIAMOND FILMS ON SURFACES WITH INTRICATE SHAPE

V. Ralchenko, I. Vlasov, V. Frolov, D. Sovyk, A. Karabutov
General Physics Institute of Russian Academy of Sciences, Vavilov str. 38, Moscow 119991, Russia
K. Gogolinsky
Technological Institute of Superhard and Novel Carbon Materials, 7a Centralnaya str., 142190 Troitsk, Moscow region, Russia
V. Yunkin
Institute of Microelecronics Technology RAS, 142432 Chernogolovka, Moscow Region, Russia

Diamond deposition on patterned substrate surfaces is considered, primarily as the step to produce CVD diamond components by molding technique. The importance of substrate seeding process to obtain high diamond nucleation density and conformal coatings is underlined. In case of free-standing films the morphology, defects and impurities of fine-grained nucleation (substrate) side are different from those on growth side. Examples of the molding technique for manufacturing of various diamond shapes, from micron-scale pyramidal arrays for optics and field electron emitters to centimeter size X-ray refractive lenses, are presented.

Keywords: diamond film, nucleation, patterning, molding, cantilever, field emission, X-ray lens.

1. INTRODUCTION

A remarkable progress in past two decades in CVD diamond technologies resulted in emergence of new engineering materials with outstanding properties, many of them approaching to those known for best quality single crystal diamonds [1,2]. Nowadays thin polycrystalline diamond coatings on large area, up to half square meter [3], and thick (above 2 mm) wafers [4] are produced commercially. The CVD diamond synthesis methods are based on a decomposition of hydrogen-hydrocarbon (usually methane) gas mixtures followed by diamond growth on a hot (typically 600-1000ºC) substrate. The gas activation can be realized with many ways, in particular, using hot filament DC discharge, microwave plasma, DC arc-jet, laser-sustained plasma, acetylene-oxygen flame and some other techniques [1,2]. Different substrate materials, such as Si, WC-Co alloy, Cu, Ti, steel, SiO_2, composites, and many others, which are able to withstand the high deposition

A. A. Voevodin et al. (eds.),
Nanostructured Thin Films and Nanodispersion Strengthened Coatings, 209–220.

temperatures, can be coated with the diamond directly or with an appropriate interlayer. In the present paper we focus on peculiarities of the diamond nucleation and growth not on flat surfaces but on substrates patterned on micro- and macroscale. The conformal growth is important not only for the production of thin (particularly, nanocrystalline [5]) coatings, but also for the fabrication of various diamond components of pre-designed shape using a transfer molding technique [6].

2. DIAMOND NUCLEATION AND GROWTH

Since the spontaneous diamond nucleation on foreign substrate materials is poor, the effective nucleation centers must be artificially introduced on the surface to stimulate the diamond growth and to form a continuous film. Probably the simplest and the oldest approach, i.e. the mechanical abrasion of substrate with a diamond grit, can enhance the nucleation density to 10^7 – 10^8 cm^{-2} (Fig. 1a). However, a formation of defects (scratches) on the substrate surface is undesirable in some cases, e.g. in electronic or optical applications. Very high nucleation density, up to 10^{12} cm^{-1}, can be achieved by the substrate biasing, when a negative DC voltage (typically –150 - 200V) is applied to the substrate at the beginning of diamond deposition [7]. The nuclei are formed in a carbonaceous layer under bombardment with accelerated ions. Moreover, the bias pretreatment allows the growth of highly oriented textured diamond films on flat surfaces [8]. However, because of shadow effect the nucleation will not be complete for patterned surfaces, for example, having undercut structures (Fig. 1b). The seeding technique compatible with a standard microelectronic technology is the spin-off coating on the substrate with a photoresist containing diamond particles [9] (Fig. 1c), which, again, requires the flat surface. The same restriction on the surface relief seems to be valid for the seeding by printing [10] or laser-induced particles transfer [11,12]. For seeding of surfaces with intricate shape as well as the porous materials an ultrasonic pretreatment in suspension of nanomentric diamond grit in acetone, alcohol or other liquid (Fig. 1d) seems to be the most efficient [13,14]. The particles attach the surface and promote heteroepitaxial diamond growth. The ultrasonic treatment with 5 nm detonation synthesised diamond powder provides the nucleation density higher than 10^{10} cm^{-2} without any surface damage [13-15]. The seeds can penetrates in small, several tens nanometres, pores going deep into the substrate, allowing a filling the porous network, sometimes very complex, with the CVD diamond [16].

We note that the seeding is not required for diamond-based composite substrates in which diamond particles are always present on the surface as well as in the bulk, so the diamond growth starts without any pre-treatment. For example, diamond coatings with good adhesion due to anchor effect were realized on nanocomposites containing 5 nm diamond particles embedded in a matrix of pyrolytic carbon [17], and on composites with diamond in TiC+SiC+Si or TiB_2+$TiSi_2$+Si matrix produced by self-propagating (combustion) high-temperature synthesis [18].

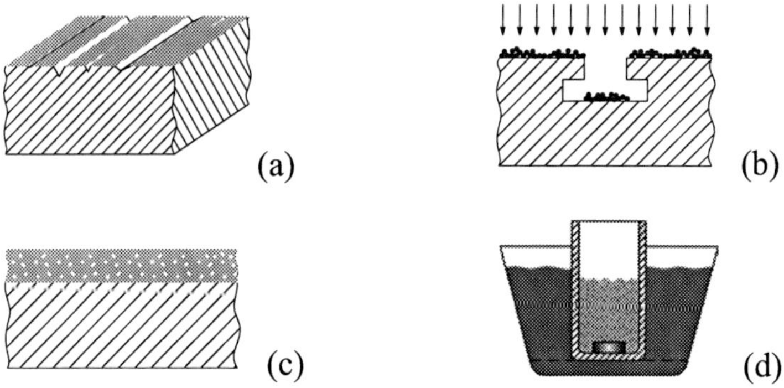

Figure 1. Seeding the substrate surface with diamond nucleation centers: (a) scratching with diamond grit, (b) biasing, (c) spin-off coating, and (d) ultrasonic treatment.

The diamond films described below were grown on silicon using a DC plasma deposition system [13-15] and a microwave plasma CVD reactor [19,20]. Fig. 2 shows a thin diamond film uniformly deposited on Si grating by DC plasma CVD. The Si bars with 10 μm spacing were formed by lithographic patterning followed by reactive ion etching.

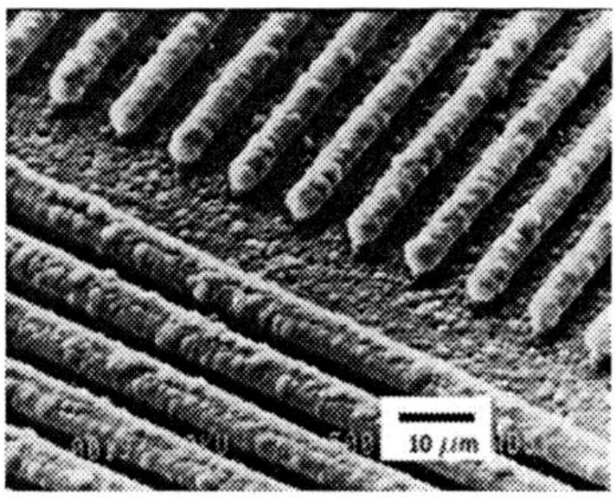

Figure 2. Diamond coating on Si bars deposited in DC plasma CVD reactor. The grating period is 10 μm.

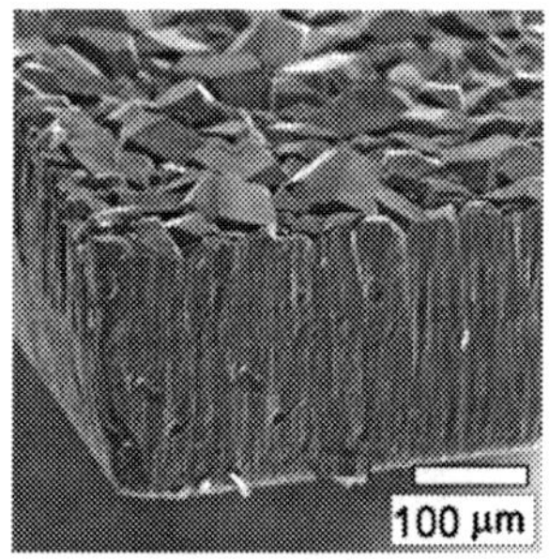

Figure 3. MPCVD free-standing diamond film of 200 μm thickness grown on mirror-polished silicon.

The diamond film can be separated from the substrate, e.g. Si can be removed chemically in HF-HNO_3 acid, to produce diamond plate or membrane. Fig. 3 demonstrates the free-standing film grown in a MPCVD reactor (frequency 2.45 GHz, power 5 kW) under the following conditions: source gas 2.5% CH_4 in H_2, gas pressure 100 Torr, hydrogen flow rate 400 sccm, substrate temperature 715°C, substrate diameter 57 mm, deposition rate 3 μm/hour. To provide a high nucleation rate elevated methane concentration (4%) was kept for the first 30 minutes of deposition. The top (growth) surface displays well-faceted crystallites with lateral size of 30-60 μm, while the bottom (nucleation) side adjacent to the substrate consists of small, micron-scale grains. The diamond film is inherently a gradient material with grain size, defect abundance and physical properties changing across the film. The nucleation side is smooth being the replica of the mirror-polished Si substrate. Obviously, the replica of a patterned substrate can also be produced, providing a high nucleation density is achieved.

3. PROPERTIES OF NUCLEATION SIDE

Nucleation side normally is smooth enough and does not need to be polished if an enhanced concentration of defects is tolerable. AFM images of nucleation side are shown in Fig. 4. The surface topography is mapped in tapping and phase imaging modes. The latter mode enhances the contrast, the information signal is the product MAG×*sin*φ of the cantilever vibration amplitude MAG and the phase shift φ. It visualizes ~300 nm grains, and the average surface roughness R_a=7 nm being measured for this particular sample. The inspection of a large number of MPCVD films showed the surface roughness R_a < 15 nm on nucleation side, and the average grain size ranging from 70 nm to 2 μm depending on the deposition conditions.

Hydrogen and nitrogen are most common impurities in CVD diamond. Since the diamond growth proceeds in a hydrogen-reach environment, the hydrogen is a major impurity in the CVD diamond, its concentration ranges in the range from a few tens ppm for high quality, large grain size films to 20,000 ppm for thin nanocrystalline films [21]. Hydrogen is found to be distributed non-uniformly across the film thickness, a 20-50 μm thick fine-grained layer adjacent to substrate is enriched by an order of the magnitude

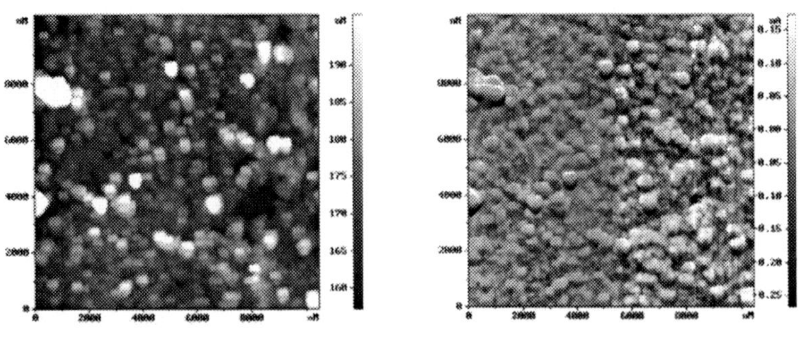

Figure 4. Comparative AFM maps of nucleation surface: relief (left); phase image (right). The scanned area is 10×10 μm^2.

with H compared to a bulk of the film [22,23]. This is in agreement with higher density of defects and grain boundaries (which are decorated by hydrogen) on the nucleation side. Therefore, the hydrogen concentration can be an indicator of defect abundance in the film [24]. In particular, a clear inverse correlation between the total bonded (C-H bonding) hydrogen content in CVD diamond and its thermal conductivity was observed [21,24]. Local thermal conductivity $k_{\|}$ (in the film plane) at growth and nucleation sides of thick (0.3-0.5 mm) diamond films was measured [25] by a transient thermal grating (TTG) technique using a quasi-surface TTG excitation in VUV spectral range (213 nm wavelength), where the diamond has a strong intrinsic optical absorption. At the growth side of translucent films the $k_{\|}$ values of thin (ca. 30 microns) sub-surface layer reach 18-22 W/cmK, while much lower values (~7 W/cmK) were measured at the nucleation side. Due to a specific columnar growth of diamond grains the thermal conductivity $k_{\perp}$ in direction perpendicular to the film plane is by 10-20% lower than $k_{\|}$ values. The thermal conductivity of thin (~10 microns) diamond films typically does not exceed 10 W/cmK [26] that is one half of that of the best single crystals or coarse-grained CVD diamonds.

Upon heating in vacuum the diamond films preserve their structure, optical and mechanical properties up to 1200°C [27]. At higher temperatures, the break of the C-H bonds on grain boundaries and defects triggers a graphitization of the domains which are primarily decorated (and stabilized) with hydrogen. The fine-grained nucleation side is involved in this process more intensively compared to a less defective growth side.

Since some nitrogen is a contamination in the source gas, the incorporated nitrogen in a substitutional form can be detected in the film, for instance, by ESR and UV optical absorption spectroscopy. If not doped intentionally, CVD diamond contains less than 10-20 ppm nitrogen [28] that appears to be uniformly distributed across the film [29].

In addition, silicon can be incorporated in diamond from the Si substrate often used for the diamond deposition. The presence of Si in diamond is associated with a zero-phonon line at 737 nm in the photoluminescence (PL) spectrum, which presumably belongs to a substitutional silicon atom and a vacancy in the nearest lattice site according to Clark and Dickerson [30]. A typical PL spectrum (excitation with an Ar^+ laser at 514 nm wavelength) taken on nucleation side of the diamond film grown on a Si substrate is shown in Fig. 5. Besides the 737 nm Si-V center, two nitrogen-vacancy (N-V) optical centers at 575 nm and 638 nm are also observed in the PL spectrum. The PL cross-section profiling on fractured surfaces revealed that the nucleation side is enriched with Si impurity [29]. The depth profiles of the 737 nm line intensity for two films deposited on Si substrates at temperatures ≈700°C and 800°C, and one grown on a molybdenum substrate are shown in Fig. 6. For a correct comparison, the integral PL intensity was normalised to the Raman diamond peak area. The probing volume was minimized to 2 μm in the film plane and 5 μm in the film depth using a confocal optical scheme. The laser light was always focused on the surface.

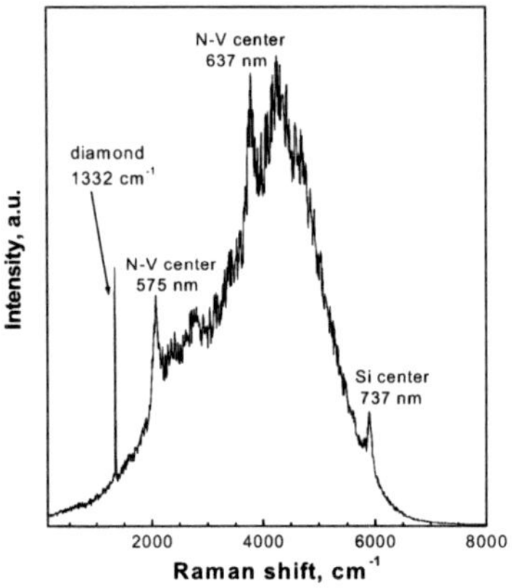

Figure 5. PL spectrum of a MPCVD diamond film recorded on nucleation side at 90°K. The line at 737 nm belongs to Si center, while the lines at 575 and 637 nm are due to N-V centers. The line at 1332 cm^{-1} is diamond Raman peak.

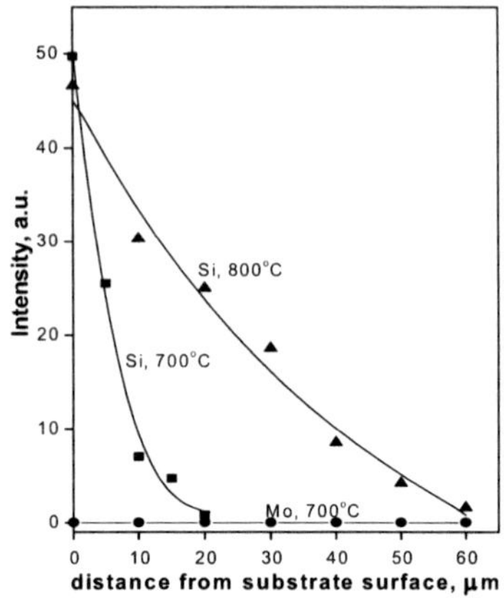

Figure 6. Depth profiling of PL intensities of a Si–related defect center at 737 nm for diamond films deposited on Si at temperature 700°C (squares) and 800°C (triangles), and on Mo substrate at 700° (circles).

No PL signal from the 737 nm center has been detected in the film grown on Mo substrate. This confirms the role of Si substrate as a main impurity source. The contamination proceeds via the substrate etching in hydrogen plasma at an initial stage of the deposition process, followed by the Si incorporation from the plasma into the growing film. When the substrate is completely covered with thin diamond film the Si etching stops, and the intensity of Si-V centres reduces to zero at the depth of 20 μm for 700°C sample. The Si penetration depth extends to 60 μm for the sample prepared at 800°C, due to a lower nucleation density (and, consequently, a longer exposure time of naked Si to plasma) in this case. Note that the diffusion in diamond is negligible for all elements except hydrogen, so the direct diffusion of Si from the substrate is excluded.

The Si contamination, partly in form of SiC, influences the wettability of nucleation side. The growth side becomes hydrophobic with a contact angle of water θ=93° after hydrogenation, and hydrophilic one (θ=32°) after oxidation. The nucleation side shows a similar tendency under these treatments, but the contact angle varies in a more narrow range, between 65° and 53° [31]. The latter phenomenon was ascribed to the presence of Si contamination (4 at% according to XPS analysis) on the nucleation side.

In boron-doped diamond films the boron content as well as a total acceptor concentration on the nucleation side can be significantly, by order of magnitude, higher compared to that on the growth side [32].

4. DIAMOND COMPONENTS PRODUCED BY MOLDING TECHNIQUE

For real applications in optics, micromechanics, electronics and other fields the diamond films must be shaped in a desired geometry both on micro- and macroscales. For example, the fabrication of the diffractive diamond optics [33] or cantilevers for scanning probe microscopy (SPM) [34] requires a special relief formation with a micron level definition, or even better. Because of extreme chemical stability and hardness of the diamond its processing is difficult. The molding technique for a direct growth of CVD diamond components (see [6] for a review) is attractive since it allows to avoid or minimize the further surface treatment, and sometimes to produce the patterns which could not be obtained even with the sophisticated methods of laser, plasma or ion beam etching. The transfer molding is based on the diamond deposition on a patterned substrate, the replica of those patterns being formed on nucleation side of the diamond film. The diamond nucleation density must be very high to produce the exact replica of the master structure after removal of the substrate.

The idea of the molding technique for CVD diamond has been proposed in 1982 by Belyanin and co-authors in a Russian patent [35], which probably remained unknown for other researchers in the field of CVD diamond. After the work of Okano et al. [36] on the fabrication of boron-doped diamond pyramid arrays for field emitters, this approach was employed to produce a variety of diamond structures, such as the diamond optics with antireflective "moth eye" surface structures [37-39], lenses [40], fluidic cooling systems and optical MEMS (microelectromechanical systems) [41,42], pyramid-shaped arrays for the field electron emission cathodes [43,44] and other MEMS (piezoresistors, pressure membranes, field effect transistors) [45], "pockets" and spheres on diamond substrates [6,39], cantilevers and sharp tips for atomic force microscopy [34,46], microchannels [47], “diamond opal” with regular pores which is an example of photonic crystal [16], micropatterned diamond tools for high-rate machining of ceramic materials [48], diamond flow controller micronozzles [49].

A possible sequence of processes to form the microstructures in diamond is shown in Fig. 7. Si substrate is patterned with a standard photolithography and is etched in SF_6-O_2 RF plasma using a Si_3N_4 mask. Then the patterned surface is seeded with 5 nm diamond powder in an ultrasonic bath to provide a high nucleation density for successive diamond growth. After the diamond deposition in a DC plasma the silicon was removed in a mixture of hydrogen fluoride and nitric acids to obtain the replica.

Fig. 8 illustrates the qualities of diamond replicas produced by a DC plasma CVD in case of poor and good nucleation density (the Si master patterns are not identical). This stresses again an importance of the seeding

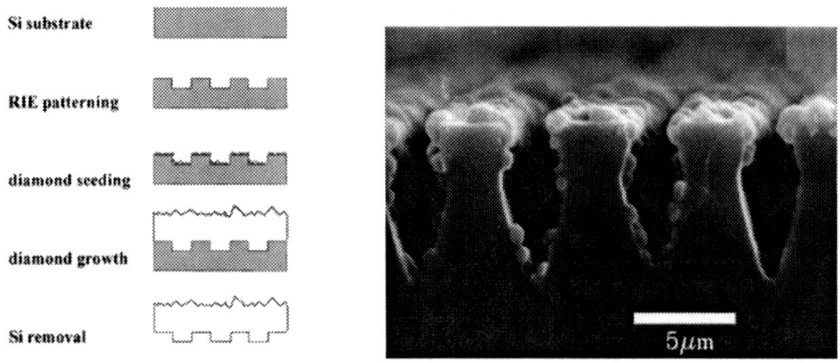

Figure 7. Fabrication process of diamond replica using a Si template (left), and diamond nucleation on a Si pattern (right).

process and high nucleation rate. Note the presence of channels inside ridges as a result of stopping the growth process *inside* the trench in Si template when the diamond film edges merge at certain moment above the trench top.

The technique to grow the regular diamond micropyramids, primarily for field electron emission (FEE) applications [43,44], became common in the last decade. The inverted pyramidal pits in Si are anisotropically etched in KOH, then they are filled with diamond (Fig. 9). The FEE onset field less than 10 V/μm was measured for the pyramid arrays when the surface conductivity was induced by the following means: ion bombardment, annealing, or boron doping [6,44]. Micron-scale pyramids (base length <10 μm), densely arranged on the surface, also make possible to essentially increase IR optical transmission of the diamond plates [37-39], being an alternative to the antireflective coatings.

Larger, but single diamond pyramids are of interest for the use in scanning probe microscopes designed for the indentation of materials, nanoengraving or evaluation of their elastic properties. The integrated diamond cantilever and the tip for SPM are shown in Fig. 10. An array of the cantilevers with pyramids of 100 μm base length was formed on a diamond wafer (60 mm diameter, 100-200 μm thick), and then was laser-cut to single

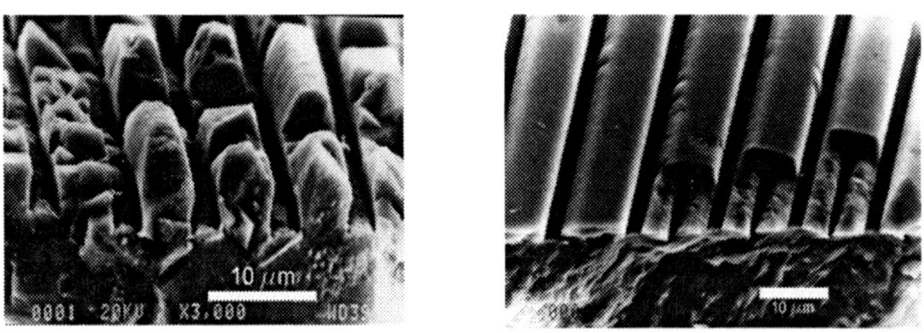

Figure 8. Diamond gratings obtained at low (left) and high (right) nucleation densities.

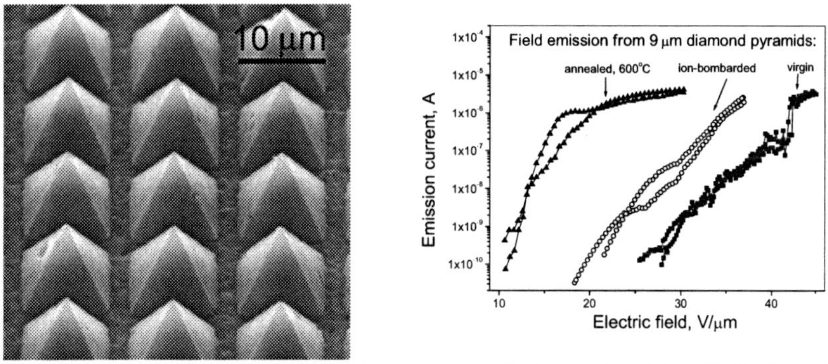

Figure 9. Array of diamond pyramids grown on anisotropically etched Si (left), and its I-V characteristics of field emission [44].

cantilevers in a way that each pyramid is at the end of the cantilever. The pyramids have smooth surfaces and sharp tips (curvature radius is < 50 nm).

Grooves (width 500 nm, depth 8 nm) on the sapphire were produced with an SPM ("Nanoscan" model) equipped with the diamond cantilever (Fig. 11). Then the formed surface relief was imaged with the same diamond tip. Similar grooves (width 250 nm, depth 15-45 nm) have been engraved on silicon, but imaged with the silicon tip (Fig. 11).

Not only the microstructures, but also rather large, a few centimeters, diamond components and articles, such as X-ray lenses [51] or coins [6], can be grown using an appropriate template. The manufacturing and tests of first diamond refractive lenses for hard X-ray focusing has been reported recently [51]. The diamond is the most attractive among other low-Z materials for X-ray transmission optics due to its low absorption, high refraction index and very high thermal conductivity, the latter being especially important for handling with high intensity synchrotron beams. The silicon mold, a plate with a deep (up to 200 µm) relief of parabolic form, has been prepared using lithography and plasma etching. The successive diamond growth in the MPCVD reactor for 57 hours produced 110 µm thick replica (Fig. 12a). The synchrotron beam focusing to 2 µm spot size at the focal distance of 0.5 m was achieved with 25% transmission at the photon energy 9 keV (Fig.12b), while for 38 keV energy the transmission increased to 80%. The lens

Figure 10. Integrated diamond cantilever with a pyramidal tip for SPM: (left) general view, (right) detail. The cantilever length is 2 mm, the pyramid base length is 100 µm.

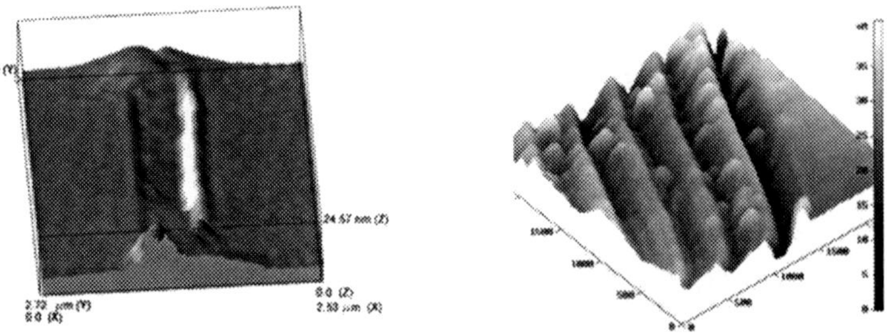

Figure 11. Engraving on sapphire (left) and silicon (right) surfaces with diamond tip.

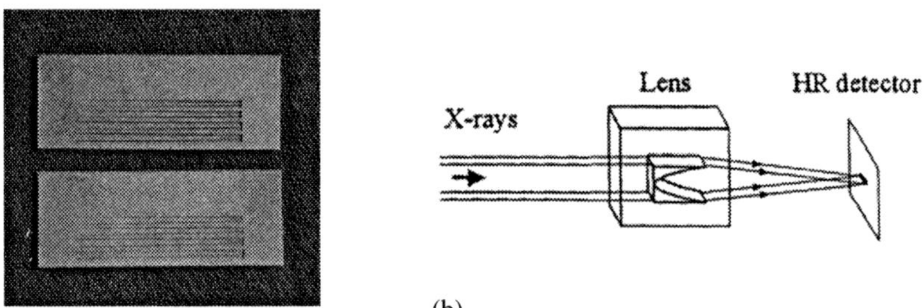

(a) (b)

Figure 12. (a). X-ray diamond lenses of 15 x 40 mm^2 size with relief depth of 100 and 200 μm as viewed from growth side. Four parabolic lenses are formed on each 110 μm thick plate. (b) Geometry of X-ray focusing test [51].

performance was quite stable at the beam power density 50 W/mm^2, what confirms the CVD diamond optics to be excellent material for extremely powerful X-ray fluxes.

5. CONCLUSIONS

The molding technique, based on the diamond film deposition on the patterned sacrificed template, which is then removed to obtain the free-standing diamond replica of a pre-designed shape, is promising for the manufacturing of micro- and macroscale CVD diamond articles. To achieve a good pattern definition, the very high diamond nucleation density is required, that can be realized using nanodiamond particles as seeds. The molding technique in some cases allows fabricating unique diamond shapes which hardly can be produced by other processing methods.

ACKNOWLEDGEMENTS

The authors are thankful to S. Voronina and I. Sychov for preparation of diamond film samples, S. Lavrischev for making SEM pictures, and A. Popovich for the help in preparation of the manuscript. This work was supported in part by the Russian Ministry of Science and Technology via "Integration" Program, under the contract No. IO 470/1188.

REFERENCES

1. Handbook of Industrial Diamonds and Diamond Films, M. Prelas, G. Popovici, L. Bigelow (Eds), Marcel Dekker, New York 1997.
2. B. Dischler, C. Wild (Eds), Low-Pressure Synthetic Diamond: Manufacturing and Applications, Springer, Berlin 1998.
3. L. Schaefer, M. Fryda, D. Herrmann, I. Troester, W. Haenni, A. Perret, Proc. 6th Applied Diamond Conference/2nd Frontier Carbon Technologies Joint Conference (ADC/FCT 2001), ed. By Y. Tzeng et al., NASA/CP-2001-210948 (2001) 158.
4. S.E. Coe, R.S. Sussmann, Diamond Relat. Mater. 9 (2000) 1726.
5. D.M. Gruen, Ann. Rev. Mater. Sci. 29 (1999) 211.
6. V. Ralchenko, in Nanostructured Carbon for Advanced Applications, G. Benedek, et al. (Eds), Kluwer, Dordrecht (2001) 27.
7. S. Yugo, T. Kanai, T. Kimura, Diamond Relat. Mater. 1 (1992) 388.
8. A. Floter, H. Guttler, G. Schulz, D. Steinbach, C. Lutz-Elsner, R. Zachai, A. Bergmaier, G. Dollinger, Diamond Relat. Mater. 7 (1998) 283.
9. V.G. Ralchenko, K.G. Korotoushenko, A.A. Smolin, V.I. Konov, in Advances in New Diamond Science and Technology, S. Saito et al. (Eds), MYU, Tokyo (1994) 493.
10. K. Higuchi, S. Noda, Diamond Relat. Mater. 1 (1992) 220.
11. S.M. Pimenov, G.A. Shafeev, A.A. Smolin, V.I. Konov, B.K. Vodolaga, in Advances in New Diamond Science and Technology, S. Saito et al. (Eds), MYU, Tokyo 1994 109.
12. M.B. Guseva, V.G. Babaev, V.V. Khvostov, Z.Kh. Valioullova, A.Yu. Bregadze, Diamond Relat. Mater. 3 (1994) 328.
13. A.A. Smolin, V.G. Ralchenko, S.M. Pimenov, T.V. Kononenko, E.N. Loubnin, Appl. Phys. Lett. 62 (1993) 3449.
14. V.I. Konov, A.A. Smolin, V.G. Ralchenko, S.M. Pimenov, E.D. Obraztsova, E.N. Loubnin, S.M. Metev, G. Sepold, Diamond Relat. Mater. 4 (1995) 1073.
15. L.C. Nistor, J. Van Landuyt, V.G. Ralchenko, A.A. Smolin, K.G. Korotushenko, E.D. Obraztsova, J. Mater. Res. 12 (1997) 2533.
16. A.A. Zakhidov, R.H. Baughman, Z. Iqbal, C. Cui, I. Khayrullin, S. Dantas, J. Marti, V.G. Ralchenko, Science, 282 (1998) 897.
17. V. Ralchenko, A. Smolin, I. Vlasov, A. Karabutov, V. Frolov, V. Konov, S. Gordeev, S. Zukov, Molecular Materials, 11 (1998) 143.
18. E.A. Levashov, V.G.Ralchenko, Advances in Science and Technology, Vol. 9, P. Vincenzini, P. Ascarelli (Eds), Proc. Forum on New Materials of CIMTEC 2002-10th, Faenza, Techna (2003) 49.
19. V.G. Ralchenko, A.A. Smolin, V.I. Konov, K.F. Sergeichev, I.A. Sychov, I.I. Vlasov, V.V. Migulin, S.V. Voronina, A.V. Khomich, Diamond Relat. Mater. 6 (1997) 417.
20. V. Ralchenko, I. Sychov, I. Vlasov, A. Vlasov, V. Konov, A. Khomich, S. Voronina, Diamond Relat. Mater. 8 (1999) 189.
21. V. Ralchenko, A. Khomich, R. Khmelnitskii and A. Vlasov, in Hydrogen Materials Science and Chemistry of Metal Hydrides, N. Veziroglu et al. (Eds), Kluwer (2002) 203.
22. J.E. Graebner, J.A. Mucha, F.A. Baiocchi, Diamond Relat. Mater. 5 (1996) 682.
23. S.M. Pimenov, V.V. Kononenko, V.G. Ralchenko, V.I. Konov, S. Gloor, W. Luthy, H.P. Weber, A.V. Khomich, Appl. Phys. A. 69 (1999) 81.
24. D.T. Twitchen, C.S.J. Pickles, S.E. Coe, R.S. Sussmann, C.E. Hall, Diamond Relat. Mater. 10 (2001) 1731.
25. V.G. Ralchenko, A.V. Vlasov, E.V. Ivakin, A.V. Sukhadolau, A.V. Khomich, Proc. 7th Applied Diamond Conference/3rd Frontier Carbon Technologies Joint Conference (ADC/FCT 2003), M. Murakawa et al. (Eds), NASA/CP-2003-212319 (2003) 309.
26. J.E. Graebner, V.G. Ralchenko, A.A. Smolin, E.D. Obraztsova, K.G. Korotushenko, V.I. Konov, Diamond Relat. Mater. 5 (1996) 693.
27. V. Ralchenko, L. Nistor, E. Pleuler, A. Khomich, I. Vlasov, R. Khmelnitskii, Diamond Relat. Mater. 12 (2003) 1964.

28. S.V. Nistor, M. Stefan, V. Ralchenko, A. Khomich, D. Schoemaker, J. Appl. Phys. 87 (2000) 8741.
29. I. Vlasov, V. Ralchenko, A. Khomich, DIAMOND'99 Conference, 12-17 September 1999, Prague, paper No. 15.616.
30. C.D. Clark, C.B. Dickerson, Surf. Coat. Technol. 47 (1991) 336.
31. L. Ostrovskaya, V. Perevertailo, V. Ralchenko, A. Dementjev, O. Loginova, Diamond Relat. Mater. 11 (2002) 845.
32. V.G. Ralchenko, Y.V. Pleskov, V.I. Polyakov, A.V. Khomich, Y.E. Evstefeeva, M.D. Krotova, E.N. Loubnin, I.I. Vlasov, Diamond Relat. Mater. 12 (2003) 531.
33. V.V. Kononenko, V.I. Konov, S.M. Pimenov, A.M. Prokhorov, V.S. Pavelyev, V.A. Soifer, New Diamond and Frontier Carbon Technology, 10 (2000) 97.
34. E. Oesterschulze, W. Scholz, Ch. Mihalcea, D. Albert, S. Sobisch, W. Kulisch, Appl. Phys. Lett. 70 (1997) 435.
35. A.F. Belyanin, A.E. Alexenko, A.A. Botev, N.A. Bulienkov, L.L. Bouilov, B.V. Spitsyn, A.A. Markaryan, Russian Patent No. 1114194, priority of 16.07.1982 (in Russian).
36. K. Okano, K. Hoshina, M. Ida, S. Koizumi, T. Inuzuka, Appl. Phys. Lett. 64 (1994) 2742.
37. V.G. Ralchenko, A.V. Khomich, A.V. Baranov, I.I Vlasov, V.I. Konov, Phys. Stat. Sol.(a), 174 (1999) 171.
38. V. Ralchenko, A. Khomich, L. Butvina, I. Vlasov, V. Konov, L. Schirone, G. Sotgiu, New Diamond and Frontier Carbon Technology, 10 (2000) 109.
39. V. Ralchenko, L. Schirone, G. Sotgiu, A. Zakhidov, R. Baughman, A. Khomich, M. Nunuparov, I. Vlasov, V. Frolov, A. Karabutov, in Proc. 6th Int. Symp.on Diamond Materials, The Electrochemical Soc., Pennington, USA, 99-32 (2000) 72.
40. E. Woerner, C. Wild, W. Mueller-Sebert, P. Koidl, Diamond Relat. Mater. 10 (2001) 557.
41. H. Bjiorkman, P. Rangsten, P. Hollman, K. Hjort, Sensors and Actuators, A73 (1999) 24.
42. H. Bjiorkman, P. Rangsten, K. Hjort, Sensors and Actuators A78 (1999) 41.
43. W.P. Kang, J.L. Davidson, M. Howell, B. Bhuva, D.L. Kinser, D.V. Kerns, Q. Li, J.F. Xu, J. Vac. Sci. Technol. B14 (1996) 2068.
44. A.V. Karabutov, V.G. Ralchenko, I.I. Vlasov, R.A. Khmelnitsky, M.A. Negodaev, V.P. Varnin, I.G. Teremetskaya, Diamond Relat. Mater. 10 (2001) 2178.
45. K.C. Holms, A. Wisitsora-at, T.G. Henderson, J.L. Davidson, W.P Kang, V. Pilugurta, in Proc. 6th Int. Symp.on Diamond Materials, The Electrochemical Soc., Pennington, USA 99-32 (2000) 360.
46. V. Ralchenko, V. Migulin, T. Kononenko, V. Kononenko, V. Konov, M. Negodaev, 9th CIMTEC'98 - Forum on New Materials, Symp. IV-Diamond Films, P. Vincenzini (Ed), Techna (1999) 109.
47. V.G. Ralchenko, E.I. Givargizov, Proc. 7th Applied Diamond Conference/3rd Frontier Carbon Technologies Joint Conference (ADC/FCT 2003), M. Murakawa et al. (Eds), NASA/CP-2003-212319 (2003) 540.
48. J. Anderson, S. Jacobson, DIAMOND'2003 – 14th European Conference on Diamond and Diamond-like Materials, Salzburg, Austria, 7-12 September 2003, Paper No. 12B.2.
49. S. Silva, M.C. Salvadori, K. Kawakita, M.T. Pereira, W. Rossi, M. Cattani, Diamond Relat. Mater. 11 (2002) 237.
50. A. Flotter, G. Schaarschmidt, B. Mainz, S. Laufer, S. Deutschmann, H.-J. Hinneberg, Diamond Relat. Mater. 4 (1995) 930.
51. A. Snigirev, V. Yunkin, I. Snigireva, M. Di Michiel, M. Drakopoulos, S. Kouznetsov, L. Shabel'nikov, M. Grigoriev, V. Ralchenko, I. Sychov, M. Hoffmann, E. Voges, Proceedings of SPIE 4783 (2002) 1.

Chapter 22

FILMS OF SOFT-MAGNETIC FE-BASED NANOCRYSTALLINE ALLOYS FOR HIGH-DENSITY MAGNETIC STORAGE APPLICATION

E.N. Sheftel', O.A. Bannykh

Baikov Institute of Metallurgy and Materials Science, Russian Academy of Sciences, Leninskii pr. 49, Moscow 119991, Russia

This paper presents a review of works related to the problem of development of soft-magnetic film Fe-based alloys exhibiting the combination of properties, such as the high saturation inductance, high level of soft-magnetic parameters over a wide range of MHz-frequencies, high wear-resistance and thermal stability, and an ability to be prepared by thin-film technologies. Magnetic cores of high-density recording heads are the principal application of these materials. Physical fundamentals of explanation of ferromagnetic behavior of a material with nano-sized structure, which have been developed for FINEMETs (Fe-Nb-Cu-Si-B) with the mixed amorphous + nanocrystalline structure, are considered. A new class of alloys Fe-Me-X (Me = Metal of III-V groups of the Periodic Table and X = C, N, O, and B) whose properties are higher that those of FINEMETs are discussed. The structure of these alloys consists of two nanocrystalline phases, such as the ferromagnetic α-Fe-based and nonmagnetic MeX phases. This structure provides the dispersion strengthening of the alloys. Metal science approaches to the selection of both chemical compositions and conditions of structure formation for these film alloys have been developed by authors and are discussed in this article.

Keywords: magnetic recording, Fe-based alloys, nanocrystalline structure, amorphous structure, soft-magnetic properties, magnetization, dispersion strengthening, phase diagram.

1. MAGNETIC RECORDING: PRESENT TENDENCY

The magnetic recording, which has more than 100-year history, is one of the principal technologies allowing one to store and transfer information.

The ever-growing demand for the increase in the body of stored information as well as the miniaturization of recording devices and increase in their speed of response determines the principal tendency of the magnetic recording development: the increase in both the recording density and recording speed on retention of the high-quality signal.

A. A. Voevodin et al. (eds.),
Nanostructured Thin Films and Nanodispersion Strengthened Coatings, 221–230.

Over a period of forty years (from the late 50s to the late 90s of the last century), the magnetic recording density on magnetic tapes and hard disks has increased by 5-6 orders of magnitude [1]. By 1997, the surface recording density was 3-4 Gbits/inch^2; the volume recording density equal to 100 Gbits/inch^3 has been reached in experiments on the perpendicular digital video recording on a magnetic tape. The hard disk recording density of ~ 50 Gbits/inch^2 has been reached by Hitachi in 2000. The recording density of 1 Tbits/inch^2 was announced can be reached in near-term outlook [2, 3].

The elementary stage of magnetic recording process consists in the magnetization of medium section (called a bit - "binary digits") with a magnetic core of a head. The lower the bit size, the higher the recording density. The size of bit is determined by head dimensions such as the nonmagnetic gap length of head core and recorded track width.

The each stage of increase in the recording density is determined by advances related to different "branches" of the complex technology of magnetic recording. In this case, technological advances related to the head-medium system play a leading part.

The revolutionary jump in the increase in the magnetic recording density took place in 90s. It was determined in particularly by the development of digital magnetic recording technology, application of high-coercivity media, development of new highly soft magnetic film materials applied in heads, and the development of a new design of recording heads and new producing technology of them. For example, over a period of last 15-20 years, parameters of video heads, such as gap length and recorded track widths, were decreased by factors of 5 and 30, respectively. As a result, over ten years (from 1988 to 1998), the 10-fold decrease in the bite size has been reached. To decrease the bit size, the higher coercive force of magnetic medium should be obtained to "hold" the so small bit. In 1972, the coercive force of magnetic media (in particular, in commercial floppy disks) was only 280 Oe, whereas in the early 90s, it reached up to 1800 Oe [4]. Now, thin-film media with a coercive force of 5000 Oe have been prepared and are under study [5].

The following combination of properties is required today for alloys used in cores of high-density magnetic recording heads: 1. a high saturation inductance (up to 2 T) providing the production of high magnetic field in the high-coercivity media during the signal recording; 2. a low coercive force (< 0,1 Oe) and a high magnetic permeability (>>1000) in a frequency range of up to 100 MHz providing a response of a head to the weak magnetic fields of input signals; 3. a high electrical resistance (at least 200 Ωcm) providing the decrease in eddy current losses at high frequencies in order to retain the high magnetic permeability at the high frequencies; 4. a high wear resistance that becomes the most important in the context of miniaturization of magnetic head core since the decrease in the gap dimensions (gap length and gap depth) results in accelerated wear of the head core; 5. a thermal stability (at least up to 600^0C) preventing the degradation of the properties upon heating during the manufacture (soldering and glueing) of heads; 6. a corrosion resistance allowing the operation of heads under different environmental conditions and 7. the ability to be prepared by thin-film technology.

2. CHEMICAL COMPOSITION AND STRUCTURE OF THE FILM ALLOYS FOR HIGH - DENSITY MAGNETIC RECORDING HEADS APPLICATION

Metal science approaches to the selection of both chemical composition and conditions of structure formation, which determine the above-mentioned special combination of properties, have been developed by us and are discussed in this paper.

It is obvious that the preset high saturation inductance determines the use of Fe-based systems since Fe exhibits the highest B_s value equal to 2.2 T; the Fe content must be maximum allowable.

According to a classic knowledge, the coercive force H_c of ferromagnetic microcrystalline bulk alloys decreases with increasing grain size. In these materials, the grain size is higher than the ferromagnetic exchange length and they are magnetized along the easy magnetization axes of individual grains. In this case, domains are formed within grains and the magnetization process is determined by the magnetocrystalline anisotropy of the individual grains.

2.1. Fe-Based Alloys with Nanocrystalline & Amorphous Structure

In [6], it was first shown experimentally that the Fe-Cu-Nb-Si-B system alloys with the nanocrystalline structure, which was obtained upon annealing of amorphous alloys, exhibit excellent magnetic-soft properties. The structure of these alloys is characterized by the presence of fine α-(FeSi) grains ~10-20 nm in size (D), which have a random texture, and amorphous phase within intergrain spacings. These data show that ferromagnetic materials with nano-sized grains should be considered in terms of other fundamental approaches in comparison with those used for bulk material.

Earlier, Hoffmann [7] theoretically showed that the effective magnetic anisotropy K_{eff} decreases with increasing number of grains (N) included in the ferromagnetic exchange interaction length. The K_{eff} value is represented by the expression $K_{eff} = K_1/N^{1/2}$, where K_1 is the magnetic anisotropy constant. Moreover, Hoffmann showed that the coercive force H_c is proportional to the square of grain size D, i.e., $H_c \sim D^2$.

Herzer [8,] performed an analysis of the data [6] using a so-called random anisotropy model that was first developed for amorphous ferromagnetic materials in [9]. Herzer derived that, in a three-dimensional material with a grain size substantially lower than the ferromagnetic exchange interaction length (L_{ex}), the effective magnetic anisotropy is represented by the expression $<K> = K_1^4D^6/A^3$, where A is the exchange interaction stiffness, K_1 is the magnetic anisotropy constant, and D is the grain size. The power function found for the K_{eff} value, $K_{eff} \sim D^6$, is also held for the coercive force and initial magnetic permeability, i.e., $H_c \sim D^6$ and $\mu_0 \sim 1/D^6$.

Fig. 1 shows the coercive forces of different soft magnetic alloys as functions of the grain size that varies over the wide range, namely, from interatomic spacings in amorphous alloys and several microns corresponding to nano-sized grains to macroscopic sizes of coarse grains. As it was shown for the $FeCu_{0-1}Nb_3(SiB)_{22.5}$ alloy, the structure containing the α-(FeSi) phase with a grain size of less than 40 nm (the L_{ex} value for α-(FeSi) was estimated to be equal to 34 nm) determines the high level of magnetic-soft properties. The basic mechanism of this effect (similar to that in amorphous alloys) consists in the averaging random magnetic anisotropies of grains whose size is less than the ferromagnetic exchange interaction length.

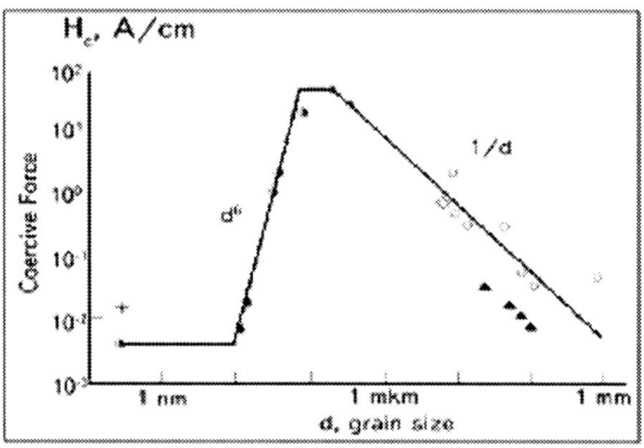

Figure 1. Coercive force *vs* grain size of soft magnetic phases [8].

These alloys annealed at a certain temperature are characterized by virtually zero magnetostriction [6] along with the low magnetic anisotropy. According to [8], the zero magnetostriction is determined by the balance between the magnetostrictions of α-(FeSi) crystallites ($\lambda_s^{FeSi} \approx -6x10^{-6}$) and amorphous matrix ($\lambda_s^{am} \approx +20x10^{-6}$), i.e. $\lambda_s \approx V_{FeSi}\lambda_s^{FeSi} + (1-V_{FeSi})\lambda_s^{am}$, where V_{FeS} is the volume fraction of α-(FeSi) crystallites.

In 90s, these results give arise to the new generation of soft magnetic Fe-based materials, such as FINEMETs (FeCuNbSiB) [10]. The coercive force

H_c of these alloys can be less than 10 mA/cm and the initial magnetic permeability can reach $\mu_0 \sim 100000$. The alloys are used in cores of miniature transformers, magnetic recording heads, etc.

However, bearing the high-density magnetic recording in mind, these alloys with the mixed amorphous + nanocrystalline structure are characterized by the inadequate saturation inductance (1.2-1.45 T) that is determined by the presence of the amorphous phase as well as by the high content of nonmagnetic alloying components. The alloys are incapable of retaining the high magnetic permeability over a wide range of MHz-frequencies; moreover, this structure is unstable upon heating to rather high temperatures (above 300°C).

2.2. Dispersion-Strengthened Fe-Based Alloys with Nano-Crystalline Structure

In the early 90s, a new class of alloys, Fe-Me-X (X = B, C, N, O and Me = Ti, Zr, Hf, Nb, Ta, etc.) with the nanocrystalline structure came to the attention of investigators. It was shown that the magnetic properties of the alloys are higher than those of FINEMETs. These alloys exhibit the following magnetic parameters: the saturation inductance B_s = 1.5-1.75 T, magnetic permeability μ_{10MHz} = 2000-7000, coercive force H_c < 10 A/m, magnetostriction $\lambda_s \sim 10^{-7}$; moreover, their structure is more stable upon heating [11-14]. The structure of these alloys consists mainly of two nanocrystalline phases, such as ferromagnetic α-Fe-based solid solution with a grain size of about 10-20 nm and thermodynamically stable nonmagnetic MeX compounds formed as rather smaller crystals and located at triple points of ferromagnetic-phase grain boundaries.

The high soft-magnetic properties of the alloys with such structure are determined by both very low magnetic anisotropy constant (like in the case of the mixed nanocrystalline + amorphous structure) and virtually zero magnetostriction. Speaking about the zero magnetostriction of such alloys and taking into account the similarity of the amorphous structure and the structure of boundary regions in nanocrystalline materials [15, 16], we assume that the effective magnetostriction of the alloys should be averaged, on the one hand, over ferromagnetic grains and, on the other hand, over developed intergrain as well as MeX/α-Fe interface regions [17].

Moreover, along with the high soft-magnetic properties and the high saturation inductance, the alloys with described structure should be characterized by the high wear resistance and the high thermal stability, which are reached at the expense of dispersion strengthening with particles of the thermodynamically stable and hard MeX compounds.

2.3. Dispersion-Strengthening Effect in Soft-Magnetic Bulk Alloys with Micron-Scaled Structure

In the late 70s and in the early 80s, the use of the dispersion strengthening effect in soft magnetic alloys allowed us to give rise a series of new bulk alloys characterized by micron-sized structure and applied in video heads [18-21]. The wear resistance of these alloys is higher than that of Sendust (Fe-Si-Al), which is conventional alloy having the highest wear resistance, by a factor of 3. Moreover, our alloys also compare favourably with Sendust in magnetic properties at frequencies at least to 30 MHz (Fig.2).

The structure and corresponding chemical composition, which provide the observed effects, can be obtained in a Fe-based system alloy that is characterized by the formation of compounds exhibiting the maximum thermal stability and highest both hardness and melting temperature. It is known, that these compounds are carbides, nitrides, oxides, and borides of metals of the IIIA, IVA, and VA groups of the Periodic Table. In this case, bcc Fe-based ferromagnetic phase should be in quasi-binary equilibrium with the MeX compounds and the quasi-binary phase equilibrium diagram Fe-MeX should be characterized by the eutectic-type solidification. Moreover, the shift of the eutectic point to the Fe-rich end of the quasi-binary phase diagram is favourable. The alloys having the hypoeutectic compositions can solidify with the formation of structure consisting of ferromagnetic-phase grains and disperse particles of the nonmagnetic MeX phase located along the grain boundaries. We have shown that, for the alloys with micron-sized structure, the higher the cooling rate upon solidification is used, the more disperse structure is formed and the higher magnetic and strength properties are reached.

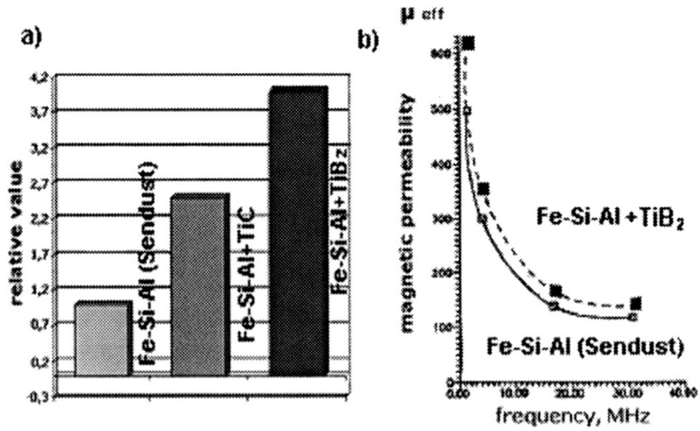

Figure 2. Properties of bulk dispersion-strengthened soft-magnetic alloys and Sendust.
a) Lifetime of commercial magnetic heads produced from the alloys.
b) Magnetic permeability *vs* frequency.

3. $Fe_{78}Zr_{10}N_{12}$ FILMS: STRUCTURE AND MAGNETIC PROPERTIES

The nanocrystalline structure of the Fe-MeX alloy can be formed by annealing of the initial amorphous structure. The amorphous state can be obtained upon rapid solidification in the alloys, whose chemical compositions are close to the invariant eutectic point of proper phase diagrams [22,23]. We showed that the Fe-Me_{III-V}X system alloys can be obtained in the amorphous state upon solidification of the gas-state alloys at cooling rates no less than 10^{11-13} K/s. So high cooling rate can be realized upon deposition of the vapour, in particular, upon magnetron sputtering.

The results of some our experimental investigations of Fe-Zr-N alloys are discussed below [24-27]. The films 1.5-2 μm in thick were obtained by reaction magnetron sputtering of a Fe target containing 10 at. % Zr under mixed argon+nitrogen atmosphere with variable nitrogen content. The films were deposited on glass or melt quartz substrates. Subsequently, the films were annealed in vacuum at 300-700°C for 1 h. According to X-ray diffraction analysis, the virtually complete amorphization upon sputtering can be reached for the film alloy containing ~15 at. % N.

Using differential microcalorimetry, it was shown that, upon heating of the film, a substantial exothermic effect is observed in a temperature range of 248-468°C. We related this effect with the nanocrystallization of the amorphous phase obtained upon sputtering.

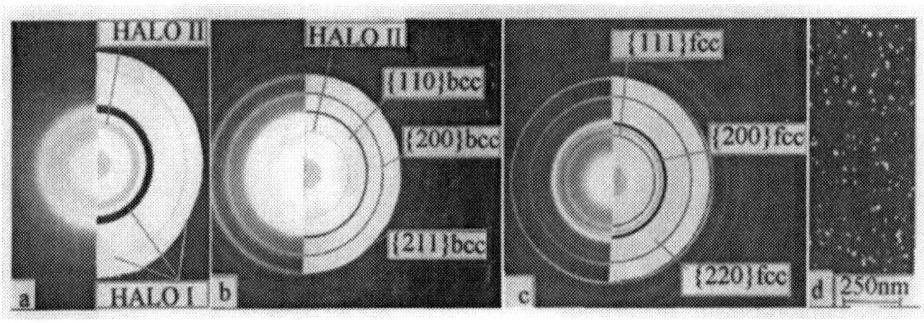

Figure 3. Electron diffraction patterns of the $Fe_{78}Zr_{10}N_{12}$ film: as-sputtered (*a*), annealed for 1 h (*b*) at 400°C and (*c*) at 550°C. Micrograph (dark-field image taken in the of the (110) ring of bcc Fe-based phase) of the film annealed at 550°C (*d*).

Electron microscopic studies confirmed that, upon sputtering, the amorphous alloy is obtained. Moreover, already upon sputtering, it is observed the separation of the alloy with the formation of two amorphous phases differing in the composition. Two series of halos are observed in the electron diffraction pattern at different angles (Fig. 3a). After annealing at 400°C the electron diffraction pattern demonstrates distinct diffraction rings

{110}, {200}, {211} belonging to the bcc Fe-based phase (Fig.3b). The annealing at 550°C for 1 h causes the subsequent development of the crystallization. Along with the diffraction reflections of the bcc Fe-based phase, there are observed weak diffraction rings corresponding to {111}, {200} and {220} reflections of an fcc phase with the lattice parameter close to that of the ZrN compound (Fig.3c). The dark-field images (Fig.3d) demonstrate the Fe-based phase crystals. The computer processing of the electron micrographs showed that, upon annealing at 400-600^{0}C, the α-Fe-based phase grains of 5-15 nm size are formed. These grains form the main volume fraction, although grains 20-30 nm in size are also present (after annealing at 400^{0}C, their volume fraction is ~4%). The increase in the annealing temperature from 400 up to 600^{0}C leads mainly to an increase in the number of grains of 5-15 nm size; in this case, virtually no growth of them is observed. This fact shows the high thermal stability of the structure.

The magnetic properties of the $Fe_{78}Zr_{10}N_{12}$ films being in the different structural states were studied using a magnetometer (Fig. 4).

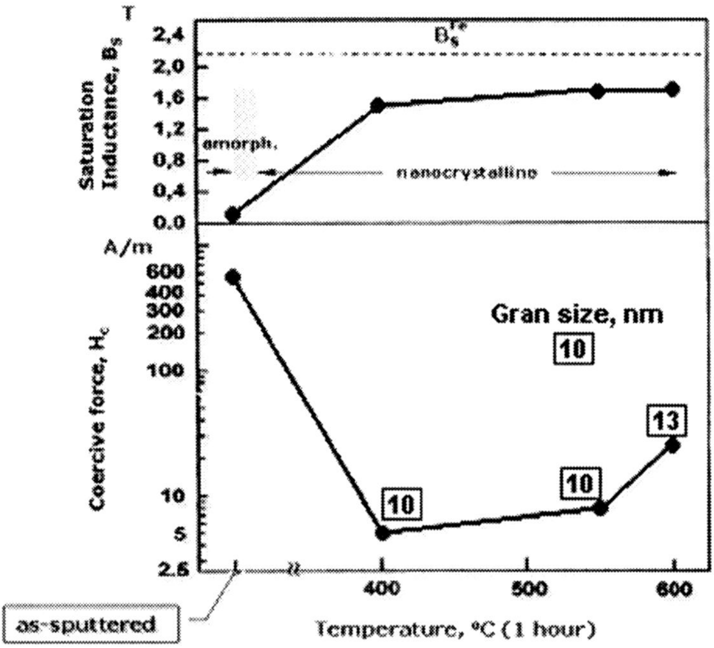

Figure 4. Saturation inductance and coercive force of the $Fe_{78}Zr_{10}N_{12}$ film *vs* the annealing temperature.

The crystallization of the amorphous alloy leads to a decrease in the coercive force and an increase in the saturation inductance. It was shown the possibility to reach the record low coercive force equal to 5-8 A/m and saturation inductance equal to 1.7-1.8 T.

Visual studies of films 3 x 8 mm in size by magnetooptics showed that their magnetic structure is realized as several domains (3) separated by straight parallel domain boundaries (Fig. 5). Magnetic measurements and magnetooptics allowed us to find that the magnetization of the films under investigation occurs via the motion of domain boundaries in the applied magnetic field and the coercive force of the alloy is determined by the coercive force of a domain boundary.

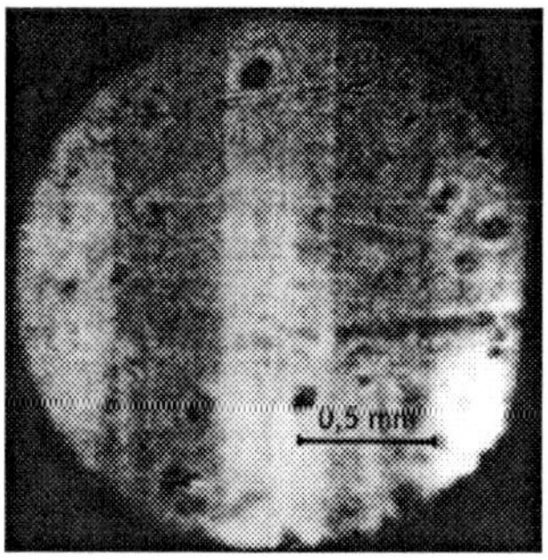

Figure 5. Domain structure (magnetooptics) of the Fe_{78} Zr_{10} N_{12} film produced by magnetron sputtering and subsequently annealed at 400°C for 1 h.

4. CONCLUSIONS

Authors had worked up a scientific approach that allowed to develop bulk super high wear-resistant soft-magnetic alloys (dispersion strengthened by thermodynamically stable compounds such as carbides, borides etc.) for magnetic video-head application. The subsequent evolution of the approach for film soft-magnetic Fe-based alloys will allow one to choice the chemical composition and the type of nanocrystalline structure that provide the special combination of properties of the material applied in high-density recording heads. In turn, the preset type of structure will give rise to the development of new film manufacturing technologies, whereas the unique combination of the properties of these materials will initiate a search for their new applications.

5. ACKNOWLEDGMENTES

This work was supported by the Russian Foundation for Basic Research, project no. 01 - 03 - 32510a, state program of supporting leading scientific schools, project no. NSh - 2079.2003.3, and program “Fundamental Problems of Physics and Chemistry of Nano-Sized Systems and Nanomaterials”.

REFERENCES

1. L. T. Romankiw, J. Magn. Soc. Jpn. 21 (S2) (1997) 429.
2. K. Ouchi, N. Honda, Abstracts of Papers, 8-th European Magnetic Materials and Applications Conference, June 7-10, 2000, Kiev, Ukraine.
3. A. Ursache, M. Bal, J. T. Golbach, R. L. Sandstrom et.al., Mat. Res. Soc. Symp. Proc. 721 (2002), Digest Toyohashi University of Technology, International Workshop on Novel Electromagnetic Functions of Nano-Scaled Materials, (2003) 84.
4. O. Kohmoto, IEEE Trans. Magn. 27 (4) (1991) 3640.
5. Dong Hyoun Km, In Tak Nam, Yang Ki Hong, Mat. Sci. 21 (1) (2003) 65.
6. Y. Yoshizawa, S. Oguma, K. Yamauchi, J. Appl. Phys. 64 (1988) 6044.
7. H. Hoffmann, Thin Solid Films. 58 (1979) 223.
8. G. Herzer, IEEE Trans.Magn. 26 (1990) 1397.
9. R. Alben, J. J. Becker, M. C. Chi, J. Appl. Phys. 49 (1978) 165.
10. K. Yamauchi, Y. Yoshizawa, Nanostuctured Mater. 6 (1995) 247.
11. M. E. McHenry, D. E. Laughlin, Acta Mater. 48 (2000) 223-238.
12. Nago, H. Sakakima, K. Ihara, IEEE Trans. Magn. Jpn. 7 (2) (1992) 119.
13. A. Hasegava, M. Saito, IEEE Trans. Magn. Jpn. 6 (1) (1991) 91.
14. K. Katori, K. Hayashi, H. Ohmori, M. Hayakawa, K. Aso, Proc. Int. Symp. on 3d Transition Semi-Metal Thin Films. Magnetism and Processing, Sendai, Japan, March 5-8, (1991) 219.
15. P. Keblinski, S. R. Phillpot, D. Wolf, H. Gleiter, Nanostructured Mater. 9 (1997) 651.
16. R. A. Andrievskii, A. M. Glezer, Fiz. Met. Metalloved. 88 (1) (1999) 50.
17. O. A. Bannykh, E. N. Sheftel, D. E. Kaputkin, V. E. Zubov, R. E. Stroug, G. Sh. Usmanova, Physics and Modelling of Intelligent Materials and their Applications (PMIMA), Proc. of the Russian-Japanese joint seminar, September 19-22, 1996, 229.
18. V. K. Grigorovich, N. I. Sheftel', E. N. Sheftel', V. I. Matorin, USSR Inventor's Certificate no. 524854, MKI S22s 38, 14.02.75, Byull. Izobret. (1976) no. 30.
19. V. K. Grigorovich, E. N. Sheftel', I. R. Polyukhova, A. S. Mkrtumov, Izv. Akad. Nauk SSSR, Met. 4 (1986) 134.
20. O. A. Bannykh, E. N. Sheftel', V. K. Grigorovich, R. E. Strug, A. S. Mkrtumov, I. R. Polyukhova, A. V. Evdokimov, USSR Inventor's Certificate no. 1765240, MKI S22s 38/14, 12.09.89.
21. V. K. Grigorovich, E. N. Sheftel', R. E. Strug, I. R. Polyukhova, Izv. Ross. Akad. Nauk, Met. 6 (1993) 173.
22. K. Sudzuki, H. Fujimori, K. Hasimoto, Amorphous Metals, Transl. Japanese, Moscow: Metallurgiya, 1987.
23. S. P. Alisova, P. B. Budberg, Yu. K. Kovneristyi, Rasplavy, 4 (1990) 8.
24. O. A. Bannykh, E. N. Sheftel, A. I. Krikunov, D. E. Kaputkin, G. Sh. Usmanova, R. E. Stroug, J. Magn. Magn. Mater. 215-216 (2000) 397.
25. E. N. Sheftel, E. N. Blinova, G. Sh. Usmanova, O. A. Bannykh, A. M. Glezer, A. I. Krikunov, Phys. Met. Metallogr. 91 (5) (2001) 482. (English translation of Fiz. Met. Metalloved. 91 (5) (2001) 56.
26. O. A. Bannykh, E. N. Sheftel, V. E. Zubov, D. E. Kaputkin, A. I. Krikunov, A. D. Kudakov, G. Sh. Usmanova, T. S. Fedulova, Pis'ma Zh. Eksp. Teor. Fiz. 27 (5) (2001) 8.
27. V.E. Zubov, A.D. Kudakov, T.S. Fedulova, E.N. Sheftel', A.I. Krikunov, Vestn. Mosk. Univ., Ser. 3: Fiz., Astron. 1 (2003) 60.

Chapter 23
PROSPECTS OF NANODISPERSIVE POWDER APPLICATIONS IN SURFACE ENGINEERING TECHNOLOGIES

E.A. Levashov, A.E. Kudryashov, P.V. Vakaev
SHS-Center of Moscow State Institute of Steel and Alloys (Technological University) and Institute of Structural Macrokinetics and Materials Science Problems of RAS, Leninsky prospect, 4, Moscow 119049, Russia

General potentials of UDD (ultra dispersive diamond), NbC, WC, W, WC-Co, ZrO_2, Al_2O_3, Si_3N_4, Co, nanosized powders in determining structure and properties of composite electrodes and coatings deposited by electrospark alloying (ESA) and thermoreactive electrospark surface strengthening (TRESS) techniques were considered. It was shown that an addition of refractory compound nanosized powder to the electrode material positively effects microstructure and tribological characteristic of ESA- coatings. Nanoparticles incorporated in the coating on grains boundaries serve as a lubricant for friction pairs. Wear resistant W-C-Co coatings with a friction coefficient below 0.15 were deposited by TRESS using nanopowders of Co and W. Nano- or microstructural coatings on the base of cemented carbides can be formed, depending on pulse discharge energy and frequency. Examples of beneficial industrial application of the coatings strengthened by nanosized particles were presented.

Keywords: surface engineering, nanosized powder, coating, structure, tribology.

1. INTRODUCTION

Basic and applied research in the field of nanodispersive material science enables a great progress in the synthesis of new composite materials and coatings. Deposition of wear resistant nanocrystalline thin films and composite coatings strengthened by nanosized particles of metals and refractory compounds provides high continuity, low roughness, tailored tribological properties, and extended service life. It can be realized using different surface engineering techniques such as: PVD, electrochemical deposition (ECD), thermal (plasma) spraying (TS), electrospark alloying (ESA), thermoreactive electrospark surface strengthening (TRESS) referred in recent papers [1-16]. These papers describe the investigation results in producing coatings strengthened by nanosized particles. The main advantages of such coatings are deposition of thin wear-resistant composite coatings with improved tribological characteristics and long-life

A. A. Voevodin et al. (eds.),
Nanostructured Thin Films and Nanodispersion Strengthened Coatings, 231–240.

performance; production of composite coatings of high adhesion to substrates of steel, titanium, nickel, aluminum alloys, and ceramics. Significant performance improvements in coating properties were achieved by changing size, shape and distribution of the phases in production ultra fine-grained materials [13].

Electrode materials, especially for ESA in Ti-C-Cr-Ni, Ti-B-Al, Ti-C-Ni-alloy, Ti-C-Steel, Ti-C-Ni-Al, Ti-B, Ti-Ta-C-steel systems produced by self-propagating high-temperature synthesis (SHS), should content nanosized refractory component for improving coating quality [2-8, 15,16]. It is shown that introduction of a minor amount (3 to 10 volume %) of refractory compound nanosized powder as a relatively inert filling material into the green exothermal mixture results in decreasing the combustion temperature and rate. Moreover nanosized powder component in the most studied systems takes an active part in primary and secondary structuring of the synthesis products due to solid solutions and new chemical compound formation. Introduction of a nanosized component also modifies grain alloy structure. For example, just a 7 vol. % addition of sol-gel zirconia powder of 30 nm size into a green mixture of Ti-C, Ti-B based compositions results in tenfold grain size reduction of titanium carbide and titanium boride [15-17].

It is necessary to summarize the influence of additives of ZrO_2, Al_2O_3, NbC, WC, WC-Co, W, Si_3N_4, and ultra dispersive diamond (UDD) nanosized powders on spark deposition process kinetics as well as on ESA-coating's composition, structure, and properties. A positive effect of nanosized metal powder Co, W, C addition on microstructure and tribological characteristic of coatings deposited by TRESS will be demonstrated. The TRESS method involves an electric discharge, which is realized using ESA installations, to initiate and support a chemical reaction between the powder components of a mixture previously deposited on a substrate-cathode or incorporated into the composition of an electrode-anode [11, 12].

2. EXPERIMENTAL PROCEDURE

Electrode materials (Ti, Cr)C-10%Ni (marked SHIM-3B), (Ti, Cr)C-10%Fe (SHIM-3V), Ti-B (SHIM-4),TiC-40%Ni-alloy (SHIM 2/40NG), TiB_2-TiAl with nanosized powder additives were synthesized by the force SHS-pressing technology according to a known procedure [15-17].

Ultra dispersive diamond (UDD) (particles size of 40-60 nm, detonation synthesis), stabilized zirconia ZrO_2 (30 nm, zol-gel), Al_2O_3 (5 nm, zol-gel), NbC (12 nm, plasma-chemical synthesis), W (170 nm, pl-chem), WC (7 nm, pl-chem), WC-8%Co (10 nm, pl-chem), Si_3N_4 (20 nm, pl-chem), Co (20 nm, zol-gel) were used as additives in ESA and main components in TRESS technologies. Novel high-frequency electrospark installation "Alier-Metal" trade mark with wide pulse discharge energy and frequency ranges and conventional lower-frequency installation "Elitron-22A" were used to produce qualitative ESA- and TRESS-coatings of a low level of roughness. The first of these installations is designed for operation in both manual and mechanized modes. The main characteristics of the installation are given in papers [6, 7]. Mass transfer kinetics of the deposition process was measured by gravimetric analysis.

TRESS-coatings in W-C-Co system were synthesized using nanosized powders of W, Co and C (carbon black). Coatings deposited by ESA, TRESS-technologies on nickel alloy, stainless steel, HSS, and titanium alloy substrates were examined with metallographic, X-ray, and tribological analyses. Effect of nanosized particles incorporation on wear characteristics and friction coefficient were studied. Metallographic analysis of coating cross sections (thickness, continuity) was made using optical microscope "Axiovert" with "Video Test" image analysis program and SEM JSM-U3 with Edison system involving WINEDS and DIPS. Microhardness measurements of the coatings across the thickness of the alloyed layer were completed with Nanohardness (CSEM Instruments) and microhardness testers. X-ray spectrums were recorded with a diffractometer DRON-4 using monochromatic Co-K_α radiation. The study of coating wear-resistance was carried out on pin-on-disk and ball-on-disk tribometers (CSM Instrument Inc., Switzerland). Pin-on-disk friction pair was composed of a diamond containing side surface of a 42 mm diameter roll and the coating [3-6]. For this test, coatings were deposited on the face surface of cylindrical specimens 8 mm diameter. Sliding linear speed was 1.5 m/s, load on the friction pair – 10 N, wear rate was continuously detected with induction transducer. In the case of ball-on-disk test, the counter-body was a 3 mm diameter ball made of 94%WC+6%Co alloy , linear speed – 0.1 m/s, load – 1-2 N. Heat resistance was studied after heat treatment at T=750°C; τ=12 h.

3. RESULTS AND DISCUSSION

3.1. Application of Nanosized Powder Additives in ESA Technology

A most complete information concerning the influence of nanosized particles additives can be obtained from studying kinetic dependences of total anode erosion and cathode (substrate) weight increment $\sum\Delta\kappa$, which makes it possible to determine electrode material transfer rate and to estimate properties of the formed coatings. $\sum\Delta\kappa$ is a function of electrode material composition, microstructure, residual porosity, thermal and electro conductivity as well as pulse discharge energy and frequency. An electrode with very small carbide grains allows increasing $\sum\Delta\kappa$ and improving coating properties. Effects of a 7 vol. % nanosized powder additive on the microstructure of SHS-electrode materials are shown on Fig. 1 with an example of a TiC+40% Ni-alloy system. Figs 2 and 3 demonstrate kinetic dependences of $\sum\Delta\kappa$ on a processing time in an optimal mode. It can be seen that introducing nanocrystalline additives into electrode composition promotes the deposition rate as compared to an electrode without additives.

Figure.1. Effect of 7 vol. % nanosized powder additives on the microstructure of electrodes with composition based on TiC - 40%Ni-alloy: a)- without additives; b)- WC additive; c)- ZrO_2; d)- Al_2O_3.

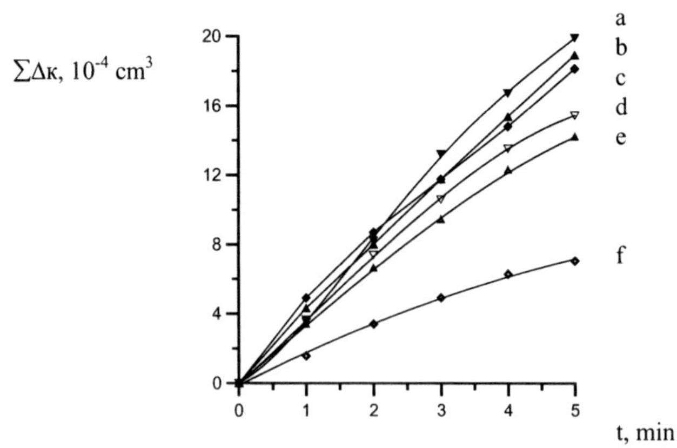

Figure 2. Effect of 7 vol. % nanosized powder additives W(a), NbC (b), Si_3N_4(c), WC-Co (d), ZrO_2(e) on kinetics of (Ti, Cr)C-Ni coatings deposition process using "Alier-Metal" set at E=0.51 J, f=500 Hz, substrate- Ni. Deposition of (Ti, Cr)C-Ni coating without additives – "f".

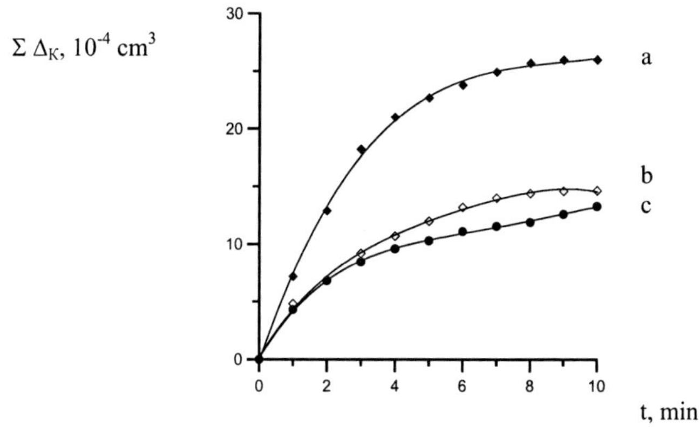

Figure 3. Effect of ultra dispersive diamond (UDD) additives (weight %): 15(a); 5 (b); 0 (c) on kinetics of (Ti, Cr)C-Ni coatings deposition using "Elitron-22A" set at E=0.3 J; F=100 Hz; substrate- stainless steel

Moreover, ESA- coatings, which are deposited on Ti-alloy substrate using SHS-electrodes modified with nanosized powders, have better properties, in particular, continuity (S), thickness (h), microhardness (H_μ), friction coefficient (K_f) and heat resistance (T) (Table 1). In these experiments coatings were obtained using an «Alier-Metal» unit, which was set to a pulse discharge energy E=0.13 J and a pulse discharge frequency F=3000 Hz. Friction tests conditions were as the following: counter-body - 3 mm diameter ball 94%WC+6%Co composition; load = 1N; linear rate - 10 cm/s.

As can be seen from Figs 4, 5 and Table 1, the TiC-Ni-alloy based coatings, which were deposited with electrodes strengthened by W, ZrO_2 and Al_2O_3 nanosized powders, are thicker as compared to the coatings produced without such powder strengthening. It was found by SEM that a maximum concentration of nanosized additives is located in the surface layer of the coating and diminishes toward the coating-substrate transition zone.

Table 1. Properties of the ESA-coatings strengthened by nanosized particles.

Coating	H_μ^*, GPa	S, %	h, μm	T, g/m^2	K_f
TiC-Ni Alloy	11.66	70	57	16.1	0.26
TiC-Ni Alloy + $ZrO_{2\ nano}$	12.64	100	104	13.8	0.19
TiC-Ni Alloy + $Al_2O_{3\ nano}$	14.91	90	105	14.1	0.18
TiC–Ni Alloy + NbC_{nano}	13.29	90	69	12.0	0.15
TiC–Ni Alloy + W_{nano}	13.50	96	102	13.5	0.17

*- H_μ (Ti-alloy) = 1.9 GPa

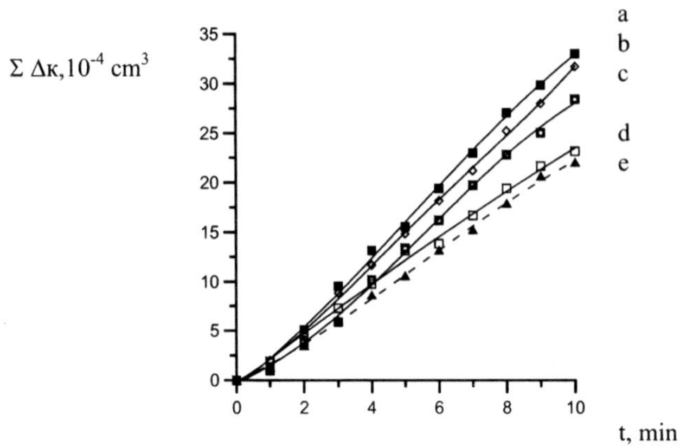

Figure 4. Effect of 7 vol. % nanosized powder additives ZrO_2 (a), Al_2O_3 (b), W (c), WC (d) on kinetics of TiC-Ni-alloy coatings deposition process using "Alier-Metal" set at E=0. 13 J, F=3000 Hz, substrate- Ti- alloy. Deposition of TiC-Ni-alloy coating without additives – "e".

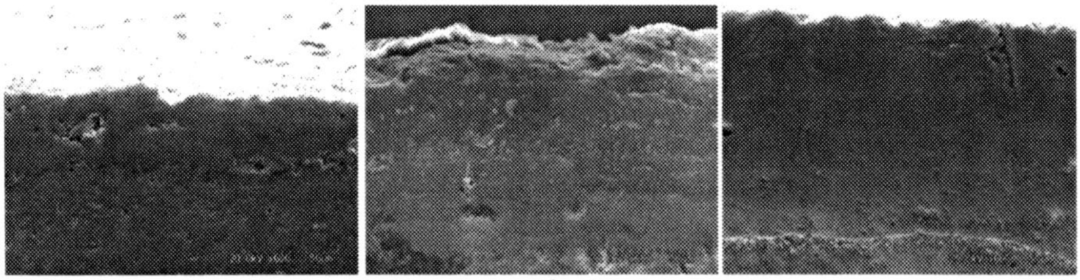

a) *b)* *c)*

Figure 5. Overview of the coatings on Ti- substrate deposited by ESA at the similar energy-frequency regime using TiC-Ni-alloy composite electrode without nanosized additives (a) and electrodes with additives of nanosized particles W (b) and ZrO_2 (c).

The positive effect of nanoparticles on the tribological characteristics (when using nanoparticle containing coatings) can be explained by the following: grain size of the wear resistant component in the coating is much less when using nanoparticles in electrode and coating composition; nanoparticles which are distributed on interfaces between carbide (or boride) grains and metallic binder are dry lubricants; coatings deposited using electrodes with nanosized additives have E=0.51 J a better continuity and microhardness, and reduced roughness.

Coating modification by nanosized additives promotes wear resistance due to the friction coefficient reduction below 0.2 (Figs 6, 7). Wear resistance of the coatings with 5% UDD is 3.5-4.0 times above that of coatings with the same composition but without UDD additives. Addition of 15% UDD to the green exothermic mixture allows a strong increase in the residual porosity (up to 20%) of the composite material of SHS- electrode and, therefore, ESA- coatings deposited with such an electrode become rougher and their wear rate - higher.

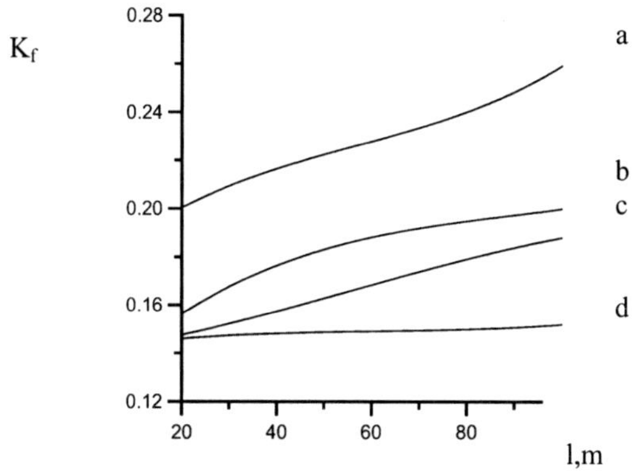

Figure 6. Dependences of friction coefficient of TiC–Ni- alloy coatings (Ti- alloy substrate) without nanosized additives (a) and coatings with additives of ZrO_2 (b), Al_2O_3 (c), NbC (d) on the length way during the tribological test.

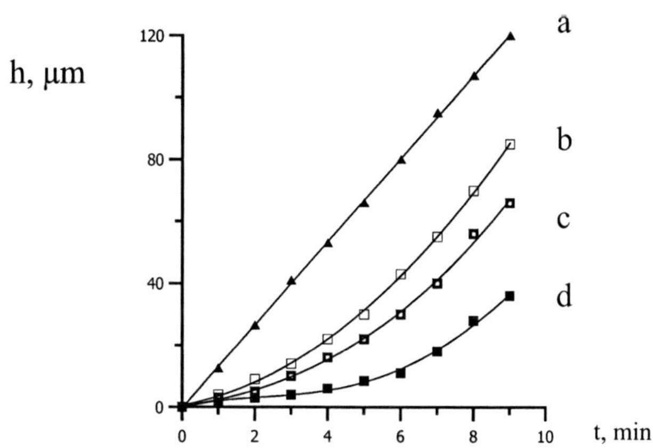

Figure 7. Dependences between wearied part of the coatings and time of pin-on-disk wear test: a- stainless steel substrate without coating; b- (Ti, Cr)C-Ni coating with 15%UDD; c- (Ti, Cr)C-Ni coating; d- (Ti, Cr)C-Ni coating with 5%UDD additives.

3.2. Application of Nanosized Powders in TRESS Technology

Pulse discharge energy and frequency are effective parameters for the coatings structure control in TRESS deposition. For example, nanocrystalline structure of coatings based on cemented carbide is formed at a medium value E and coatings with microcrystalline structure are deposited at a discharge energy above 1.6 J. Chemical reaction W + C + Co = WC + Co is assisted by a spark processes on the substrate and recrystallization of synthesized products proceeds more completely and results at higher values of E. In the case of nanosized W and Co powders, roughness of TRESS-coatings formed at a medium discharge energy is much less in comparison to the microsized powder application (see Fig. 8).

Wear resistant W-C-Co system coatings with friction coefficient below 0.15 were deposited on steel and nickel alloy substrates by TRESS. Variations of friction coefficients of WC-8%Co TRESS- coatings produced with micro- (c), nanosized (d) W and Co powders in ball-on-disk tests are shown in Fig. 9. K_f of conventional WC-8%Co ESA- coating (b) was similar at the test start but the wear life of TRESS-coating is significantly longer.

a) *b)*

Figure 8. Microstructure of WC-8%Co coatings on stainless steel substrate deposited by TRESS- method using: a)- nanosized and b)- microsized W+C+Co powders.

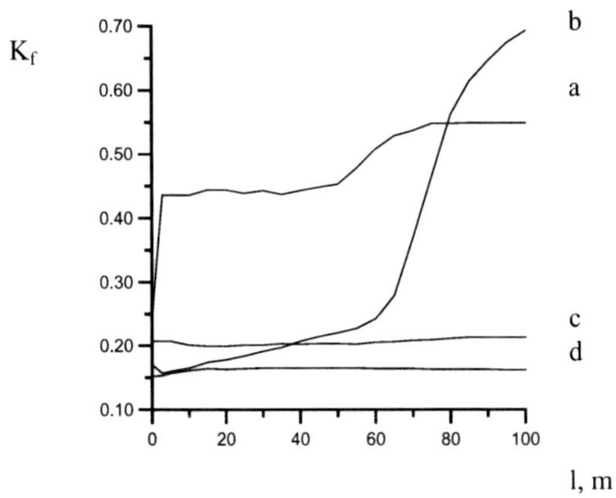

Figure 9. Friction coefficient of WC-Co coatings on Ti- alloy substrate on the length of counter body path: a- Ti- substrate (without coating); b- WC + 8% Co (ESA-coating using WC-8%Co electrode); c- WC + 8%Co TRESS – coating using W (micro) + C + 8% Co (micro); d- WC + 8% Co TRESS – coating using W (nano) + C + 8% Co (nano).

4. INDUSTRIAL APPLICATION AND TESTS

Coatings strengthened by nanosized powders and deposited by ESA, TRESS and other technologies can be successfully applied in various fields, where low roughness, high adhesion, low friction coefficient, high wear- and heat resistance are strongly required. Authors used ESA and TRESS techniques to produce tribological thin and thick films on metallurgical rollers of rolling mills for hot and cold deformation (Fig. 10 a), airospace engine parts made of Ni- Ti-alloys (Fig. 10 b), steel substrates of diamond segmented wheels (under segmented surface) (Fig. 10 c). Nanoparticles of the most refractory and hard compounds incorporated in the coating on grain boundaries serve as dry lubricants in friction pairs. According to the first industrial tests, the coated tools have extended service lives: 1.3 – 1.5 times for diamond wheels; 1.6-1.8 times for power units; 2.5-3.0 times for rollers. Optimization of coating compositions, deposition technological parameters in applications to real parts and test conditions will continue.

a) *b)* *c)*

Figure 10. ESA- and TRESS- deposition processes in mechanization modes using disk (a, c) and rod (b) electrodes: a)- surfacing the working part of roller by hard alloy coating; b)- protective film deposition on airspace power unit; c)- deposition of abrasive resistant coating on diamond segmented wheel.

5. CONCLUSIONS

Positive effects of nanosized powder additives on microstructure and properties of ESA- and TRESS- coatings were established. Low friction, wear and heat resistant coatings on the base of TiC-Ni-alloy, (Ti,Cr)C-Ni, TiC-steel, TiB_2-TiAl, TiB-Ti, and W-C-Co alloys were produced on steel, titanium and nickel alloy substrates . Nanosized powders of ZrO_2, Al_2O_3, NbC, WC, WC-Co, W, UDD, Co are recommended for applications in the surface engineering field.

ACKNOWLEDGMENTES

Work was completed within the framework of the ISTC project No. 1852.

REFERENCES

1. R. Birringer, H. Gleiter, H.-P. Klein, P. Marquard. Phys. Lett. B 102 (1984) 365.
2. E.A. Levashov, A.G. Merzhanov, I.P. Borovinskaya, G.V. Sakovich, M.G. Potapov, V.N. Shalyuta, A.M. Stolin, Russian patent 1788061, 01.02.1993.
3. A.E. Kudryashov, E.A. Levashov, V.V. Podlesov, A.M. Stolin., Int. J. of Self-Propagating High-Temp. Synth. 6 (1997) 445.
4. M.G. Potapov, E.A. Levashov, V.N. Shalyuta, P.A. Kozharskii, Weld. Ind. 13 (1997) 15.
5. E.A. Levashov, A.E. Kudryashov, M.G. Potapov, Nonferrous Metal.6 (2000) 67.
6. E.A. Levashov, O.V. Malochkin, A.E. Kudryashov, R. Suchentrunk, F. Gammel, J. Mater. Synth. Process. 9 (2001) 199.
7. E.A. Levashov, O.V. Malochkin, A.E. Kudryashov, R. Suchentrunk, F. Gammel, Galvanotechnik 4 (2002) 1052.
8. E.A. Levashov, A.E. Kudryashov, O.V. Malochkin, T.A. Sviridova, S.A. Glukhov, F. Gammel, R. Suchentrunk, Nonferrous Metal: 4 (2002) 62 (in Russian), 1 (2003) 54 (in English).
9. D.V. Shtansky, S.A. Kulinich, E.A. Levashov, A.N. Sheveiko, F.V. Kiriuhancev, J.J. Moore, Thin Solid Films 420-421 C (2002) 330.
10. D.V. Shtansky, T. Kaneko, Y. Ikuhara, E.A. Levashov, Surf.Coat.Technol.149 (2001) 206.
11. E.A. Levashov, E.I. Kharlamov, A.E. Kudryashov, A.S. Rogachev, M. Ohyanagi, M. Koizumi, S. Hosomi, J. Mater. Synth. Process. 7 (1999) 23.
12. M. Koizumi, M. Ohyanagi, S. Hosomi, E.A. Levashov, A.G. Nikolaev, A.E. Kudryashov, US Patent 6336950 B1, Jan. 8, 2002
13. B.M. Allcock, P.A. Lavin. Surf. Coat. Technol. 163-164 (2003) 62.
14. R.A. Andrievskii, Russian Chemical Journal - J. Russian Chem. Soc. named after D.I. Mendeleev XLVI 5 (2002) 50 (in Russian).
15. E.A. Levashov, E.S. Mishina, B.R. Senatulin, A.S. Kudryashov, Sviridova T.A., Blagoveschchenskii Yu.V, The Physics of Metals and Metallography 95 (2003) 1.
16. E.A. Levashov, E.S. Mishina, O.V. Malochkin, D.V. Shtansky, J.J. Moore, M.I. Fadeev, Sci. Technol. Adv. Mater. 4 (2003) 221.
17. E.A. Levashov, O.V. Malochkin, A.E. Kudryashov, S.A. Glukhov, T.A. Sviridova, F. Gammel, R. Suchntrunk, Nonferrous Metal: 3 (2002) 60 (in Russian), 1 (2003) 45 (in English).

Chapter 24
METAL MATRIX COMPOSITES PRODUCED BY ELECTROPLATING. A REVIEW ON TECHNOLOGY AND APPLICATIONS

R. Suchentrunk
Senior Manager Aerospace and Automotive (retired), Munic, Germany

In this paper selected results of research and development achieved in the Messerschmitt Bölkow Blohm (MBB) Central Laboratory (today part of EADS) within the last four decades will be presented. The focus will concentrate on investigations in the aerospace industry, mainly the production of structures consisting of fibre reinforced metals. By embedding high strength filaments (for example boron fibres) in a ductile metallic matrix (electrodeposited copper, nickel or aluminium) composite materials with excellent strength and stiffness can be produced. By this – on a pilot plant scale – cylindrical metal matrix composite structures have been fabricated and tested and lightweight pressure vessels for future space application (mainly for long time missions) have been produced. The materials, electrolytes, processes and equipment for fabrication and testing of these structures will be presented in detail and the advantages and limitations of the applied processes will be discussed. Beside of this application the technology of electroplating has been used at MBB for the fabrication of dispersion hardened coatings by embedding micro- as well as nano-sized particles in various matrix materials.

Keywords: electrodeposition, electroforming, metal matrix composites, fibre reinforced metals, space structures.

1. INTRODUCTION

The electroforming process, which we use today, has a long history. First work in this area started in Russia in the 19th century, when 1838 Professor Moritz Hermann von Jacobi ("the inventor of electroforming") presented the results of his experiments, to reproduce coins and other decorative articles by electrical current, in Riga to members of the Russian Academy of Sciences, St. Petersburg. 1840 he published his book "Galvanoplastics", which has been the first publication in the area of electrodeposition. The technology has developed since that time not continuously, but step by step. Today, more than 150 years later, it is a modern production technique suited for a constant increasing number of functional applications.

A. A. Voevodin et al. (eds.),
Nanostructured Thin Films and Nanodispersion Strengthened Coatings, 241–250.

By electrodeposition of thick metallic layers structures of various geometry can be formed (Electroforming). Examples include combustion chambers, wave guides, heat exchangers, reflectors and mirrors, erosion protection shapes and forming tools and moulds. The process can be successfully applied in many cases where standard mechanical processes (milling, drilling e.g.) will not work economically or will not work at all. The product (electroform) is produced by an electrochemical process, by electrodeposition of thick metallic layers on a pre-shaped mandrel made of stainless steel, aluminium alloys, low-melting alloys, plastics or wax. The mandrel can be removed after the deposition process or remain part of the final structure. The metallic layer can be applied with good or extremely bad adhesion, the latter to allow removal of the mandrel. The mandrel is usually made a mirror image of the internal surface of the part to be fabricated. The extremely high reproduceability of the electroforming process (peak-to-valley heights down to 0,00005 mm can be reproduced) makes this technology suitable for the production of reflectors, mirrors (1) and wave guides used in high-frequency technology (2).

A metallic composite system with high mechanical properties can be produced by electrodeposition of thick tightly adherent metallic layers on a preformed mandrel. The strength of the metal-to-metal bond can be at least as high as the tensile strength of the weaker partner, when a substrate specific pretreatment is applied (3). By this technique combustion chambers, concentrators for solar energy plants, water-cooled screens, boxes and targets for nuclear fusion experiments and compact wind tunnel models have been fabricated (4,5).

By using the same technology metal matrix composites (MMC) can be fabricated, incorporating high-strength fibres or (nano) particles into an electrochemically deposited matrix material. Electroformed structures offer some additional advantages compared with parts produced by other techniques. They can be fabricated in one piece, weak points caused by welding or brazing operations will thus be avoided in most cases. Additionally, they provide high stiffness, exact geometry of the reproduced surface and practically no danger of distortion because of a very low level of internal stresses.

The metals most frequently used for the electrodeposition of thick layers are mainly, nickel-cobalt alloys or copper. For special purposes silver, gold or even aluminium can be applied in comparable thickness. They are deposited using industrially improved electrolyte systems like the nickel-sulphamate electrolyte, the sulfuric- acid copper bath or cyanide-containing silver and gold solutions. Aluminium is deposited using a nonaqueous organic complex solution (6).

2. FABRICATION AND PROPERTIES OF FIBRE REINFORCED COMPOSITE TEST SPECIMEN

By embedding high strength filaments, e.g. boron fibres in a ductile metallic matrix, composite materials with excellent strength and stiffness properties can be produced. Investigations (7, 11) have shown that boron fibres can be embedded in copper, nickel or aluminium by means of an electrodeposition processes. Nickel and copper are deposited using the commercially available aqueous electrolytes applied in electroforming. Aluminium can be deposited by nonaqueous organic electrolytes, for example complex aluminium organic compounds (8,9,10). The advantages of electroplating for the fabrication of metal matrix composites as compared with liquid metal infiltration, braze bonding or hot isostatic pressing are the following:

- The electrodeposition process is working at a very low temperature, usually far below 100 °C. Therefore, there is no danger of extortion or degradation of the material properties by overheating.
- Electroplating makes it possible to fabricate even very complicated parts in one piece. Welding or brazing operations, which might be a danger for the composite, can thus be avoided completely.

In first investigations special formed tensile test specimen (NOL rings, see Fig. 1) have been fabricated and tested. Tear tests were performed in a standard test equipment.

Figure 1. NOL-rings and tensile testing device.

The mechanical properties of different types of fibres used and of the metal matrix composites produced are listed in Table 1 (copper, nickel matrix) and Table 2 (aluminium matrix). Based on the promising results in the next phase of the project more complex structures were fabricated.

Table 1. Tensile strength of fibres and composites (copper, nickel)

Fibre Type	Fibre Diameter (μm)	Tensile Strength (MPa)	Fibre Content (Vol%)
B	140	2800	-
B_4C-B	140	3500	-
BN-B	140	1500	-
SiC-B	100, 140	2900	-
B/Cu	140	1100	48
B_4C-B/Cu	140	1700	49
BN-B/Ni	140	1350	34
BN-B/Cu	140	1040	49
SiC-B/Cu	100	900-1100	42
SiC-B/Cu	100	610	18
SiC-B/Ni	140	1000	34

Table 2. Tensile strength of boron firer reinforced aluminium (NOL-rings) as a function of temperature

Fibre type	Fibre content (Vol.%)	Tensile strength (MPa) RT	Tensile strength (MPa) 150 °C	Tensile strength (MPa) 300 °C
B_4C-B/Al	30	870	-	-
B_4C-B/Al	30	730	-	-
B_4C-B/Al	34	970	900	535

3. FABRICATION OF STRUCTURES

3.1. Copper Matrix

For future space applications there is a need for extremely lightweight materials, yet offering high stiffness properties. Fibre-reinforced resins are a well known way to reach this goal. In some cases however, especially for long-term space missions, a metal matrix instead of an organic resin would provide much more promising results, since it is less sensitive to degradation caused by the extremely high intensity of cosmic radiation.

Therefore as the next step of the program prototypes of fibre-reinforced pressure vessels for space applications haven been produced on a pilot-plant scale by applying the following technique (2, 7, 11): High-strength fibre-material, for example boron filaments, are wound around a cylindrical mandrel and after this embedded in a metallic matrix by electroplating. The winding machine used is shown in Fig. 2.

Figure 2. Equipment for winding fibers onto mandrel.

When the fibres are completely embedded, the metal surface is reworked by mechanical milling in order to remove rough and imperfect surface structures. Then the next fibre layer is applied and the final structure is built up step by step. For deposition of the copper matri a commercially available electrolyte system based on sulphuric- acid with a copper sulphate content of 200 g/l was used.

Fig. 3 shows a boron- fibre reinforced copper pressure vessel produced by applying the technique described above. As it was not possible, to fabricate the complete structure by winding, only the cylindrical part of the vessel was strengthened wit 8 fibre layers positioned alternating +/- 45° to the cylinder axis.

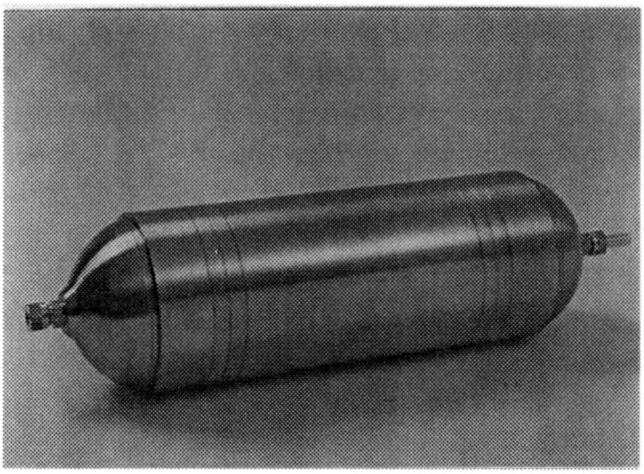

Figure 3. Pressure vessel with copper matrix.

This pressure vessel can be seen in Fig. 4 after burst pressure testing. The composite structure failed at a pressure of 130 bars (13 MPa). Because of the high specific weight of copper (density 8,9 g/cm^3) the stress-to-weight ratio and the specific properties may be a limitation for space application. An aluminium matrix would provide much more favourable specific properties due to a density of 2,7 g/cm^3.

Figure 4. Pressure vessel with copper matrix after burst pressure testing.

3.2. Aluminium Matrix

Therefore the following investigations concentrated on the replacement of copper by aluminium. This became possible by a process developed and patented by Siemens, Erlangen between 1970 and 1975, which uses a nonaqueous electrolyte system for the electrodeposition of very pure and thick aluminium layers (8, 9, 10). The organic electrolyte contains complex aluminium organic compounds and offers advantages like a process temperature below 100 °C, a micro- crystalline structure with a grain size far below 1 micrometer, a dense and pore-free morphology and the possibility to deposit very thick layers. Some features of this electrolyte are listed in Table 3. The equipment used for aluminium deposition is shown in Fig. 5.

The electroforming technique described in 3.1. was used to produce boron – fibre reinforced aluminium structures (7). The cylindrical part shown in Fig. 7 a was electroformed using a mild alloy steel mandrel which was removed by etching with nitric acid. The mechanical properties of the composite are quite promising. By burst pressure tests using the equipment shown in Fig. 6 an average tensile strength of 900 MPa (theoretically 1200 MPa) was measured. The density was 2,5 $g.cm^{-3}$ and the boron- fibre content 33 percent per volume. The composite cylinder after the test is shown in Fig. 7 b.

The micrograph of the boron-fibre reinforced cylinder wall shows, that the composite has a comparable dense and pore-free structure (Fig. 8).

Based on these fundamental investigations a prototype of a larger lightweight pressure vessel was fabricated, consisting of a cylindrical boron-fibre reinforced aluminium cylinder and two shells made of titanium alloy.

Table 3. Chemical composition and process parameters of Siemens electrolyte (9)

Chemical composition	$2\ Al\ (C_2H_5)_3\ .NaF + 3{,}35\ m\ C_6H_5CH_3$
Electrical conductivity	$(0{,}5 - 1{,}5)\ .10^{-2}\ \Omega^{-1}\ cm^{-1}$
Process temperature	80 – 95 °C
Current density	$0{,}5 - 5.0\ A.dm^{-2}$
Deposition rate	10 – 20 µm per hour
Layer thickness	µm – mm
Current form	DC or DC + AC
Impulse form	Rectangular pulses, periodic reverse
Stirring of bath	Continuos filtration or movement of parts
Anode material	Pure aluminium
Pretreatment	Nonaqueous or aqueous
Posttreatment	Anodizing, colouring, glass peening, chromate treatment, electropolishing
Substrate materials	All conductive materials

Figure 5. Equipment for aluminium electrodeposition.

Figure 6. Equipment for burst pressure testing.

7 *a* 7 *b*

Figure 7. Boron-fibre reinforced aluminium cylinder before (a) and after (b) burst pressure testing.

Figure 8. Micrograph of boron-fibre reinforced aluminium cylinder (5).

Figure 9. Pressure vessel components

Figure 10. Pressure vessel

The shells were not strengthened with fibres, because at that time a technology for winding and fixing the boron fibres on a spherical geometry was not available. In Fig. 9 the components of the pressure vessel are shown. The complete structure can be seen in Fig. 10.

4. CONCLUSIONS

Electroforming is a production technique for the fabrication of even very complicated structures. It can be successfully applied in many cases, where standard mechanical processes will not work economically or will not work at all. In the investigations described in this article it has been used for the fabrication of metal matrix composites. Boron fibres have been embedded by electrodeposition in matrix materials like copper and aluminium and by this composite materials with high strength and stiffness have been produced. Based on first fundamental investigations a step-by-step process (winding of fibres and deposition of matrix material alternatively) was developed and used for the fabrication of partially fibre reinforced pressure vessels for space application.

Caused by new priorities in the European space program after 1985 the project was not continued. Long-time space missions were not expected as a realistic goal in the near future. Additionally, the boron fibres available on the market at that time suffered considerably from low quality (fibre breakage, limited reproduceability e.g.), which would make production on a larger scale extremely expensive But beside of special space applications, metal matrix composites revealed a high application potential in other industrial areas, using particles instead of filaments for strengthening (Fig. 11).

By incorporating nano-sized particles in a metallic matrix it should be possible to fabricate composite structures or coatings with good mechanical,

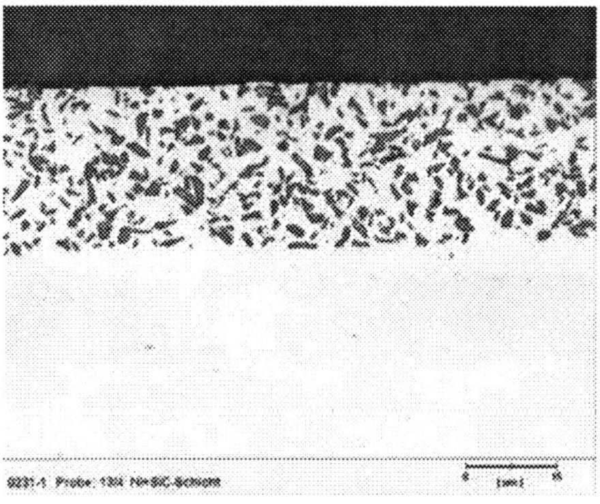

Figure 11. Nickel-Silicon carbide composite layer for tribological application.

thermal or tribological properties at an economic price. Some recent applications will be presented by EADS on this conference.

REFERENCES

1. R. Suchentrunk and O. Tuscher, Galvanotechnik 70 (1979) 1178.
2. R. Suchentrunk, Proc. Interfinish 80, Kyoto, Japan, 1980,180.
3. O. Tuscher and R. Suchentrunk, Galvanotechnik 69 (1978) 674.
4. R. Suchentrunk, Trans. IMF 64 (1986) 19.
5. R. Suchentrunk and F. Gammel, Z. Werkstofftech.14 (1983) 298.
6. R. Suchentrunk, AES Annual Technical Conference, June 1981, Boston, Mass., USA.
7. R. Suchentrunk, Metall 6 (1981) 539.
8. W. Wittich, R. Suchentrunk, H. Kellerer, Metall 10 (1976) 943.
9. R. Suchentrunk, AGARD LS 106, Lisboa-Athens-Ankara, March-April 1980.
10. R. Dötzer, VDI- Berichte 309 (1978) 95.
11. W. Wittich, Metall 10 (1973), 696.

Chapter 25
STRENGTHENING OF HEAT-INSULATING SURFACES

V.I. Bogdanovich and V.A. Barvinok
Samara State Aerospace University, Samara 443086, Russia

The methods of pulsed-plasma dynamic, laser strengthening of the coating surfaces, and coating formation by thermogaseous plasma deposition of finely-dispersed particles are used to form fine and nanocrystalline structures on the outer surfaces of the heat-insulating coatings.

Keywords: thermogaseous plasma, deposition, heat-insulating coatings, pulsed plasma, laser strengthening, finely-dispersed particles.

I. Introduction

The known effects of fine dispersion and nano-crystalline phases on the functional properties of objects are stimulating the search for the technologies of formation of such phases, particularly in the near-surface layers. In our previous studies, we observed an important role of finely dispersed phases and inclusions in improving the functional quality the coated objects [1, 2, 3].

One of the most complicated problems of the toady's machine-building is that of forming highly efficient systems of heat-insulating coatings (HICs) along the hot path of gas-turbine engines (GTEs). Although an impressing progress has been made by now [3, 4], the potential of enhancing the HIC efficiency is far form being exhausted. One of the obvious approaches to achieve this goal is to maintain a viable control over the HIC structure during the stages of the coating formation and its further processing on the final product.

In the present study, the HIC efficiency has been enhanced by creating a special dense fine-grained layer on the substrata surface by the method of plasma dynamic pulse and laser modification of the surface and plasma thermogaseous deposition of a powder material containing finely dispersed particles. The theoretical aspects of the methods used, the results of the materials examination, and their application as coatings on the element of the GTE hot routes are discussed.

A. A. Voevodin et al. (eds.),
Nanostructured Thin Films and Nanodispersion Strengthened Coatings, 251–260.

2. RESULTS

2.1. Analysis of the Trends to Enhancing the HIC Lifetime

The blades of the first-stage GTE are known [3, 4] to face the most severe conditions as compared to the machine-building parts operating in other fields and practically determine the engine lifetime. The functionality loss by the blades made of constructional refractory alloys during their cooling is due to high-temperature gaseous corrosion of the surface followed by the development of inter-grain corrosion and fracture wear under the conditions of high-temperature mechanical activation. The attempts of using oxide films to protect the alloys revealed a number of problems, the most important being those of thermal stability.

The search for the ways to solve this problem via adjusting the gradient of the coefficient of thermal expansion (CTE) across the system thickness due to the use of intermediate matching layers led to the design of a M-Cr-Al-Y composition (M = Ni, Co, Fe or their alloys). For a number of years this material (called a cermet coating, CMC) was the main HIC for the elements operating under moderate-temperature conditions. By now more than 3 mln. turbine blades covered with the above-mentioned coatings have been manufactured and put to use [3].

The material is based on the composition of a refractory Ni-Cr alloy, which minimizes the changes in the CTE values, and the additional components Al-Y that are used to form stable high-melting oxides on the surface after recrystallization annealing. The processes of diffusion, formation and reduction of aluminum oxide stabilized with yttrium oxide are going on at its erosion wear from the surface.

Thus, inside the system the CTEs are smoothly changing from the values close to refractory alloys to those of oxides. The CMC lifetime can be terminate at the decrease in the Al and Y concentrations below some critical value, at which a continuous oxide coating is not formed and an intense high-temperature corrosion of the remaining components is observed. Therefore, the CMC lifetime is practically determined by the coating thickness, since the increase in the Al and Y concentrations above 13 and 0.6%, respectively appeared to be ineffective. However, thickening of a CMC above some critical value results in lowering of the durability limit of the coating-substrate system. Although by using some specified technological approaches one can achieve partial reduction of this effect, nevertheless, the introduction of new-generation engines with a high-temperature combustion chamber stimulated the attempts of designing multilayer systems of HICs.

Among multilayer HICs, the highest efficiency was exhibited by the system containing an outer layer of ZrO_2-Y_2O_3 (a ceramic coating, CC) and an intermediate one- or two-layer matching CMC of M-Cr-Al-Y with various composition. In the present study, the system of coatings was deposited by an electron beam method (CC, CMC), a high-energy vacuum plasma technology (CMC), and a plasma thermogaseous method (CC, CMC) [3].

The experiments revealed a significant effect of HIC production technology on its lifetime and some contradictory requirements as regards the structure of such coatings. On one hand, the increase in the coating density and cohesive strength enhanced the system resistance to high-temperature gaseous corrosion, on the other, it lowered down its thermal stability and durability limit. Besides, it was found that the outer oxide layer obtained by various technologies was (although to a different extent) 'transparent' to oxygen ions and did not prevent exchange processes at the CC-CMC interface. This was due to the fact that the coating microstructure formed by the electron beam method was of a columnar character. Each particular column consisted of several crystallites (the column and crystallite boundaries are shown by solid and dashed lines, respectively Fig. 1), which had rather tight basal adhesion to CMC and were weakly adherent to the other columns. The boundaries between the columns and crystallites that began at the outer surface of CC and run to their boundary with the metal were transport channels for oxygen ions. The origin of such boundaries is technologically inherited from the coating formation by atomic fluxes [2]. It should be noted that any technological approach facilitating elimination of through boundaries in CC (Fig. 1b) may strengthen the system resistance against high-temperature gaseous corrosion and at a high thickness of the coating they cause a considerable decrease in the system durability limit [2, 3]. In plasma thermogaseous coatings, the intergrain boundaries have a quite different structure (Fig. 2), which slows down the oxygen ion penetration to the oxide-metal interface. An intermediate CMC is characterized by similar positioning of the intergrain and intercrystallite boundaries (Figs 1,2). Moreover, it was found that an increase in the coating density and the strength of the intergrain boundaries over the entire system bulk resulted in a decrease in the material heat-resistance and durability limit because of the decrease in the matching properties of the HIC layers. Consequently, a certain compromise should be kept in the system structure to enhance its inner resistance to the development of high-temperature gaseous corrosion and to strengthen its stability during multicyclic alternation of thermal and mechanical loading.

Taking into account good prospects for the applicability of the proposed system of HIC for aeroengines of new generations with significantly higher temperatures in the combustion chambers and the described mechanisms of the HIC lifetime decreasing, we continued investigations on the further improvement of HICs. They were based on an idea of a possibility of a significant increase in the HIC stability against high-temperature gaseous corrosion due to formation of a denser fine-grained surface layer without through boundaries (Figs 1b,c). At the same time, the structure of the untreated part of the CC and CMC layers should not undergo any essential changes in order to retain their matching properties.

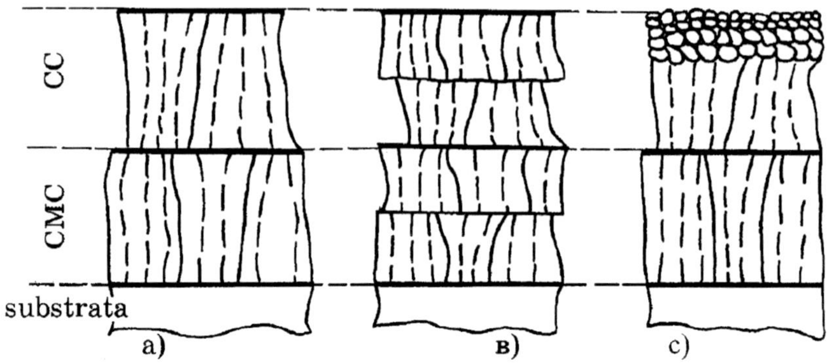

Figure 1. The HIC structure formed at the atomic flow deposition (a) and its modifications at elimination of through intercrystallite interfaces (b) and pulse thermal modification of the outer surface (c).

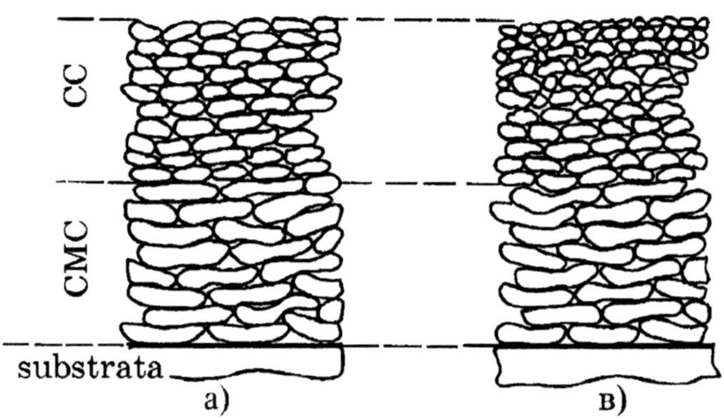

Figure 2. The HIC structure formed at plasma thermogaseous deposition of a powder material (a) and its modification after addition of finely dispersed powders at the final stage of the outer layer formation (b).

In principle, such an HIC structure can be formed on the available coating under the action of high-power thermal pulses (Fig. 1c), whereas simultaneous formation of the coating and an HIC layer on its surface can be achieved by using finer-grained powders at plasma thermogaseous sputtering (Fig. 2b) or ions with a high kinetic energy to reduce negative technological consequences of the coating deposition by the atomic fluxes (Fig. 1b) [2, 3].

Investigations are curried in these three directions. (1) Pulsed plasma-dynamic strengthening of the surface has been developed as far as the stage of pilot testing; the experiments on laser-induced surface strengthening are on the laboratory scale. (2) The activity on introduction of finely dispersed powders at the final stage of the oxide layer formation is also on the laboratory scale. As for direction 3, the experiments are at a stage of planning, designing and arrangement. Below the results which we have already obtained are considered.

2.2. Plasma Dynamic Strengthening of Heat-insulating Coatings

Upon the success in developing a dense fine-grained layer in the HIC formed on substrata there has appeared the problem of finding a possibility of employing this technology under the industrial conditions. That is because the temperature of the outer molten surface of CC should be higher than its melting point, $T_{M-1} = 2910$ K. At the same time the temperature at the CC-CMC interface must not exceed the value of an order of $T_{C-2} = 0.8 \cdot T_{M-2} \cong 1500$ K (T_{M-2} is the initial temperature of the CMC material melting). This limitation on the T_{C-2} value is determined by the fact that the processes of gas release and sublimation generated by higher temperatures could result in cohesion disruption of the CC continuity. Since there are no analytical data on such a situation in literature, we have developed a special mathematical model and used the results of numerical simulation to analyze the possibility of choosing the appropriate parameters of the thermal pulse (t_i is the pulse duration, q is the density of the energy flow), at which the described above situation could occur.

The mathematical model was formulated within a one-dimensional planar nonlinear problem of thermal diffusivity in a three-layer structure. Its surface is treated with thermal pulses and cooled due to heat conduction, radiation and the material evaporation. The stages of heating, melting, crystallization, and cooling were successively calculated.

Numerical simulation shows that the time of the oxide surface heating t_M to the melting temperature T_{M-1} is approximately proportional to q^{-2} (Fig. 3). At $t = t_M$, the thermal pulse effect is spread within a narrow undersurface layer. Beginning with this moment, the front of the melt formed on the surface is moving into the material. The oxide surface temperature is growing rather rapidly and when reaching the oxide boiling point $T_B = 4573$ K, it slows down and remains constant. The temperature distribution during the pulse action along the system thickness is characterized by a gradually decreasing function (Fig. 3).

As seen from Fig. 3, at q = *const* an increase in the pulse duration t_i is followed by a growth of the melt zone thickness and the temperature at the CC-CMC interface. Note that at $t_i = t_i(q) > t_{i,c}(q)$, the interfacial temperature becomes as high as the temperature of CMC melting. The melt depth increases with the pulse duration, at a high pulse power the oxide melt depth being two times deeper then at a low power although in both cases the temperatures at the CC-CMC interface are close to each other (Figs 3a, b).

Numerical analysis of the process proceeding at a given thickness of the system allows one to choose the q and t_i parameters, at which the maximum or preset melting of CC could occur and the temperature at the CC-CMC interface could not be higher than the given critical value.

After the pulse termination, the external surface temperature lowers down because of evaporation and heat conduction and radiation. However, the melting front motion $\xi(t)$ into the material goes on at the expense of heat stored in the overheated layer until approaching its maximum value. Calculations show that duration of the stage is no longer than $2t_i$ and the zone of melting expands no more than 20% of the $\xi(t_i)$ value. The time of the melt crystallization is proportional to ~ ξ^2_{max} and its rate depends on the degree of the melt overheating. The CC-CMC interfacial temperature is also growing at this stage.

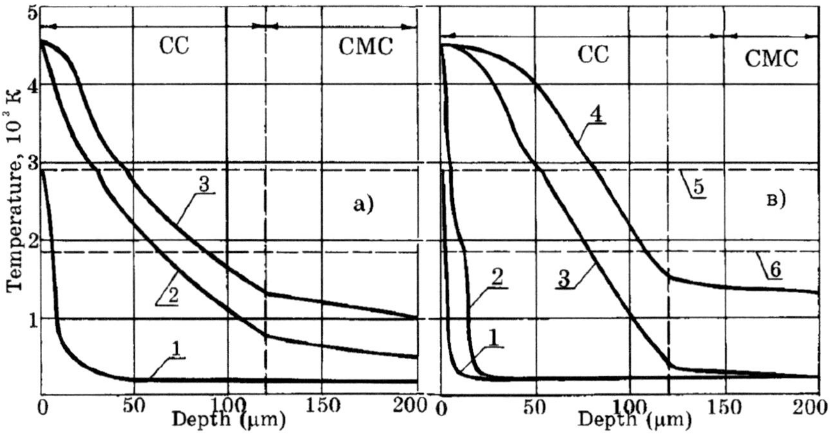

Figure 3. The temperature distribution across the thickness of the system $(ZrO_2+Y_2O_3)$ - (NiCoCrAlY) - (the substrate) for $t_i > 9.2 \cdot 10^{-3}$ s. (a) $q = 10^8$ W/m^2 at t = $4.2 \cdot 10^{-3}$ (1), $7.8 \cdot 10^{-3}$ (2), $9.4 \cdot 10^{-3}$ s (3); (b) $q = 10^9$W/m^2 at t = $4.2 \cdot 10^{-5}$ (1), $5 \cdot 10^{-4}$ (2), $4.2 \cdot 10^{-3}$ (3), $9.4 \cdot 10^{-3}$ s (4), 5 – T_{M-1} = 2910 K, 6 – T_{M-2} = 1875 K.

Moreover, the HIC was treated with several successive thermal pulses because the grain size in the crystallization zone diminished with an increase of the crystallization rate and narrowing of the range of melting. It was expected that the first pulse would provide 20-50% deep melting of CC, whereas lower melting depth resulting from further pulses would provide the formation of finer-grained structures near the HIC surface without a significant effect on the under-surface structures.

The proposed technological method [3] was performed on a pulsed dynamic accelerator of gaseous plasma. The accelerator consisted of a two-electrode collecting bar connected to two coaxial isolated copper electrodes. A plasma supplier with diameter 0.3 m is linked with the accelerator and a vacuum chamber, wherein technological gears and units are kept. The plasma-forming gas is injected into the accelerator volume (the injector chamber volume 20 cm^3 and the inside gas pressure up to 120 atm.) and ignited with a high-voltage pulse discharge from the capacitor storage bank. The section of the capacitor bank allows generation of a pulse with an up to 90 J energy and a voltage of up to 50 kV.

Specially packed turbine blades, their simulators, and participating samples covered with HICs were loaded into the vacuum chamber and treated with a gaseous plasma flow with the energy density from $2 \cdot 10^8$ to $2 \cdot 10^9$ W/m^2. The blade surfaces were successively treated with two pulses with a 12 min.-gap in-between.

Investigations of the samples with HIC obtained by the electron beam method demonstrated the validity of the proposed suppositions on the structural changes in the outer CC layer induced by pulsed plasma dynamic treatment. The adequacy of the mathematical model was evaluated by comparison of the data of theoretical calculations of the melting depth to the thickness of the zone of structural changes in the transverse cross-sections.

As shown by the experiments, the sample processing results in smoothing of the surface roughness from R_a=1.25 μm to R_a=0.8...0.63 μm, disappearance of open channels along the columnar structure boundaries, and partial reduction of zirconium oxide. The surface dull-white color changes to a dark metallic lustre. The presence of structural changes dependent on the regime of processing in the near-surface 20-60 μm-thick layer was revealed on the sample cross-sections. Across the depth of the oxide coating melt zone there are no normal-to-the-surface boundaries of the columns and crystallites and the layer structure itself looks like that in Figure 1c and is characterized by fine and equi-axial grains without dendrite formations and clearly defined directions of crystallization. The maximum grain size ranges within 1...3 μm. The presence of such a structure is an evidence for predominantly bulk character of crystallization in the melt zone. However, the structure analysis is far from being complete, since it requires more sophisticated analytical examination.

Thermal resistance testing of the samples and the turbine blades with a thermal shock (heating up to 1404 K for 3 min., 10 min. exposition, and cooling down to 473 K for 40 s) involved 2107 cycles of heat loading (T_{max} = 1233 K, T_{min} = 473 K, the cycle duration of 20 min.) and analysis of structural changes and defects formed in the as-treated surfaces. Investigations showed a no less than 1.8 times increase in the number of cycles prior to the defect formation as compared to the non-strengthened surfaces of similar or other items. After 1300 cycles of heat loading, no micro- and macrodefects was detected in the examined strengthened blades. The heat resistance of the samples with strengthened coatings increased by 1.5-2.2 times as compared to the non-treated coatings. Long-term high-temperature stability in air (T = 1248 K, σ_O = 270 MPa) surpassed that of the substrate material and the durability limit at the ambient temperatures was within the range of the durability limit of the substrate materials.

The obtained results demonstrate the perspectives of this technology and the possibility of its application to production of new and improved GTE. However, it was stated that under industrial conditions exploitation and technical maintenance of the plasma dynamic accelerator was rather complicated. Therefore, investigations were started to find the possibility of the CC strengthening with laser using dynamic laser-beam adapters.

2.3. Plasma Thermogaseous Heat-Insulating Coatings

At present plasma thermogaseous multifunctional coatings are widely employed in production of gas-turbine engines (GTE), in particular, as heat-insulating coatings (HICs) [1, 3]. Available equipment, particularities of the powder material production and storage, specific features of their heating and transportation in a plasma flow, as well as those of their interaction with the substrate surface predetermine the feasible range of the powder particle size as 40...200 μm. In general, most of technical problems being solved with the use of this method do not require powders of finer dispersion. Therefore the problem formulated in Section 2.1. is rather specific and very important from the practical viewpoint, since it can bring about an essential economic effect during the technology implementation into industry due to its simplicity, low energy consumption, and high productivity.

Presently, HICs on the elements of the GTE hot path consist of 100...150 μm-thick cermet layers deposited from the M-Cr-Al-Y powder (the particle size 60-100 μm) and an outer 150...350 μm-thick ceramic layer of the ZrO_2-Y_2O_3 powder (40...60 μm) [3]. Investigation of the plasma-deposited coatings for various dispersion used (40...200 μm) shows that deposition of powders with finer dispersion results in an improvement of the coating filling, its higher density, lower porosity, and a more uniform structure [1]. However, when stored under humid conditions, due to more developed surfaces and forces of molecular adhesion, the powder materials of 10...20 μm and less dispersion can form conglomerates consisting of different number of fine particles. Therefore, the powder material consumption is unstable during its feeding into the heating zone. Since the conglomerates exposure to the plasma action is very short, the conglomerate components attain rather different speeds and temperatures and, consequently, they form coatings with nonuniformly distributed functional properties. Besides, due to their low inertia, fine particles are swayed with a gaseous flow near the surface to either miss hitting it or to hit it under a very high angle. The as-obtained coatings exhibit lower adhesive and cohesive strength, higher porosity, and small coefficients of the material consumption efficiency.

Therefore, the problem of compacting the upper layer of CC (see Section 2.1.) required development of the technology of fine powder preparation before spraying, optimization of spraying regimes (particularly, those of plasma-forming gases consumption and the distance of spraying) and employment of specially designed ultrasonic feeders and plasmatrons.

The (ZrO_2 + Y_2O_3) powder fractions of less than 40 μm, 20...40 μm, less than 20 μm, and 10...20 μm were investigated.

They were used both independently and in mechanical mixtures (the standard fraction of 40...60 μm). Better results were observed at deposition of an outer 40-60 μm-thick layer from the fraction of 10...20 μm. Employment of the latter fraction reduced the maximum geometric deviation of the surface profile from 30 to 10 μm and gave an almost 3-fold decrease in the open porosity. The value of the total porosity determined from metallographic cross-sections appeared to be as small as 4%. The coatings were subjected to a series of standard testing described in Section 2.2. When compared to the coatings obtained via earlier used technologies, they exhibited a 50%-higher heat-resistance without diminishing of their thermal stability, long-term strength, and durability limit. However, the coefficient of the material efficiency at spraying with this fraction appeared to be ca. 2 times lower than at spraying with the 40...60 μm fraction.

3. DISCUSSION AND CONCLUSIONS

As shown by the experiments with heat-insulating coatings on the elements of the hot route of gas-turbine engines, the materials with a dense fine- or nanocrystalline structure can significantly contribute enhancing the material service life. The main field of improvement for HICs is the quality of the outer surface of the ceramic layer.

Development of specific technological methods is required for the formation of fine and nanocrystalline structures on the outer surfaces of the ceramic layers. The present authors used the methods of pulsed-plasma dynamic and laser strengthening of the coating surfaces and the method of the coating formation by thermogaseous plasma deposition of finely-dispersed particles.

The data obtained in the present study show the perspectives and usefulness of investigating and developing the proposed directions of nanostructural strengthening of heat-insulating surfaces.

REFERENCES

1. Barvinok V.A., Regulation of Tensile Stress and Properties of Plasma Coatings, Mashinostroenie, Moscow, 1990, 384 p.

2. Barvinok V.A. and Bogdanovich V.I. Physical Fundamentals and Mathematical Simulation of the Processes of Vacuum Ionic Plasma Sputter, Mashinostroenie, Moscow, 1999, 309 p.
3. Barvinok V.A., Bogdanovich V.I., and Dokukina I.A. Highly Efficient Technological Processes in Production of Turbine Blades and GTE Combustion Chambers, Nauka i Tehknika, Moscow, 2003, 640 p.
4. Contemporary Technologies of Gas-turbine Engine Production (A.G. Bratukhin, G.K. Yazov, and B.E. Karasev, Eds.), Mashinostroenie, Moscow, 1997, 408 p.

Chapter 26
NANOPOWDERS - AN APPROACH TO ENHANCED SURFACE COATINGS

F.J. Gammel, D.P. Jonke, O. Rohr
EADS Corporate Research Center, Munich, Germany

There are a lot of attempts to improve mechanical properties of coatings by using nanoscaled powder materials. Similar to the strengthening effect in oxide dispersion strengthened bulk materials a hindrance of dislocation movement as well as a synergetic effect of matrix material and strengthening particles should improve coating performance. Thus more elevated surface loads can be tolerated in service - light-weight materials or more cost effective base materials are acceptable due to enhanced coatings. Two ways of coating formation will be presented in this paper: A compaction of particles on the surface and an incorporation of particles in a metallic matrix. The compaction of powders was done using a new thermal spray process "Cold Kinetic Compaction" based on the Cold Gas Dynamic Spray Process developed by Papyrin [1, 2]. The incorporation of powders was performed by means of co-deposition of nano-particles during metal deposition by electroplating.

Keywords: coating, nanosized powder, thermal spray, cold gas compaction, dispersion coating, electroplating, co-deposition.

1. INTRODUCTION

Finely dispersed nanometer-sized oxide particles improve mechanical properties effectively in oxide-dispersion-strengthened (ODS) bulk materials. The improvement is due to the hindrance of dislocation movements within the bulk material leading e.g. to enhanced creep resistance as well as to increased oxidation and corrosion stability. There is a great interest to take advantage of this strengthening effect for advanced coatings.

There are different ways to use particles for production of coatings. The two mostly used are: A compaction of particles on substrate surfaces e.g. by means of thermal spray processes producing surface layers totally built up from particle material. Or an incorporation of particles during coating processes like electroplating resulting in particle reinforced dispersion coatings. Both technologies are already in industrial application using micron-sized powders, but there is only limited information about processing of nano-sized powders and their influence on resulting coating properties [9 - 11].

A. A. Voevodin et al. (eds.),
Nanostructured Thin Films and Nanodispersion Strengthened Coatings, 261–272.

2. COMPACTION OF PARTICLES

Thermal spray processes such as plasma spraying or HVOF are established processes within surface technology for generating of coatings. The characteristic here is that totally or partially melted powders are accelerated and sprayed onto substrates. Recent developments in HVOF show that with increasing velocity and decreasing temperature the quality of coatings can substantially be increased [3]. The innovative process of kinetic cold-gas compaction (cold spray process, K3) carries this trend on. Particle temperatures are significantly below the melting point at particle velocities up to 1200 m/sec. The coating structure is achieved through the high deformation of particles when impacting on the substrate.

2.1. Cold Spray Process - K3

Fig. 1 shows a schematic sketch of the cold spray process technology. Gas subjected to high pressure > 20 bar is heated to temperatures of 300-500°C, led to a Laval nozzle and accelerated there to a supersonic speed. Powder particles are fed into the supersonic gas jet and accelerated. Particle velocity is to be influenced by the geometry of the nozzle, the type of process gas and gas temperature.

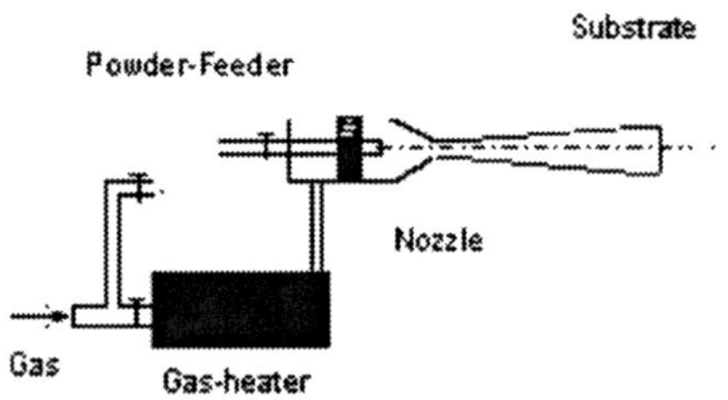

Figure 1. Principle of Cold Spray Process K3.

Particles strike the substrate at a high velocity and are deformed and compacted to create a coating while being well below the melting point of the coating material. No diffusion or grain coarsening effects will appear during deposition of the nano-particles.

The high impact of particles at moderate temperatures provides a low degree of porosity as well as a negligible chemical reaction of coating material with ambient air or process gas. In Fig. 2 the microstructure of plasma sprayed and K3 coatings are illustrated. The strong deformation of the particles in the K3 process is clearly visible in the deformed silicon precipitates that are oriented in parallel to the substrate surface.

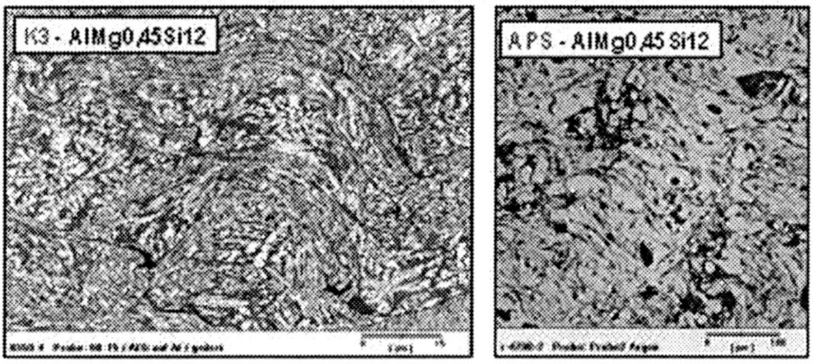

Figure 2. Metallographical analysis of aluminum alloys: plasma sprayed (APS), right and cold gas kinetically compacted (K3), left; (micron sized powders)

One of the main issues in the use of nano-sized powder material in thermal spray processes is the feeding of the powder to the coating nozzle. Electrostatic charge or strong agglomeration of particles due to high surface energy can led to a blocking of the feeding device and feeder tubes. This was the reason why nano-powders in the beginning were reported as not applicable in spray processes.

Best suited for feeding should be powders which exhibit a weak agglomeration to aggregates of some microns in diameter. This weak agglomeration would allow a de-agglomeration during acceleration and while striking the surface. So compact coatings should be accessible.

2.2. Analysis of Nano-Powders

To estimate particle size and agglomeration tendency metallic and ceramic nano-powders were analysed using SEM. The material (see Table 1) was prepared by MSAI [3] and IMET [4] using plasma and sol-gel techniques.

All powders had a primary particle size of 10 – 50 nm. The agglomeration rate however differed significantly. Some materials were agglomerated to cubic like compact structures, while others showed only limited cohesion between particles. Particles prepared in plasma processes sometimes were melted together to bone-like structures. Secondary particle size (size of aggregates) was in the range of some microns up to 50 microns. Fig. 3 gives an impression of the appearance of aggregates.

As all powders showed some degree of cohesion of particles, feeding was possible with appropriate modification of feeding equipment.

Table 1. Nano-Powders

Material	Code	Preparation
Tungsten	W-0.01/15O	Plasma
Tungsten	W-0,04/3O	Plasma
Molybdenum	Mo-0.03/17O	Plasma
Molybdenum	Mo-0.05/3O	Plasma
Iron/Cobalt	Fe/Co-42/58	Sol-Gel
Iron/Molybdenum	Fe/Mo-78/22	Sol-Gel
Iron	FeMUCuC	Sol-Gel
Tungstencarbid	WC	Plasma
Molybdenumoxide	MoO_x	Plasma
Molybdenumoxide	MoO_2	Sol-Gel
Zirconiumoxide	ZrO_2	Sol-Gel
Nickel/Iron	Ni/Fe-50/50	Sol-Gel
Copper/Silver	Cu/Ag-97/3	Sol-Gel
Tin/Copper	Sn/Cu-10/90	Sol-Gel

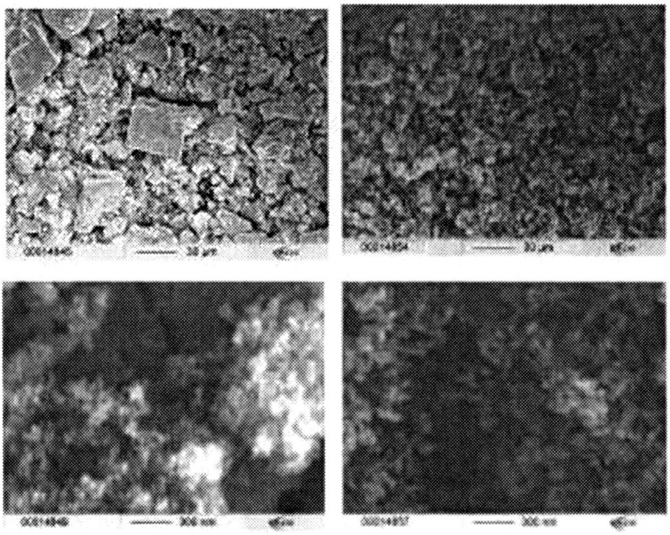

Figure 3. SEM-Analysis of powders: Size and agglomeration rate: W-0.01/15O left, Mo- 0.03/17O right.

2.3. Characterization of Coatings

2.3.1. Metal Coatings

Cohesion of particles and ductility of coating material had a significant influence on results. Strongly agglomerated material lead to a strong roughening of the base material. Only small quantities of coating material were implanted into the surface. No closed surface layers were accessible. This phenomenon was especially visible with tungsten. With loosely agglomerated material closed homogeneous coatings with smooth surfaces could be produced. Material with intermediate cohesion lead to closed coatings with bulky surfaces. Those bulks resulted from clusters which had not been de-agglomerated during acceleration like most of the other particles (Fig. 4).

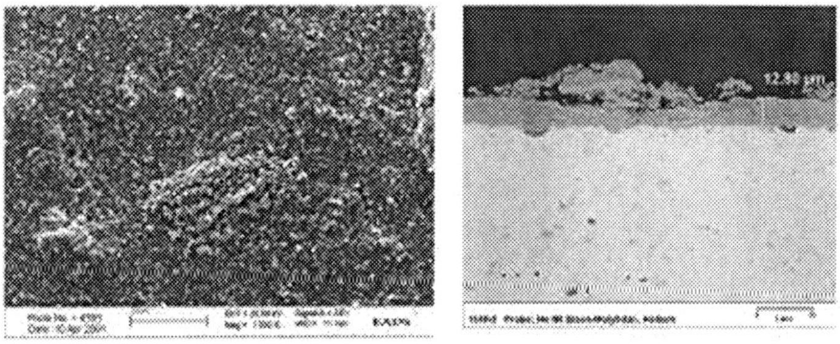

Figure 4. Topography and thickness of coatings (Fe/Mo).

Coating thickness was limited to microns. Ductile powders like copper generated layers up to 50 microns while less ductile material lead to coatings with a thickness of 5 to 10 microns .The compaction during coating increases the hardness of the deposited material. So the velocity and hence deformation of additional nano particles is not high enough any more to adhere to the initial coating layer. Nevertheless due to the homogeneity and lack of voids of the thin coatings better functional properties compared to coatings from micron-sized powders are expected. Fig. 5 gives an impression of the difference in coating build-up of nano-sized and micron-sized powders, both processed with K3 technology.

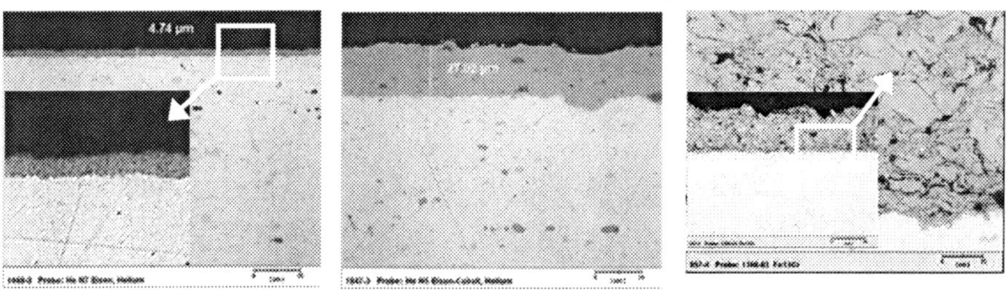

Figure 5. Cross section of coatings processed with nano (Fe, left), (Fe/Co, middle) and micron-sized (Fe, right) powders.

The process gas had also a significant impact on layer formation. With nitrogen and air roughly the same gas velocities and consequently comparable particle velocities are reached. Helium offers the potential of a distinct increase in particle velocity due to differing gas-dynamic properties.

While with helium as process gas total surface coverage was accessible, nitrogen lead to incomplete coating. The lower particle velocity when using nitrogen causes insufficient dissipation of agglomerates and locally too low impact for adherence. Fig. 6 e.g. shows the surface coverage for Fe/Co with identical coating parameters for He and N_2 as process gases.

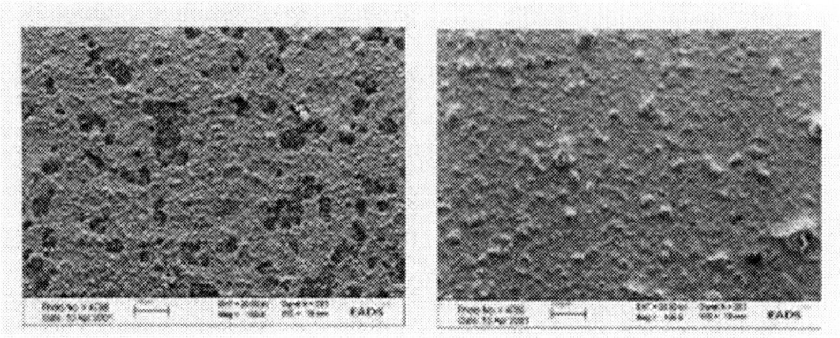

Figure 6. Surface coverage depending on process gas (N_2 left, He right): powder Fe/Co.

2.3.2. Ceramic Coatings

Ceramic particles are hard and brittle. So sufficient deformation for adherence of particles with one another cannot be expected. This is true for the WC powder investigated. Only roughening of the substrate and implantation of particles into the surface to a limited degree can be noticed. With ZrO_2 however, thin compact coatings up to 4 – 7 μm could be produced (see Fig. 7).

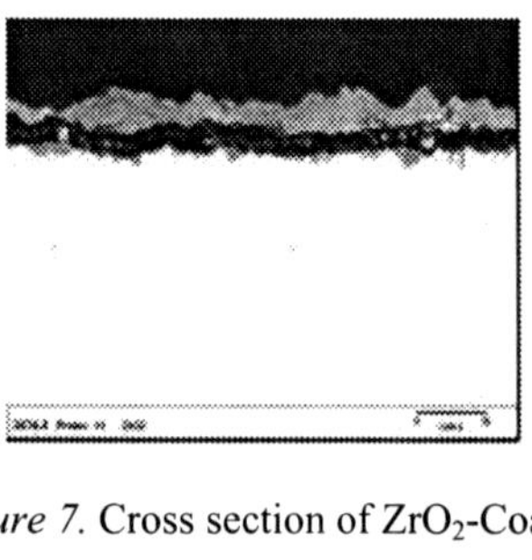

Figure 7. Cross section of ZrO_2-Coating.

2.4. Tribological Properties of Coatings

Tribological behavior was investigated with a ball-on-disc tester using 100Cr6 steel balls as counter-part without lubrication. The load applied was 1 N, the velocity of the ball 6.3 m/sec. Coatings had been applied on aluminum discs with no post treatment after deposition i.e. the sprayed layers were tested in the topography as plated with no polishing or surface smoothening. Table 2 gives the mean value for the coefficient of friction (CoF) after a sliding distance of 200 – 400 m. For some materials the CoF stabilized after a short induction time. Some however exhibited a continuous increase of CoF. Best results in terms of low value and stability of CoF showed Mo and Fe/Co (Fig. 8).

Table 2. Coefficient of Friction CoF, mean value after 300 m sliding distance
Ball: 100Cr6 - ∅ 6 mm, v = 6.3 m/sec, load 1 N, rel. humidity 50 %

Material	Code	CoF
Aluminium-Reference	AlSi9Cu3	0.9
Steel (hardened)		0.6
Molybdenum	Mo-0.05/3O	0.4 – 0.5*
Iron/Cobalt	Fe/Co-42/58	0.5
Iron/Molybdenum	Fe/Mo-78/22	0.5 – 0.7*
Iron	FeMUCuC	0.8 – 0.9*
Tungstencarbid	WC	0.6
Molybdenumoxide	MoO_x	0.6 – 0.9*
Molybdenumoxide	MoO_2	0.7 – 0.9*
Zirconiumoxide	ZrO_2	0.6 –0.7*
Nickel/Iron	Ni/Fe-50/50	0.7
Copper/Silver	Cu/Ag-97/3	0.5
Tin/Copper	Sn/Cu-10/90	0.5/0.9

* band indicates a continuous increase during test – no stabilisation of CoF

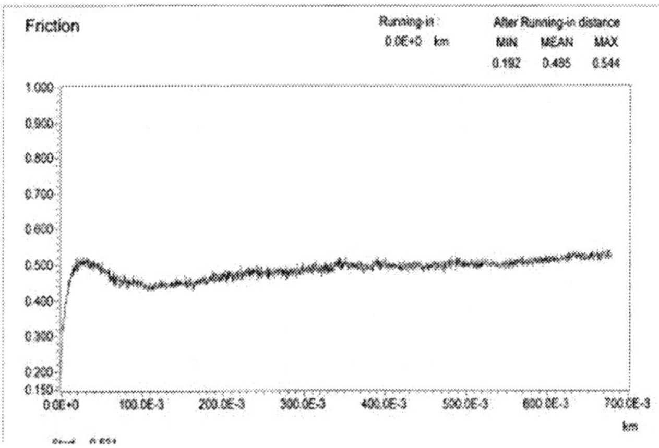

Figure 8. Coefficient of Friction vs sliding distance
Material: Fe/Co (parameters see Table 2).

3. DISPERSION COATINGS

Like thermal spray processes, dispersion coating with micron-sized particles is an already industrially applied technique. Applications are wear resistant and low friction surfaces in automotive and textile industries [6, 7, 8].

3.1. Coating Process

During metal deposition in an electroplating or electroless plating process, particles, suspended in the electrolyte, stick to the surface and are embedded into the metallic matrix. To prepare void-less coatings only not conductive particles are used. Most critical aspect using nano-powders instead of micron-sized material is the high surface area of the particles. This can lead to excessive adsorption of electrolyte constituents to the particles interfering with the metal plating process and/or agglomeration of nano-particles in the aqueous surrounding with a high concentration of ions.

3.1.1. Nano-Powders

Nano-particles used for incorporation were Alumina (Al_2O_3), Siliconcarbide (SiC) and Diamond (n-Dia) with a particle size < 30 nm. Concentration of particles in the electrolyte had been 20 g/l. All nano-particles showed a strong tendency for agglomeration in the electrolyte. This agglomeration increased with time. So various attempts to stabilize the particles in the plating bath were taken: Modification of particle surfaces by means of plasma or wet chemical treatment and use of surfactants (cationic, anionic and amphoterous). Surfactants showed increase in stability of particle suspension in pure water, but no long-term stability of suspension in electrolyte. Only wet chemical treatment of SiC had a noticeable positive effect on suspension stability.

3.1.2. Electrolyte

A nickel electrolyte, Watts type, was used (Table 3). For preparation of the dispersion electrolyte, the particle containing solution was treated with Ultraturrax (Type T50, 6000 rpm) to de-agglomerate the powder due to the high occurring shear stresses. This was the best method to prepare a homogeneous solution. Additional tests were run with electrolytes that were ultra-sonic treated before and partially during electrodeposition.

Table 3. Electrolyte composition and plating parameters

Electrolyte		Parameters	
Nickel Sulphate	250 g/l	Anode	S-Nickel
Nickel Chloride	35 g/l	pH	3.9 ±0.1
Boric Acid	40 g/l	Temperature	50 ± 5 °C
Additive A5	40 ml/l	Current density	4 A/dm²
Additive NPA	1,5 ml/l		

3.2. Characterization of Coatings

3.2.1. Nickel + nano-Al_2O_3

During electroplating of nickel agglomerates of nano-Al_2O_3 are adsorbed on the surface and subsequently embedded into the coating (Fig. 9). The structure of the nickel deposit is thereby not changed, we notice the expected pyramidal growth. The agglomerates are fairly regularly distributed in the nickel matrix. By periodic ultrasonic (US) treatment of the plating solution during plating (Cycle 0.1), agglomerates were dissipated to smaller shapes, which were more evenly incorporated into the coating (Fig. 10). Ultrasonic treatment however increased the tendency to rough surfaces especially in areas with high current densities like edges and peaks.

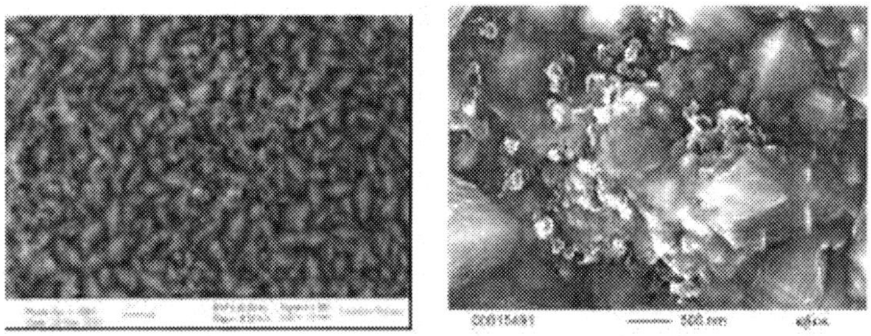

(a) (b)

Figure 9. Surface of dispersion coating: Ni + nano-Al_2O_3: (a) adsorbed agglomerates of nano-particles; (b) partially embedded agglomerates.

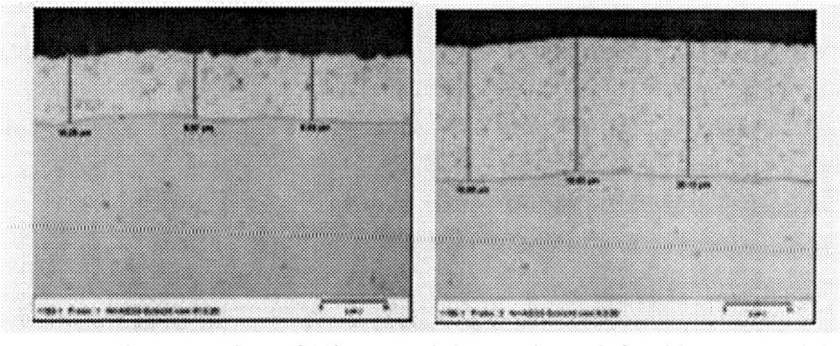

Figure 10. Cross section of Ni-nano-Al_2O_3 coatings: left without US; right with US treatment during deposition.

3.2.2. Nickel + Nano-SiC

As delivered SiC powder showed low incorporation rate. Plasma treatment in various atmospheres and plasma intensities had no beneficial effect. Only wet acid treatment rendered the particle surface in such a way that noticeable and homogeneous incorporation of particle was possible.

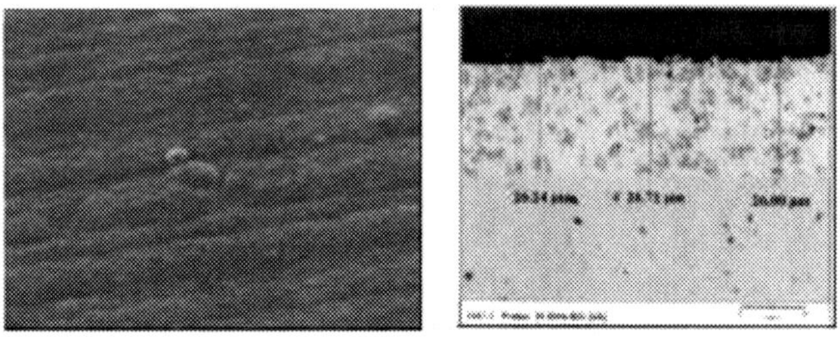

Figure 11. Surface morphology and cross section of Ni-nano-SiC coating

As already seen with Alumina, the nano-particles were agglomerated to micron-sized aggregates (Fig. 11). In contrary to Alumina, ultrasonic treatment had no additional beneficial influence on agglomeration or incorporation rate.

After 2-3 weeks the incorporation rate dropped. Even with ultrasonic treatment – neither internally nor externally in bye-pass - the incorporation rate of a new bath was not accessible (Table 4).

Table 4. Incorporation rate of nano-SiC in electrodeposited Nickel (4 A/dm^2)

Code	Powder	Age of bath	US treatment	Incorp. Rate Vol%
19	SiC-m/B1	New	No	4.23
20	SiC-m/B1	New	Internal	4.35
22	SiC-m/B1	2 weeks	Internal	3.42

3.2.3. Nickel + Nano-Diamond

Nano-Diamond could be co-deposited as delivered. As experienced with the two other materials, nano-Diamond was embedded as agglomerates. In contrary to nano-Alumina and nano-SiC however ageing of the bath did not change the incorporation rate (Table 5).

Ultrasonic treatment led to a reduction of agglomeration and at the same time to an increase in incorporation rate.

3.3. Tribological Properties

Properties were investigated with ball-on-disc test as described under 2.4. Samples (coatings on steel substrates) had been tested as plated without smoothening of surfaces. The CoF of electrodeposited Nickel coatings is in the range of 0.8. Incorporation of particles did not change the CoF. However wear of coatings (disc) and counterpart (ball) is noticeable reduced.

Table 5. Incorporation rate of nano-Diamond in electrodeposited Nickel (4 A/dm²)

Code	Powder	Age of bath	US treatment	Incorp. Rate weight%
23	n-Dia "G"	New	No	0.56
52	n-Dia "G"	New	External	0.70
25	n-Dia "G"	2 weeks	No	0.59
26	n-Dia "G"	2 weeks	Internal	0.69
53	n-Dia "G"	3 weeks	External	0.65

Wear of coating (volume loss) is reduced for all particle-containing layers compared to the pure Nickel coating (Fig. 12). US treatment during deposition had no pronounced effect on wear. The summarized volume loss of disc and ball, indicating the stability of the tribological system, demonstrates the beneficial effect of nano-particle incorporation. Increased wear of samples deposited during ultrasonic treatment might be due to increased roughness as mentioned earlier.

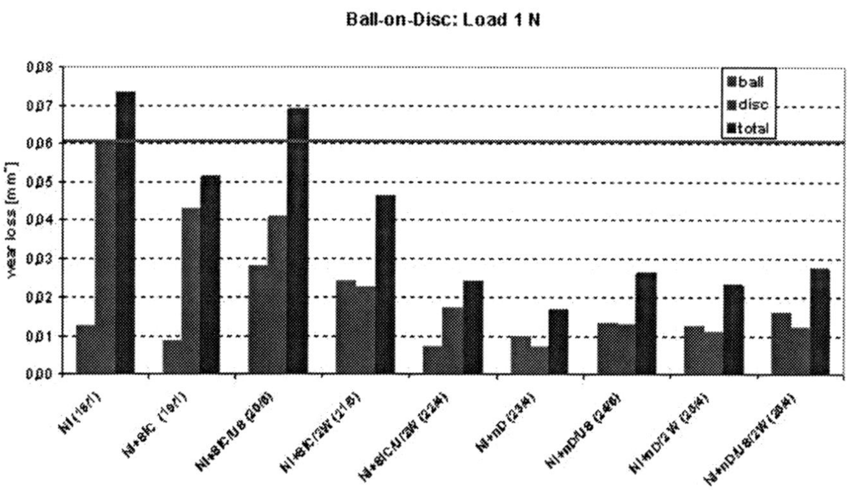

Figure 12. Ball-on-Disc: Wear loss; (2 w = deposition from an electrolyte 2 weeks in use) Ball: 100Cr6 - ∅ 6 mm, v = 6.3 m/sec, load 1 N, rel. humidity 50 %.

4 . CONCLUSIONS

Nanoscaled powder material (metals, ceramics) was used to prepare coatings – either as compact layers by means of cold gas spray process (K3) or as dispersion coatings by means of codeposition during electroplating.

Depending on agglomeration rate and ductility of particles layers up to 50 μm were accessible with K3. Due to the small size of particles a coating thickness of some microns was sufficient to get compact pore-free layers. Coatings showed good adherence to the substrate and resistance to destruction in tribological ball-on-disc tests (dry, load 1 N).

During electrodeposition of Nickel from a Watts type electrolyte electrical non conductive nano particles (Al_2O_3, SiC, Diamond) were embedded into the metal matrix. The co-deposition rate was limited to 4 Vol% (SiC) and 0.7 Weight% (Diamond). The incorporation of the nano-particles had no effect on the coefficient of friction, but wear of particle containing coatings as well as total wear (coating and counterpart) was significantly reduced (dry, load 1 N).

ACKNOWLEDGMENTS

We gratefully acknowledge the funding of thermal spraying of nano-powders within “Nadicoat” by German Ministry of Education and Research (FKZ: 13EU0152/8) as well as Russian Ministry of Industry, Science and Technologies within the umbrella of EUREKA (2260). Work on dispersion coatings using nano-powders was funded within “Inman” by the European Commission (Brite Euram BE4722).

REFERENCES

1. A. Papyrin, Advanced Materials Processes, Sept. 2001, p. 49.
2. A.P. Alkimov, V.F. Kosarev, A.N. Papyrin; Sov. Phys. Dokl. 35(12)1990, American Institute of Physics 1991.
3. H. Voggenreiter, Dissertation Bergakademie Freiberg, Herbert Utz Verlag, München 1996.
4. MSAI: Moscow Steel and Alloys Institute (Technological University Moscow).
5. IMET: Institute of Metallurgy and Material Science named after Baikov, Russian Academy of Science, Moscow.
6. K. Maier, Galvanotechnik 87 (1996) 1566.
7. J. Lukschandel, Metalloberfläche 4 (1984).
8. S. Kalitzki, Galvanotechnik 87(1996) 765.
9. JP 6196690 900313; JP 8869387 870413.
10. SU 4382417 880211.
11. G.V. Sakowich et al, Zh. Vses. Khim. O-va. Im. D.I. Mendeleeva 35 (1990) 600.
12. N.V. Mandich, J.K. Dennis, Metal Finish. 6 (2001) 117.
13. Y. Xiang, J. Zhang, C. Jin, Plating and Surface Finishing 2 (2001) 64.

Chapter 27

PHYSICOCHEMICAL PROPERTIES OF NANO-SIZED SYNTHETIC DIAMOND AND ITS DISPERSION-STRENGTHENING INFLUENCE ON DIFFERENT MEDIA

V.F. Komarov, G.V. Sakovitch, and M.G. Potapov
Institute of Problems of Chemical and Energetic Technologies,
Siberian Branch of Russian Academy of Sciences,
Federal Research and Production Center "Altai", Biysk, Russia

Development of the production technology of nano-sized diamonds from the carbon atoms of explosive molecules during their decomposition in the detonation regime and some properties of the products as-synthesized were reported in [1-3]. The further investigation of interaction of the products in various media led to the conclusion that their properties were determined by the size and aggregate state of the particles involved. The present communication will be mainly devoted to the properties of nano-sized synthetic diamonds and the effect they exert on the media they are in.

Keywords: nano-sized synthetic diamond.

When a TNT-RDX alloy is used as an explosive, the synthesis products may contain condensed carbon (CC) with an up to ~ 50% content of diamond-like carbon (DLC) (the cubic lattice constant a = 0.3562÷0.3573 nm calculated on reflections from different crystallo-graphic planes [3, 4]) enclosed into the graphite carbon matrix (Fig. 1) with the periphery consisting of disordered carbon in the proportion of an order of 5:2:3 [5].

The non-diamond carbon phase is extracted with chemical reagents by selective oxidation. As-obtained nano-sized particles of diamond (Fig. 2) with the average size of 4-5 nm measured by small angle X-ray scattering have a tendency to aggregation and formation of three-dimensional weak fractal structures [5].

The CC and DLC products have practical importance for the problem of interest, therefore the below discussed data and figures may be equally referred to both carbon products.

An X-ray structure analysis of nano-sized DLC has revealed five hkl lines on their diffractograms [3]. They correspond to reflections from the {111}, {220}, {311}, {400} and {331} planes of the cubic lattice with the respective intensity distribution of 85(44), 14 (22), 0.5(18), 0.3(4) and 0.2

A. A. Voevodin et al. (eds.),
Nanostructured Thin Films and Nanodispersion Strengthened Coatings, 273–278.

(12) % (the bracketed values present the intensity distribution for a reference pattern of the macrocrystalline diamond). It indicates redistribution of contribution of the lattice structural constituents to the particle surface energy. The DLC cubic lattice has been registered in the ^{13}C NMR spectrum and spectra of combined scattering [3-5]. The specific surface area determined with the BET method is within the limits of $(2.9 \div 3.6) \cdot 10^5$ m^2/kg for different powder batches. The enthalpy of formation estimated from the data on the combustion heat ranges within the limits of $\Delta H_f^o = (2560 \div 2950)$ kJ/kg that is a multiply higher value than that for macrocrystalline diamond (158 kJ/kg). The high ΔH_f^o value cannot be the result of overestimation of the calorimetric determination of the combustion heat by virtue of some physicochemical properties of such powders [5] but it is the consequence of an energy-saturated state of the substance.

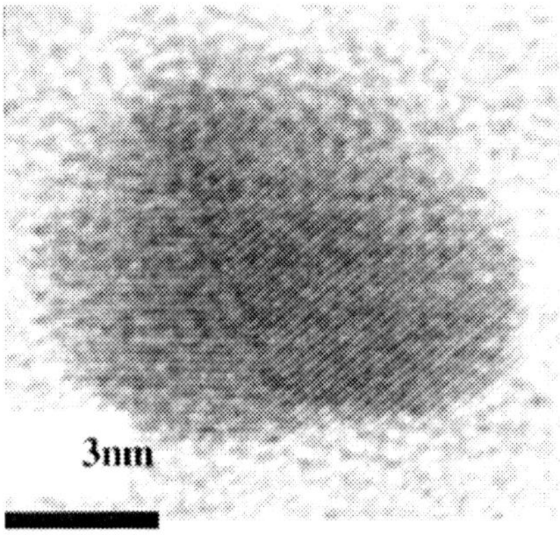

Figure 1. A TEM image of two DLC particles in a CC matrix.

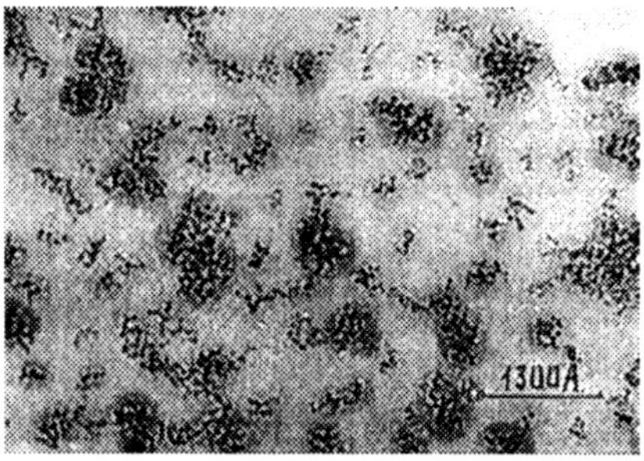

Figure 2. A TEM image of DLC.

The formation enthalpy of nano-sized DLC particles can be evaluated with the use of the standard thermodynamic values and the account made of their surface area [5]. At the constant values of temperature T and pressure P, variation of the surface area A of the given quantity of substance is accompanied by the changes in the Gibbs energy

$$dG_{T,P} = \sigma \cdot dA, \qquad (1)$$

where σ is the surface energy that leads to the gain of enthalpy ΔH and entropy ΔS as

$$\Delta H_{T,P} = \int_A \left[\sigma - T\left(\frac{d\sigma}{dT}\right)_p \right] \cdot dA \;\; ; \;\; \Delta S_{T,P} = -\int_A \left[\left(\frac{d\sigma}{dT}\right)_P \right] \cdot dA \qquad (2)$$

Assume that the surface energy of the crystalline diamond is temperature-independent, i.e. $\left(\frac{d\sigma}{dT}\right)_P = 0$ and $\Delta S_P = f(A) = 0$

$$\Delta H_P = \int_A \sigma \cdot dA \qquad (3)$$

Using expression (3), the reference value σ=10.6 J/m^2 for the plane {111} and its predominant contribution to the DLC particle structure, and the typical value of A, one quantifies the ΔH_f^o value equal to (3300÷3900) kJ/kg. The given ΔH_f^o estimate and its experimental value are indicative of a high energy content the material has obtained at the transition to the nano-sized state. If so, the increase in the formation enthalpy caused by the energy of the chemical bond cleavage in the surface layer of the substance q_0, changes in the redox potential ε^0, the specific adsorption heat of gases and liquids Q_{ad} must be reflected on the temperature boundaries and the kinetic parameters of physicochemical processes with such materials. Thus, both CC and DLC nano-sized particles are characterized by a high ability to structuration in various media such as solutions of salts, rubbers, plastics and metals. Consider some of them in detail.

A large quantity (≈2.5% of the DLC mass) of the retained by DLC gas is a common feature of interaction with the environment. The DLC injection into an electrolyte during chemical and electrochemical metal deposition causes changes in its graininess and microhardness and, as a consequence, enhances its wear-resistance.

Table 1. The influence of the DLC addition into the electrolyte on the properties of different metal deposits

Type of deposition	Grain size, μm		Microhardeness		Multiplicity of wear decreasing
	initial	With DLC	initial	With DLC	
Chrome electrochemical	0.013	0.009	9÷11	11÷14	2÷4
Nickel electrochemical	3.5÷5 .0	0.1÷0.5	2.8÷3.0	4.0÷5.8	2÷5
Copper electrochemical	2.0÷8 .0	0.2÷0.5	2.0÷2.1	3.2	–
Cobalt-Phosphorous electrochemical	–	–	5.25	6.8	4
Copper chemical	0.5÷1 .0	0.1÷0.5	2.0	6.5	–

In this case, the changes in the macropattern of the bending-induced fracture of the deposited film (Fig. 3) are also determined by the disordered structure of the deposit (see Table 2).

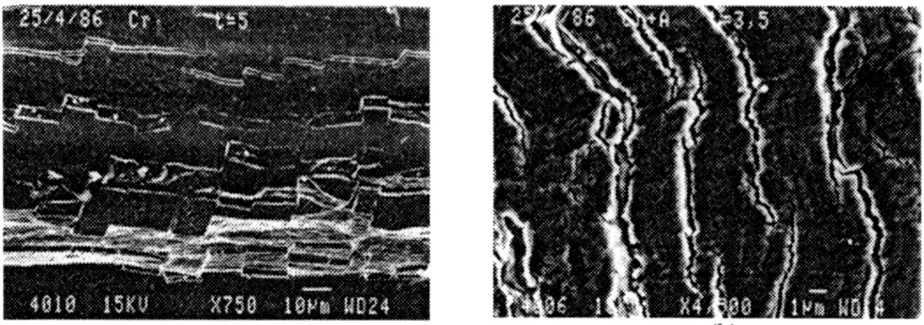

(a) (b)

Figure 3. The macropattern of the fracture of the chrome (a) and chrome with DLC (b) deposits on the plate after its bending under the angle of 45°.

The resultant data on the commercial efficiency of various tools coated with chromium-plated film with DLC (Table 3) have been summarized upon their industrial testing.

Table 2. The influence of the DLC concentration in chromium-plating electrolytes on the grain size and the deposit structure

[DLC], kg/m^3	Relative intensity of chrome reflections (nkL)						Grain size, μm	Micro-hardness GPa
	(110)	(200)	(211)	(220)	(222)	(310)		
0	1	22	9	12	100	5	0.013	11.1
3	1	22	5	4	100	1	0.010	12.4
8	28	82	39	25	100	6	0.010	12.7
15	66	100	46	28	45	10	0.009	13.8
25	1	0	0	1	100	0	0.008	12.0

Table 3. The data on the commercial efficiency of tools chromium-plated with DLC.

Tools	Working material	Multiplicity of wear decreasing
Pressing tool	Iron and stainless steel powders	8–10
	Radio ceramics	4–5
Stamping tool for cold extrusion	Steel, copper, aluminium	2–3
for cold draw-forming	Brass steel	2–3
Milling cutters, Ø = 8÷12 mm	Stainless steel, brass	1.5–3
	Steel 45	2
Drills: Ø=7÷10 mm	Steel 45	1.5–3
Ø=0.8÷1.0 mm	Glass-fiber material	5

A similar behavior is exhibited by the metals obtained by other methods, for example, aluminium compacted from the aluminium-DLC powder mixture by melting under vacuum (Table 4); compacts of hot-pressed copper or aluminium powders; magneto-soft amorphous alloys used in magnetic heads of the digital magnetic recording systems [2,5].

Structuration of composite materials based on plastics and standard resins can be noticeably affected by addition of small quantities of nano-sized CC (Tables 5 and 6).

Table 4. Characteristics of aluminium with DLC.

DC Contents, %	Microhardeness, GPa	Density, kg/m^3	Thermal conductivity, W/m·K
0	0.76	2730	162
2	1.78	2710	141
10	1.93	2690	101
20	2.63	2700	64
50	5.12	2600	9

Table 5. Characteristics of polytetrafluorethylene added with carbon and CC.

Additive and its content	Modulus of compression elasticity, MPa	Friction constant	Multiplicity of wear decreasing
0	4.2	0.21	1
20% C	11.5	0.32	25
5% CC	4.9	0.21	70

Table 6. The influence of the CC content on the resin quality coefficients.

Rubber and its basic content	CC content, %	Tear resistance, kN/m	Abradability, m^3/(kW·h)
7 IRP-1315 based on isoprene rubber	0	81.5	0.071
	5	96.1	0.017
7-404 based on the nitrile rubber	0	54.0	0.162
	2	55.9	0.101
	5	54.9	0.106

Addition of 0.5÷1.0% CC to the epoxy-resin base of the filament-wound glass-fiber material increases its tensile strength by 1.3-1.6 times.

In the presence of nano-sized carbon particles, the material tendency to changing its properties is observed in a though out vast range of earlier studied systems [5] formed by various methods (deposition, melting, pressing, mixing). Once the structural changes have arisen in the system, they lead to enhancement of the mechanical characteristics that in turn enhance the material wear-resistance. Considering this regularity one may come to the conclusion that it is determined by the same driving factor, a high surface energy of nano-materials. In our opinion, one more important conclusion can be made: any nano-sized material when introduced into a chemically agreeable medium can strengthen it. This conclusion does not contradict to the available (although infrequent so far) literature data.

REFERENCES

1. Lyamkin A.I., Petrov E.A., Yershov A.L., et al., Rpts. USSR Acad.Sci., 302 (1988) 611.
2. Sakovitch G.V., Petrov E.A., Komarov V.F. et al., Kluwer Acad. Publisher, Series I, 14 (1997) 55.
3. Vereshchagin A.L., Sakovitch G.V., Komarov V.F., et al., Diam. Relat. Mater. 3 (1993) 160.
4. Kuznetsov V.L., Aleksandrov M.N., Zadoruiko I.V., et al., Carbon 29 (1992) 665.
5. Sakovitch G.V., Petrov E.A., and Komarov V.F., Superhard Mater. 3 (2002) 3; ibid., 4 (2002) 3.

Chapter 28
SINGLE CRYSTAL SBN:Yb / OPAL MATRIX (SiO_2):Er COMPOSITE AS A NANOPHOTONIC STRUCTURE

M.I. Samoilovich[1], L.I. Ivleva[2], M.Yu. Tsvetkov[1], S.M. Kleshcheva[1], A.V. Gur'yanov[3]

[1] JSC Central Research Technological Institute "Technomash", 4 I.Franko Street, Moscow 121108, Russia

[2] Laser Materials and Technologies Research Center of GPI, 38, Bld. "D", Vavilov Street, Moscow, 119991, Russia

[3] JSC "OPALON", 1/2 4th Syromyatnichesky pas., Moscow 107120, Russia

Thin layer of opal matrix on SBN:Yb crystal substrate of <001> orientation was prepared via sedimentation method and post-growth annealing treatment. The film thickness was 25 cubic package monolayers of SiO_2 nanospheres (240 nm in diameter). Erbium oxide (Er_2O_3) was introduced into the interspherical space of opal matrix (about 35 – 40 % relative to the pore volume), using the sol-gel method. The specific features of photoluminescence in the nanocomposites are discussed.

Keywords: nanocomposite, opal, strontium barium niobate, rare earth, luminescence.

1. INTRODUCTION

Erbium lasers and amplifiers attract the essential interest due to the radiation in the range of 1.55 μm wavelength (third communications window) [1]. At the same time, erbium has a small absorption cross-section and, as a result, a considerable part of the pump energy remains not absorbed in the case of realization of compact systems. Laser transitions Er^{3+} form a 3-level system and, hence, it is necessary to realize at least 50% population inversion to receive the amplification. Careful selection of the pumping laser is required to match it with narrow pump band at 800 nm, 940 nm or 1480 nm bands in the Er^{3+} media. Reabsorption from $^4I_{13/2}$ level can lead to upconversion (especially for high dopant concentrations) and worsen the device quality.

Laser crystals doped with Yb^{3+} ions have some advantages in comparison with other widespread laser media. In particular, for Yb^{3+} (electronic configuration $4f^{13}$) only two electronic states are possible – ground $^2F_{7/2}$ and excited $^2F_{5/2}$. Hence, there is no parasitic absorption from the ground or

A. A. Voevodin et al. (eds.),
Nanostructured Thin Films and Nanodispersion Strengthened Coatings, 279–284.

excited levels in IR or visible spectral regions that considerably reduce the crystal heating during the excitation process. It prevents the thermal instability of the output radiation. The quantum efficiency is practically close to unity.

The fluorescence lifetime is 0.2 – 0.4 ms, which provides an effective accumulation of laser energy. Besides, the small ionic radius of Yb^{3+} (in comparison with other rare-earth ions) gives essential advantages when incorporated into the crystal matrix, which allows highly concentrated active media. Note, that Yb^{3+} ions have a broad absorption band that provides the possibility of tuning in IR spectral region (and also ultrafast pulses generation). As a drawback, the 3-level generation scheme leads to losses due to reabsorption and temperature-dependent parasitic effects.

The energy of the $^2F_{5/2}$ level of Yb^{3+} almost coincides with the energy of the $^4I_{11/2}$ level of Er^{3+} in the Er-Yb system. The energy exchange between closely spaced ions of Er and Yb is possible in cooperative cross-relaxation process [3]. In particular, electrons from the excited $^2F_{5/2}$ state of Yb^{3+} can pass to the ground state, providing the Er ions transition from the ground state to the $^4I_{11/2}$ excited state. In the case of effective cross-relaxation process, considerable part of adsorbed pump energy is transferred to the erbium system. Thus, Er-Yb co-doping can considerably increase the absorption of pump energy at 980 nm and provide the effective laser generation at wavelength of 1.55 μm (even in devices of centimeter length).

2. NANOCOMPOSITE PREPARATION

The development of optical systems for communication and information processing to some extent repeats a way of radiofrequency electronics from discrete elements and devices to integral optoelectronics. The latter assumes that the similar systems should work with optical signals in tiny devices located on a uniform substrate. There can be sources of optical energy, optical waveguides, connectors, detectors of optical signals and devices for the appropriate transformations of optical signals. The present paper is concerned with the opportunities of "hybridization" of optical crystals (with piezoelectric, electrooptical and acoustooptical properties) and photonic crystals (materials with photonic bandgap). Ferroelectric crystals of strontium barium niobate $Sr_{0.61}Ba_{0.39}Nb_2O_6$ (SBN:61) were used as optical medium [2]; an opal matrix (cubic package of SiO_2 nanospheres) was chosen as photonic crystal [4]. The nanophotonic structure was created based on a single crystal plate of SBN:61 doped with Yb, covered by a thin layer of opal matrix. Erbium oxide was introduced (Er_2O_3) into interspherical space of opal matrix.

Single crystal SBN:Yb^{3+} was grown from the melt by the Czochralski method [2]. The Yb_2O_3 dopant was introduced into the melt in a concentration of 2.6 wt. %. The growth parameters were as follows: pulling rate – 4mm/h, rotation rate – 15 rpm, the vertical thermal gradient at the crystallization front – 85 degree/cm. According to X-ray luminescence data, the Yb^{3+} concentration in a crystal was 1.1 x 10^{20} ions/cm^3. In our experiments the crystal plates of sizes 5 x 3 x 1 mm^3 and thin plates less than 500 μm thin were used. The plane of a plate was perpendicular to the crystal optical axis. Similar samples were also prepared from nominally pure SBN:61. The preparation of SBN plates included short-time etching in HF acid, washing in flowing water, processing by ethanol and drying in a thermostat at 100 °C during 30 – 40 minutes.

Self-ordering packing of spherical colloidal particles (SiO_2 nanospheres) is one of attractive technologies for manufacturing 3D- and 2D-photonic crystals with photonic bandgap (PBG) in an optical wavelengths, which can be used for development of optoelectronic devices. Colloidal crystals available now have not complete PBG because of small differences in the refractive indices of the periodic structure. Nevertheless, such materials provide an ideal basis for creation of structures with complete PBG by introducing substances with high refractive indices into the opal matrix and subsequent removal of the initial SiO_2 nanospheres. There are various methods to obtain the colloidal crystals [5]. The gravitational sedimentation of SiO_2 colloids from various suspensions is the most frequently used method for creation of colloidal crystals.

The realization of perfect photonic crystals remains a significant technological problem [5]. The large concentration of various internal defects, such as dislocations, vacancies, packing defects, etc, is one of the fundamental limiting factors of self-ordering. Thermodynamically stable phases are characteristic for face centered cubic packing, but the free Gibbs energies for phases with hexagonal dense packing and face centered cubic packing differ not so great, that leads, as a rule, in generation of a plenty of packing fault. As a result, the colloidal crystal is a polycrystalline mixture of misoriented domains. In a case of slowly growing samples the face centered phase prevails above a hexagonal phase. There are some technological possibilities improving quality of self-ordering colloidal crystals. We modified the methods of sedimentation by using tetraethoxysilane (TEOS) suspensions and obtained the planar structures of homogeneous multilayers of SiO_2 nanospheres (240 nm in diameter). The high quality of planar structures was provided by using suspensions with high monodispersity (deviation in diameters of nanospheres did not exceed 5 %) and special regimes of evaporation, drying and annealing. The special temperatures of evaporation provided small speeds of the growth of colloidal films. The obtained samples had small concentrations of dislocations and vacancies which was controlled by optical and electron microscopy methods. The SBN – opal matrix and SBN:Yb^{3+} – opal matrix nanocomposites were obtained. The Er^{3+} ions (erbium nitrate) were introduced into interspherical voids using standard sol-gel technique. A series of samples of various film thickness and Er^{3+} concentration were fabricated. We used the films to 20 – 80 layers with polycrystalline domains to 0.01 mm^3. Increase in Er^{3+} ions concentration leads to shift of PBG to red wavelengths which was measured a visually on the reflection.

3. EXPERIMENT

The features of behavior of Er^{3+} and Yb^{3+} ions in different matrices were investigated earlier, for example in glasses [3, 6], where cross-relaxation effects were observed (Table 1).

Table 1. Laser transitions in ionic system Er^{3+} - Yb^{3+}.

Yb		**Er**	
Transition	*Wavelength*	*Transition*	*Wavelength*
Absorption			
		$^4I_{15/2} - ^4I_{9/2}$	**800 nm**
$^2F_{7/2} - ^2F_{5/2}$	**980 nm**	$^4I_{15/2} - ^4I_{11/2}$	**980 nm**
Radiation			
$^2F_{5/2} - ^2F_{7/2}$	**1075 nm**	$^4I_{13/2} - ^4I_{15/2}$	**1530 nm**
Cross-relaxation			
$^2F_{5/2}$ **(Yb) –** $^4I_{11/2}$ **(Er)**			

The experimental setup is shown in Fig. 1. The photoluminescence spectra were investigated at room temperature. The semiconductor lasers (with emission wavelengths of 800 nm and 980 nm) were used for luminescence excitation under identical irradiation geometry [7]. The laser beam was focused to the sample at the angle of 60 degree. The luminescence radiation passed through the monochromator MDR-23 and the spectra were registered by non-cooled germanium detector DPD 2000 (DILAS Co.).

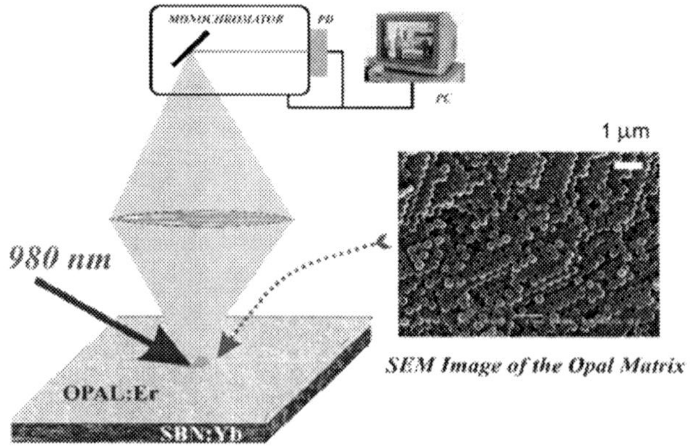

Figure 1. Experimental setup. Inset – SEM image of opal matrix.

The silicon filter cut off the pump radiation. The measurements were carried out in a lock-in mode with modulation frequency 30Hz.

Luminescence spectrum of the SBN:Yb^{3+} crystal obtained with excitation radiation at wavelength 980 nm is presented in Fig. 2. The superluminescence is observed at wavelength 1075 nm. For comparison, a low temperature luminescence spectra of SBN:Yb^{3+} is shown in the insert.

The luminescence spectra of SBN:Yb^{3+} – opal:Er^{3+} nanocomposite (curve 1) and the nominally pure SBN – opal:Er^{3+} nanocomposite (curves 2 and 3) with 980 nm pump wavelength are presented in Fig. 3. Curves 2 and 3 correspond to samples with "red" and "green" reflection, respectively. The difference spectra at around 1.5 μm for nanocomposites of nominally pure SBN:61 (curve 4) and SBN:Yb^{3+} (curve 5) is shown in the insert. The results show the presence of exchange processes between electronic subsystems of Yb^{3+} ions in the SBN crystal medium and Er^{3+} ions in the opal matrix.

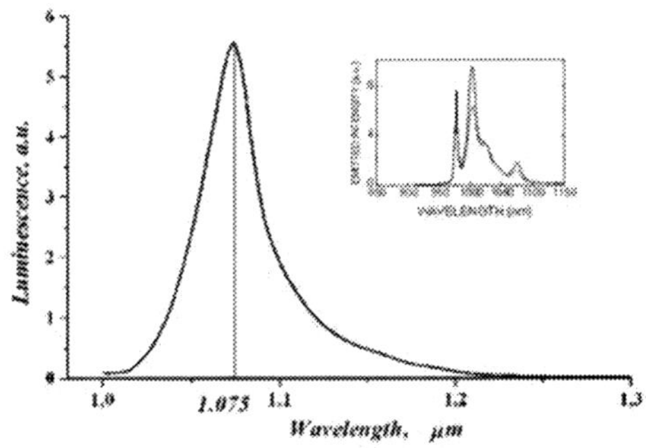

Figure 2. The luminescence spectrum of SBN:Yb^{3+} crystal in the wavelengths of 1 – 1.3 μm. In the inset: the spectra of low-temperature luminescence for different polarizations in the same area [2].

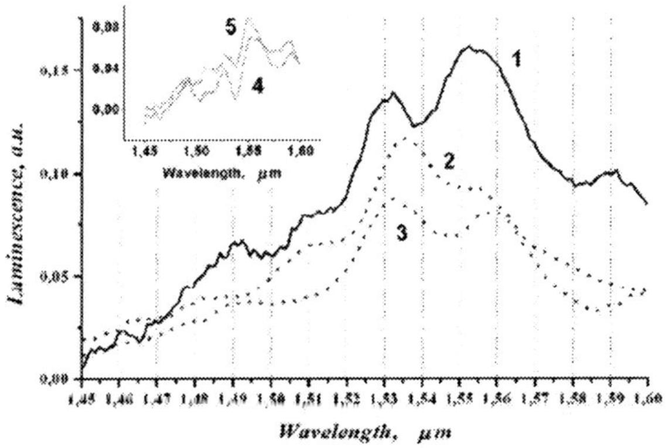

Figure 3. The luminescence spectrum of SBN:Yb^{3+} – opal:Er^{3+} nanocomposite (1) in comparison with the spectra of SBN – opal:Er^{3+} nanocomposite (curve 2 and 3 – "red" and "green" reflection). In the inset: difference spectra between curves 1 and 2 and between curves 1 and 3. Pump at wavelength of 980 nm.

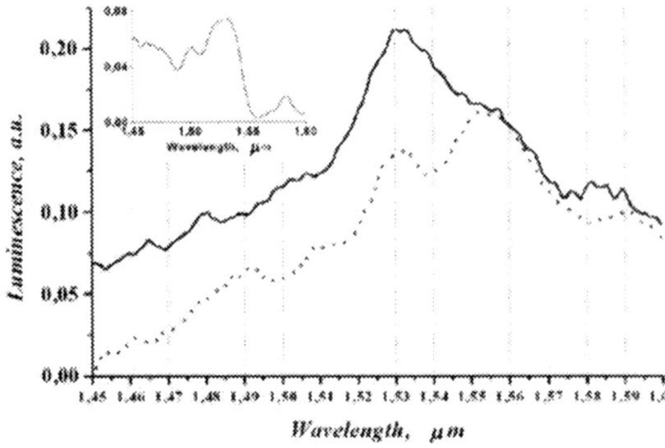

Figure 4. The luminescence spectrum of SBN:Yb^{3+} – opal:Er^{3+} nanocomposite under pump at wavelength of 800 nm (solid line) and 980 nm (dashed line). In the inset: difference spectrum.

The luminescence spectra of the SBN:Yb^{3+} – opal:Er^{3+} nanocoposite at pump wavelengths of 800 nm and 980 nm were studied (Fig. 4). The strong redistribution of luminescence intensity between Stark components of 1.53 μm and 1.55 μm was observed.

4. CONCLUSIONS

Until now various optoelectronic devices are mostly realized on optically active media, especially on the crystals having piezoelectric, electrooptical and acoustooptical properties. A new type of optical media– "photonic crystals" gives opportunity for development of new optoelectronic devices. The first experiments on hybridization of traditional optical crystals with photonic crystals are presented in this work. The luminescence data obtained for SBN:Yb^{3+} – opal:Er^{3+} nanocomposite show one of the ways to create new optical elements.

REFERENCES

1. E.Desurvire, Erbium Doped Fiber Amplifiers. Wiley, 1994.
2. T.Volk, L.Ivleva, P.Lykov, N.Polozkov, V,Salobutin, R.Pankrath, M.Wohlecke, Optical Materials 18 (2001) 179.
3. K.A.Winick, Rare-earth-doped waveguide lasers in glass and $LiNbO_3$: a review, in: S.Honkanen and S.Jiang (Eds.), Rare-Earth-Doped Devices II, Proceedings of SPIE, 3280 (1998) 88.
4. V.N.Astratov, V.N.Bogomolov, A.A.Kaplyanskii, A.V.Prokofiev, L.A.Samoilovich, S.M.Samoilovich, Yu.A.Vlasov, Nuovo Cimento D17 (1995) 1349.
5. S.M.Samoilovich, Ph.D thesis, MSU, Moscow, 1999 (in Russian).
6. R.Francini, F.Giovenale, U.M.Grassano, P.Laporta, S.Taccheo, Opt.Mater. 13 (2000) 417.
7. A.V.Gur'yanov, M.I. Samoilovich,, M.Yu.Tsvetkov, E.B.Intyushin, Yu.I.Chigirinsky, J. Appl. Spectrosc. 70 (2003) 323.

Chapter 29
MICRO- AND NANO-STRUCTURED THIN FILMS FOR BIOLOGICAL INTERFACES

F. Rossi, A.Valsesia, M. Manso, G. Ceccone, P. Colpo
European Commission Joint Research Centre, TP 203, 21020 Ispra (VA) Italy

The biological properties of a materials surface depend strongly on the way proteins adsorb on the synthetic surface. The development of the new generation biomaterials is based on surfaces that stimulate specific cellular response at the molecular level, for instance by immobilisation of biomolecules such as cell adhesion proteins, directly on the surface. Moreover, micro and nano patterning of the surface can be used for cell guidance, or for the design of parallel analysis domains on biosensors and biochips. The present work deals with the development of micro and nanostructured plasma polymer surfaces that covalently bind proteins. We present the different polymer surfaces that can be obtained by plasma deposition or by plasma treatment and relate the modification to protein attachment kinetics.

Keywords: surface patterning, protein adsorption, polymer modification, plasma discharge, ion beam.

1. INTRODUCTION

The immobilization of biomolecules on surfaces with contrasted properties is a fundamental step towards the development of electronic and optical biosensors [1]. The contrast consists actually in a patterning of the surface with a material that induces specific binding, surrounded by another material which rejects both specific and unspecific binding of the target biomolecule. Another possibility is to produce chemical contrast with 2 different types of functional polymers. Low-pressure plasma processes have been widely recognized as technology of choice for the deposition of functional polymers. One of the advantages that plasma techniques offer is the possibility to modify very first layers of the surface, keeping the bulk material properties unchanged [2]. Thus, substrate materials can be engineered with different surface properties in order to improve and control their compatibility with the biological system. Plasma treatments and plasma deposition of films are easily controllable by choosing suitable precursors and appropriate plasma conditions. Moreover, in case of polymers, the selectivity (or density) and the mobility of the functional groups on the surface can be varied to some extent by changing the degree of cross-linking.

A. A. Voevodin et al. (eds.),
Nanostructured Thin Films and Nanodispersion Strengthened Coatings, 285–295.

This article presents results obtained by plasma assisted chemical vapour deposition of 2 different polymers and their association in a patterned surface. The first polymer is characterized by a controlled density of carboxylic (COOH) functions and was deposited by plasma polymerisation of Acrylic acid (pAA). The second presents amine (NH_2) functions and was deposited by plasma polymersiation of allylamine (pAL). The same functions are present in proteins and are used as binding sites. Both pAA and pAL are suitable either to promote and control cellular growth [3] and for immobilisation of bioactive molecules [4].

Plasma diagnostics, surface analysis and physical-chemical characterisation have been performed in order to explore possible relation between the deposition processes and the film properties. Contrasted chemical functions are illustrated by Force distance curves obtained by Atomic force microscopy using functionalised tips.

2. EXPERIMENTAL

2.1. Plasma Assisted Deposition of Thin Films

The source used for the deposition is a Capacitively Coupled Plasma source illustrated in Fig. 1. The plasma chamber consists in a cylindrical stainless steel vessel (diameter = 210 mm, h = 350 mm) in which the part to be processed is placed and the process gases introduced at reduced pressure (10^{-3} to 1 torr). The plasma is ignited between two internal stainless steel parallel plate electrodes and sustained by a RF unit operating at 13.56 MHz working in continuous or pulsed mode. Samples are placed on the powered electrode, while the upper one is grounded. A vapor delivery system is connected to the chamber to vaporize the liquid precursor (Allylamine and Acrylic Acid of > 99% purity from Aldrich Chemical Co). The monomer flow rate (ϕ) is kept constant at 5 sccm, and it is injected through the upper electrode (shower head geometry). The working pressure is fixed at 50 mTorr, whilst the RF power range explored is between 5 and 60 Watts.

2.2. Thin Films Characterization

The characterization of the coatings has been mainly carried out by using X-ray photoelectron spectroscopy. A XPS-ultra photoelectron spectrometer, KRATOS UK, with monochromatic Al Kα emission at 150 W and high resolution pass energy of 20 eV. Operating pressure was 2.10^{-9} mTorr. The analyzed area was fixed at 110 μm diameter in spectral mode, whilst in imaging mode a spot area of 27 μm was used for the spatially resolved analysis. All analyses were performed at a take off angle of 90° (referred to sample surface). Both survey (0-1150 eV) and high resolution spectra of C 1s, N 1s, and O 1s regions were recorded at 160eV and 20 eV of pass energy, respectively. A flood gun was used to compensate the electrostatic surface charge. Peak fittings have been performed with VISION 2 software, whilst quantitative analysis from the survey spectra was carried out using a linear background subtraction and sensitivity factors provided by KRATOS. The COOH concentration has been evaluated after derivatisation by trifluoroethanol. The morphology of the coatings was evaluated by using atomic force microscopy (AFM) in a commercial NT-MDT Solver equipped with a Smena head. Standard silicon cantilevers with force constant of 5 N/m and working first harmonic frequency of 158 KHz were used. The surface free energy of the films were measured (Digidrop DGD Fast/60) with 3 liquids (formamide, water, glycerol, and α-bromonaphthalene), with a drop volume of 3 μl. The measurements are performed in a static mode at room temperature (23°c).

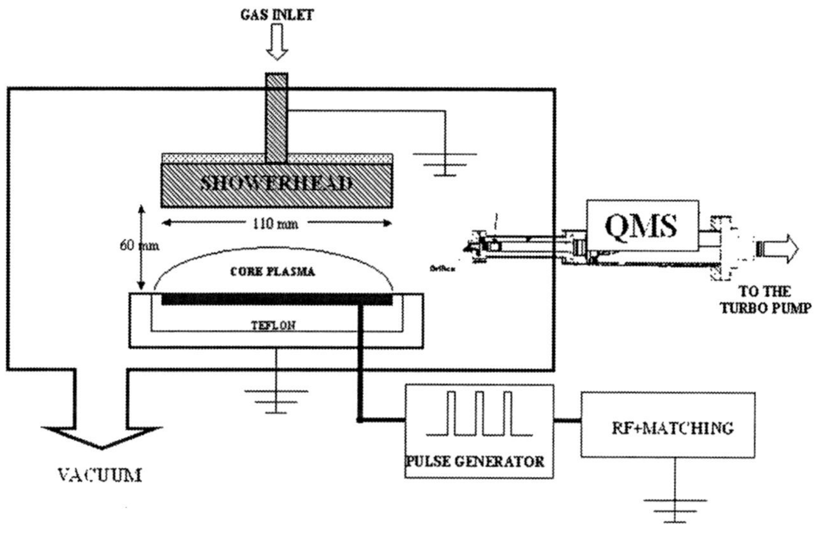

Figure 1. Schematic of the plasma reactor.

A Bi-Eka electrokynetic analyser (Anton Paar, Austria) was used to monitor the surface Z-potential values in the 4 to 10 pH range. The system allows measurements onto flat surfaces by using a capillary patterned PMMA surface inserted in an hermetically pressed head. The contribution of the PMMA surface to the Z-potential of the samples can be estimated by performing measurements onto reference samples. The Fairbrother-Mastin algorithm was used to calculate the Z-potential from the streaming curves (max pressure 600 mbar, two measurements in two flowing directions between the platinum electrodes) and the PMMA contribution was conveniently subtracted.

3. RESULTS

In order to study the rf power influence on the pAA deposited films structures, different films have been deposited at different powers between 5 and 60 watts. From a typical XPS survey spectrum obtained for a pAA film, only carbon and oxygen are detected. The C1s core level spectrum obtained for each sample has been peak-fitted based on chemical basis (see Fig. 2); starting with the hydrocarbon signal at 285 eV other four peaks are found, respectively: C-CO_2H/R (+0.5 eV); C-O (+1.9 eV); C=O (+2.9 eV); CO_2H/R (+4.35 eV) (the values in brackets correspond to the shifts relative to the hydrocarbon positions). The C1s core level spectra analysis shows that the percentage of the peak due to COOH/R carbon in the spectra decreases with the power increase.

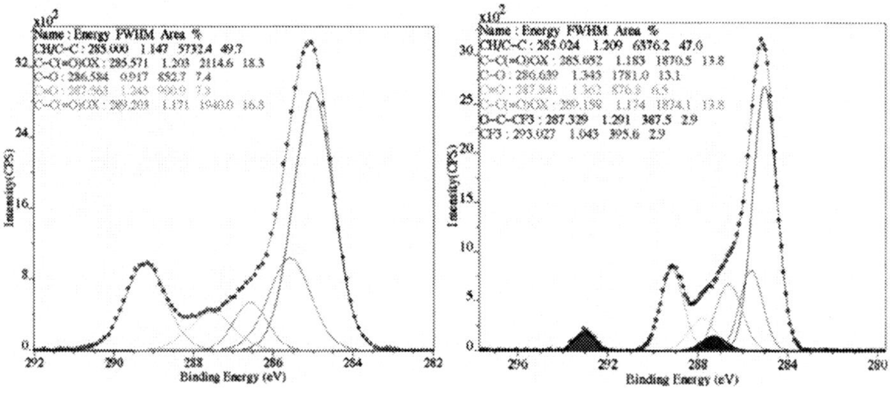

Figure 2. XPS spectra of pAA deposited at 80W before and after derivatization.

Fig. 3 shows the evolution of the film composition with power, in continuous mode. As already observed [5], it can be seen that increasing the RF power increases the fragmentation of the precursor, which leads to an increase of C and a decrease of O in the films, as well as a decrease of the COOX functionalities.

Fig. 4 shows the chemical composition of the films obtained in pulsed RF mode (time on t_{on}=3ms) varying the duty cycle. It can be seen that the COOH functionalities decrease strongly as the duty cycle is increased. Optimum COOH functionalities can be obtained at 80W with a t_{on} of 3ms and duty cycle of 10%. Other experiments show that in these conditions, the pAA films obtained are stable in water.

Contact angle measurements show that the hydrophylicity of the films decreases by increasing the power [5]. As expected with the previous results, this effect is a direct consequence of the lost of the COOH groups in the structure.

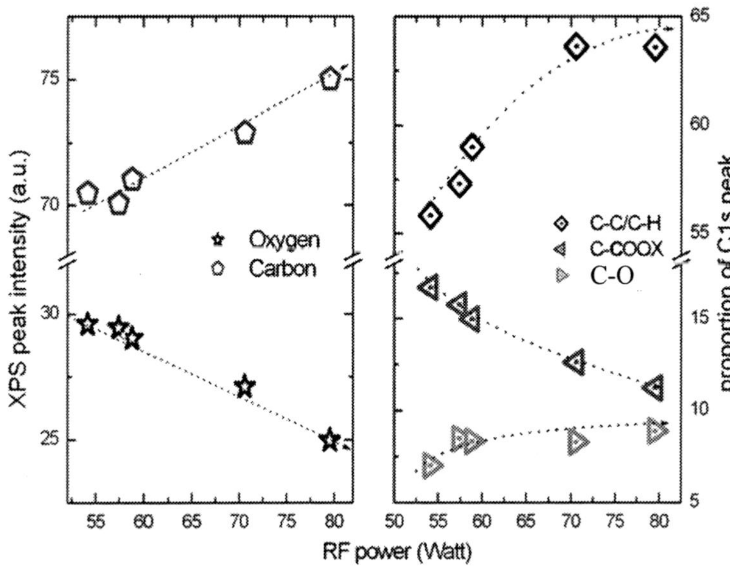

Figure 3. Effect of power on the chemical composition of pAA thin films

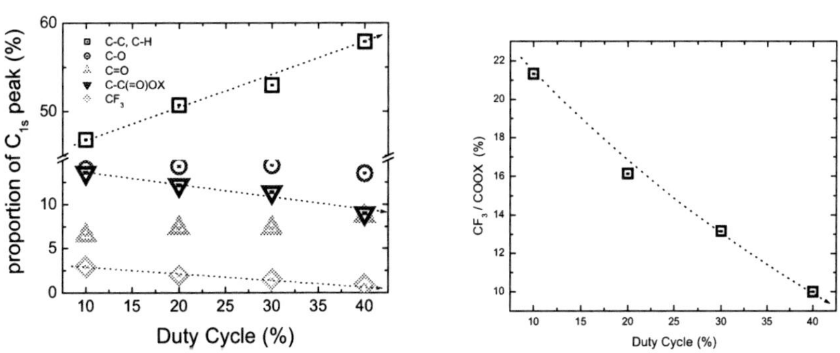

Figure 4: Composition of the pAA obtained in pulsed mode at 80 W 50 mTorr.

Then the different contribution to the total surface free energy γ_{tot} has been evaluated, and the polar, dispersive, hydrogen bonding and acid-base components have been calculated following the different approaches of Fowkes, van Oss and Owens Wendt [6, 7, 8]. Fig. 5 shows the different components of the surface energy as a function of RF power. As expected, the basic character of the films decrease as the power increases, whilst the acid components remains unchanged and the dispersive component of the surface free energies increases.

Similar experiments have been performed for the deposition of pAL films. Fig. 6 shows the C1s and N1s XPS peaks of pAL films deposited at 50W, 50mTorr and continuous mode. The complete study shows that these conditions lead to an optimum power for deposition of pAL films, ensuring a maximum of amine I functionalities and an N/C ratio close to 3. It is also found that beyond 40W, the films are stable in water after rinsing.

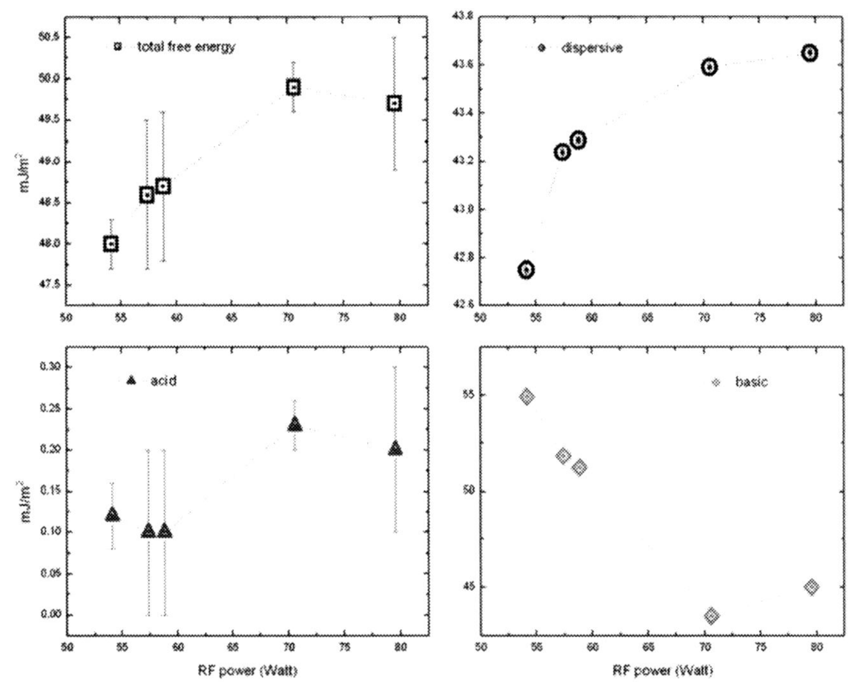

Figure 5. Surface free energy, dispersive and Lewis acid base components of pAA as a function of power.

Fig. 7 shows the AFM images of the surfaces of pAA and pAL. It is observed that pAA has very flat surface (roughness < 2 nm) while pAL has a very flat background surface (roughness < 3 nm) with circular islands (average height = 15 nm, typical diameter = 200 nm). Fig. 8 shows the Zeta potential of Si (substrate) pAA and pAL surfaces as a function of pH. It can be seen that the pAA has a very high negative Z potential for basic pH. The surface remains negatively charged also in acidic environment. In the case of pAL, Zpot=0 at pH = 5, and the positive potential saturates for pH < 4. For basic pH, the surfaces have a weak negative potential as compared to silicon.

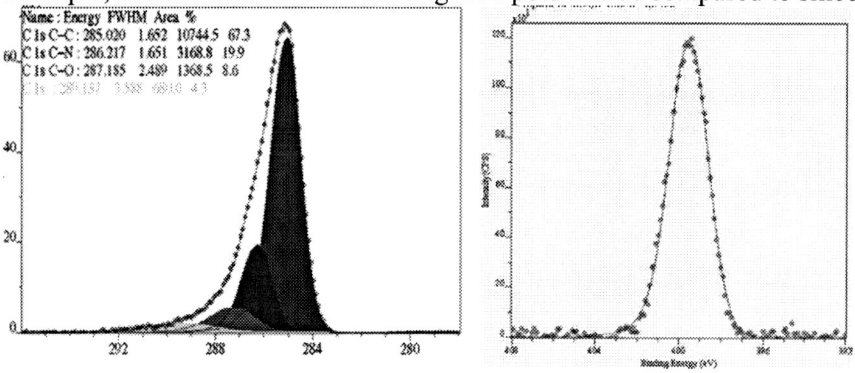

Figure 6. C1s and N1s peaks of pAL films deposited at 50mTorr and 50W in continuous RF plasma.

Micropatterned surfaces of pAL/pAA surfaces have been prepared as described in Fig. 9. The pAA film has been deposited first, then etched through a Ni mask by ion beam. The mask dimensions where typically 30um-100um holes, with a pitch of 100-500um. Then the pAA is deposited through the mask, in order to obtain a surface with negatively/positively charged domains.

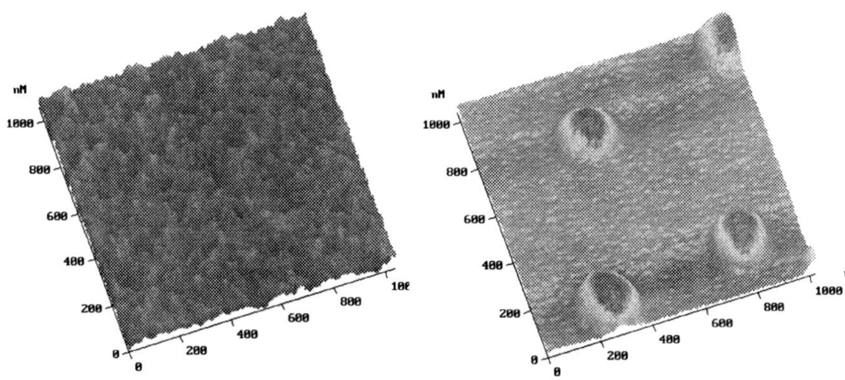

(a) (b)

Figure 7. AFM images of surface of (a) pAA (vertical scale 2nm) and (b) pAL (vertical scale 20nm).

The chemical contrast can be illustrated by Force/distance mapping of the surface over the patterns. The generation of the force images is made by measurement of the oscillation amplitude of a COOH functionalised AFM tip. The contrast in the image arises from differences of the force felt by the tip at a given distance z of the surface. Depending on the distance at which the amplitude is measured, values of the electrostatic interactions can be isolated.

Fig. 10 shows the AFM height and force curve made on pAA/pAL micropatterned surfaces at 240nm tip-surface distance. The force contrast can be clearly seen between the COOH functionalised tip and the patterned surface.

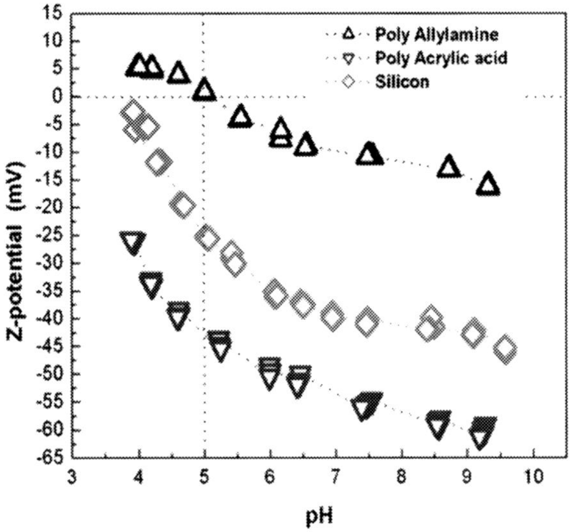

Figure 8. Zeta potential of pAA, Si and pAL surfaces as a function of pH.

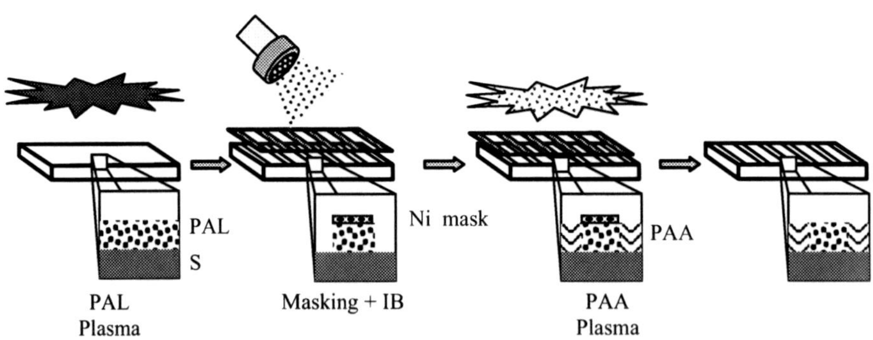

Figure 9. Preparation steps of pAA and pAL micro-patterned surfaces. Etching of PAL is made by ion beam.

The biological response of the patterned device has been evaluated by the adsorption of Bovine Serum Albumin (BSA) in a buffer solution (pH = 7.2 , BSA concentration 40 μg/ml, incubation time = 30 min at room temperature). After the incubation the samples have been rinsed in Milli-Q water and characterized by the XPS.

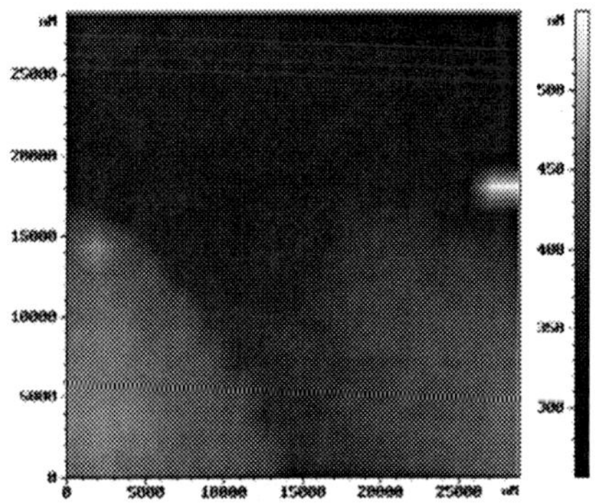

(a) 25 x 25 μm height image

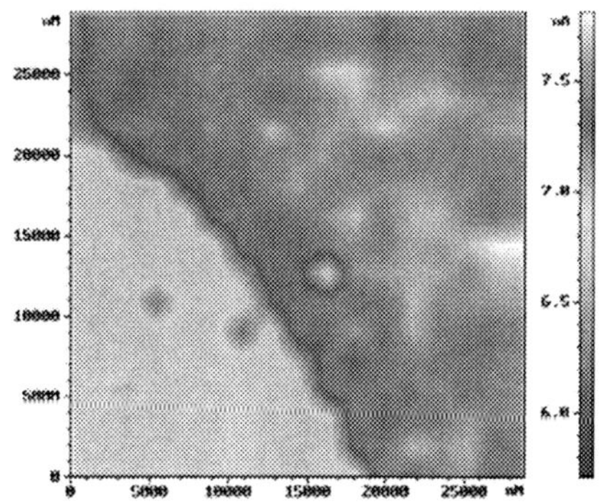

(b) 25 x 25 μm Chemical force image

Figure 10. Height and chemical force images of the pAA/pAL micropatterned surfaces. Observation mad with COOH functionalised tips in KCl 100 mM solution, pH 6. Chemical mapping made at a tip- surface distance of 240nm.

Fig. 11 shows that after the BSA solution incubation on pAA zone the signal from Nitrogen peak appears, this signal is attributed to the adsorption of a stable layer of protein that is rich of Nitrogen containing chemical groups.

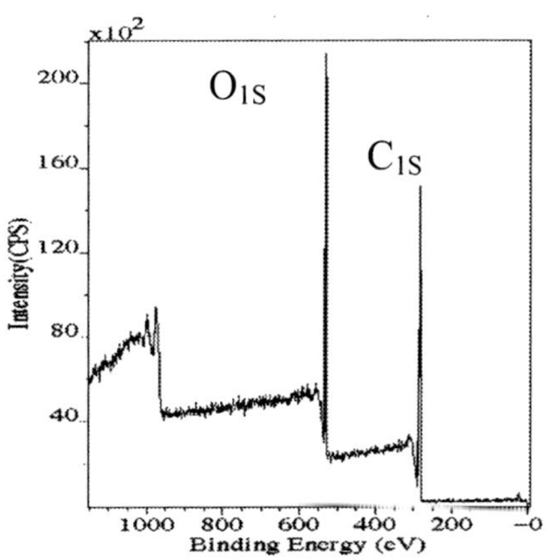

as deposited

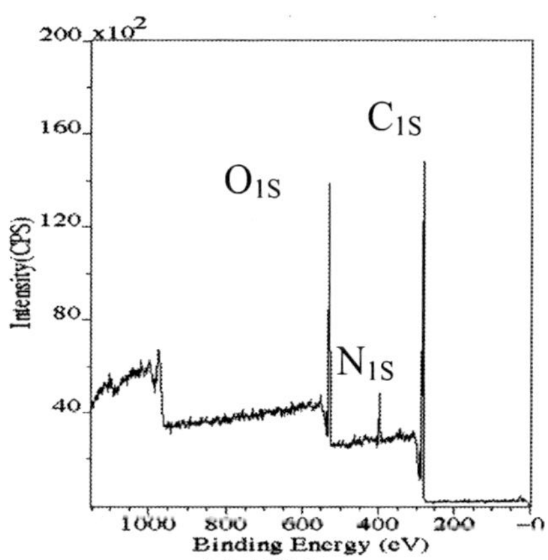

with BSA

Figure 11. XPS wide spectra of PolyAcrylic Acid as deposited and after BSA adsorption.

On the other hand (Fig. 12), on the pAL zone the total content of Nitrogen does not change considerably on the surface (from the 5 % of the as deposited to the 7 % after protein incubation) but the dominant process is the oxidation of the surface.

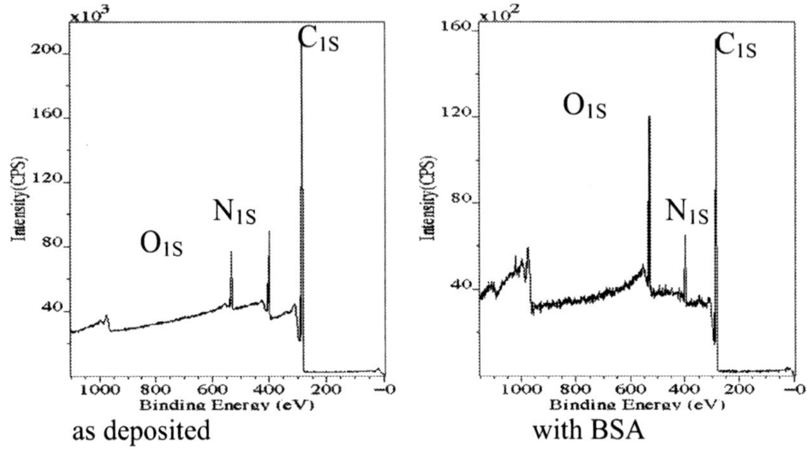

Figure 12. XPS wide spectra of PolyAllylamine as deposited and after BSA adsorption.

4. CONCLUSIONS

In this work a device with contrasted charge properties on a micro-patterned surface has been produced combining PE-CVD and Ion Beam techniques. The physico-chemical properties and the electrostatic response of the device have been evaluated by different surface analysis techniques. In particular, the capability of PE-CVD to produce polymeric films with stable chemical functionalities has been demonstrated; in this frame the modulation of plasma deposition parameters allows the possibility to tailor the density of surface active COOH and NH_2 groups. Z-potential measurements confirm that, in the whole 4 to 10 pH range, COOH and NH_2 chemical groups induce opposite charge shifts with respect to reference passive surfaces, and while the pAA surface maintains a strong negative potential in the whole pH range, pAL surface acquires a positive potential for acidic pH values ($pH < 5$) and has a weak negative potential with respect to a passive surface for basic pH. After the separated surfaces characterization, micro-patterned devices with contrasted charge properties have been produced. The interaction of the patterned device with negatively charged molecules attached on an AFM tip in a controlled environment with a fixed pH, allowed the direct measurement of the different forces acting on charged species at the boundaries of the designed motives. Preliminary BSA adsorption tests further suggest the existence of a contrasting interaction with biomolecules.

ACKNOWLEDGMENTS

The technical support of A. Hoffmann is gratefully acknowledged. This work was supported by the JRC Institutional programme FP6 4221 "NanoBiotechnologies for Life Science".

REFERENCES

1. B. Kasemo, Surface Science 500 (2002) 656.
2. H. Yasuda, M. Gazicky, Biomaterials 3 (1982) 68.
3. B. Gupta, C. Plummer, I. Bisson, P. Frey, J. Hilborn, Biomaterials 23 (2002) 863.
4. P. Favia, F. PALumbo, R. d'Agostino, S. Lamponi, A. Magnani, R. Barbucci, Plasma and Polymers 3 (1998).
5. P. Rossini, P. Colpo , G. Ceccone , K.D. Jandt and , F. Rossi, Materials Science and Engineering C 1050 (2002) 1.
6. D. K. Owens, R. C: Wendt, J. Appl. Polym. Sci. 13 (1969) 1741.
7. F. M. Fowkes, M. A. Mostafa, Ind. Eng. Chem. Prod. Res. Dev. 17 (1978) 3.
8. C. J. van Oss, R. J. Good, M. K. Chaudury, Langmuir 4 (1988) 884.

Chapter 30

APPLICATIONS OF ION-ASSISTED DEPOSITION TO SURFACE MODIFICATION ON THE MICRO- AND NANO-SCALE

J.S. Colligon

Manchester Metropolitan University, Chemistry and Materials Department, Chester Street, Manchester M1 5GD, UK

The use of ion beams to produce modified surfaces on the micro- and nano-scale is described and the advantages of using ion beams to gain a better fundamental understanding of the growth process are outlined. Examples of current projects are given and include recent results on nc-TiN/a-Si_3N_4 nanocomposites, Ti-Cr-N coatings, the control of stress in oxide coatings, production of surfaces with getter pumping properties and the effects of surface modification on biocompatibility. For the nanocomposite films it is shown that hardness values increase with substrate temperature and increased levels of ion-assistance. Ti-Zr and Ti Zr-V coatings have been produced to evaluate their getter performance and preliminary results show that these films can be re-activated at temperatures of 350°C. Ion-assisted deposition of SiO_2 films has been shown to reduce the stress in these films by a factor of 2. Biological cell-surface interaction studies on surfaces modified using ion beams are reported and show that cell proliferation is influenced both by chemical changes and sometimes by the surface topographical features.

Keywords: ion-assisted deposition, nanocomposite, getters, film stress, biocompatibility.

1. INTRODUCTION

Ion-assisted deposition of materials can produce coatings with superior properties since the ion-assistance provides incident atoms with additional energy. This added energy can modify the nucleation process, improve film adhesion, increase film density, stimulate mixing of alloy materials to form metastable compounds, trigger phase changes, influence film stress and change film microstructure [1]. All these factors can be utilised to produce coatings with improved corrosion-resistance, hardness and optical, physical and wear properties. As the coating industry for tribological and wear improvement alone is worth some 10 billion dollars p.a. in the UK alone [2] there is a clear advantage in developing and optimising the ion-assisted coating processes.

A. A. Voevodin et al. (eds.),
Nanostructured Thin Films and Nanodispersion Strengthened Coatings, 297–306.

Unbalanced magnetron sputtering [3], deflected arc [4] and plasma-immersion systems [5] can all be used to provide the ion-assisted coating conditions. These have commercial potential in that they can produce fairly large coated areas (order of square metres) at reasonably high deposition rates. Unfortunately, in all these systems, it is very difficult to separate the deposition parameters for a controlled optimisation study. Ion beam systems of the form shown in Figure 1 do offer this fine control.

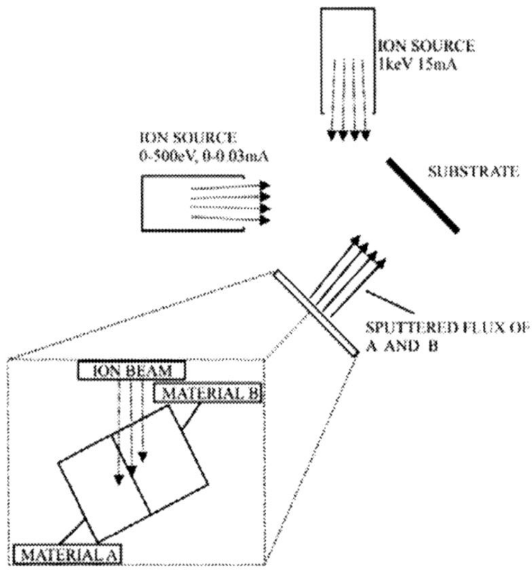

Figure 1. Dual beam IAD system.

In the system shown the vertical ion source is used to sputter atoms from a target (which may be a single material or have strips of different materials) onto the substrate. The atom arrival rate at this substrate is therefore simply controlled by increasing, or, decreasing the vertical source ion beam current. The horizontal ion source is then used to control the current and energy of the ion-assist beam. The substrate temperature can be independently controlled. If the substrate is clamped at one end, and has a polished reverse side, its flexure, and hence the stress in the coating, may be determined by monitoring the deflection of a laser beam from this back surface. If a patterned deposit is required a fine mesh grid can be placed onto the substrate. The sputtered material will now form a deposit with features of micron size. If the mesh is removed these can be free-standing pillars whereas, if the mesh is buried, regular depressions form. The system can thus not only form coatings with controlled composition and micro-, or nano-structure, but also, produce regular topographical features.

Some applications of this dual beam system with its unique capabilities to produce and study nanocomposite nc-TiN/a-Si_3N_4 hard coatings, getter coatings, film stress in optical coatings and cell-surface interactions are described in this paper.

2. RESULTS AND DISCUSSION

2.1. nc-TiN/a-Si_3N_4 Hard Coatings

Much work has already been reported concerning these coatings which have hardness values exceeding 40 GPa. [6-9]. The existing data indicates that the hardest coatings will be formed when the atomic percentage of Si in the film is about 9 percent, corresponding to a molecular Si_3N_4 content of 7 percent. High resolution TEM studies indicate that the hardest films have a crystal size of about 4-6nm and XPS data shows that the Si_3N_4 appears to form a monolayer thick percolation network around these TiN crystals. However, most of the films were produced using PECVD, or, magnetron sputtering systems where close control of deposition parameters was difficult.

In the present study, using the dual beam system illustrated in Fig. 1, a Ti target with Si strips placed on its surface was sputtered onto a (100) Si substrate using either an argon or a nitrogen ion beam. The substrate could be maintained at temperatures between 70 and 800°C. Test runs with varying total areas of Si on the Ti produced the desired range of Si concentrations of 7-12 atomic percent (measured using Rutherford Backscattering Spectroscopy). The horizontal ion beam used Nitrogen and this bombarded the growing film and provided additional nitrogen to form the nitride layers. Hardness values (H) were measured using a Micro Materials Ltd nanoindenter

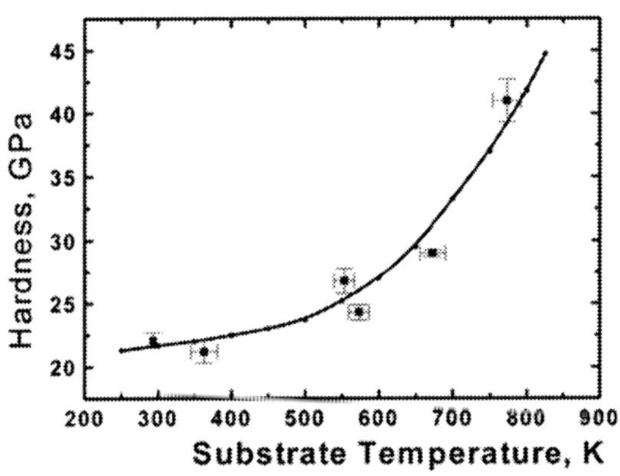

Figure 2. Hardness vs substrate temperature for Ti-Si-N bombarded with 400eV N_2 and ion/atom ratio of 1.5/1.

Results showing the influence of substrate temperature on H for an ion/atom ratio of 1.5 and a nitrogen molecular ion energy of 400eV (200eV per atom) are shown in Fig. 2. There is clearly a strong influence of temperature on the process with hardness values increasing by a factor of 2 over the range of temperatures studied. The films were highly stressed owing to the mismatch with the Si substrate and it was not possible to produce samples at higher temperatures owing to delamination. Even at the lower temperatures a short period of deposition without ion assist was needed at the beginning of film growth to prevent this delamination.

The effect of the added energy on hardness is shown for deposition at room temperature (in fact the sample heated to about 70°C) and 400°C in Fig. 3.

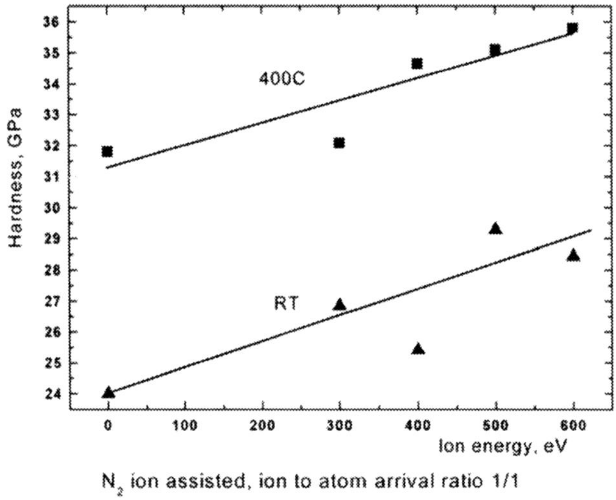

Figure 3. Hardness vs deposition temperature for Ti-Si-N films (RT has 9% Si and 400°C has 11% Si).

In terms of hardness there is clearly a trade-off between T and added energy since the RT 600eV/atom film hardness is almost as high as that for the 400°C non-ion-assist value. High resolution TEM pictures of a sectioned ion-assisted film show crystalline TiN regions clearly visible and of size about 3-5nm. There is no such microstructure for films prepared without ion-assistance where the films show columnar growth.

It is possible that the hardness changes are entirely related to changes in stress level in these films. An annealing period of 30 min in vacuo at temperatures up to 1000°C caused no significant change in hardness value. The impurity content in these coatings was seen to be very low (less than 0.1%) according to RBS and XPS data.

Similar studies are being been made for T-Cr-N films which, although less hard, have better wear and corrosion resistance properties. Results showing the effect of film composition on hardness are presented in Figure 4 and, again, show a significant increase in hardness when compared with the non-ion-assisted sample, Cr_2N. Again the ion-assistance produces nano-size structure.

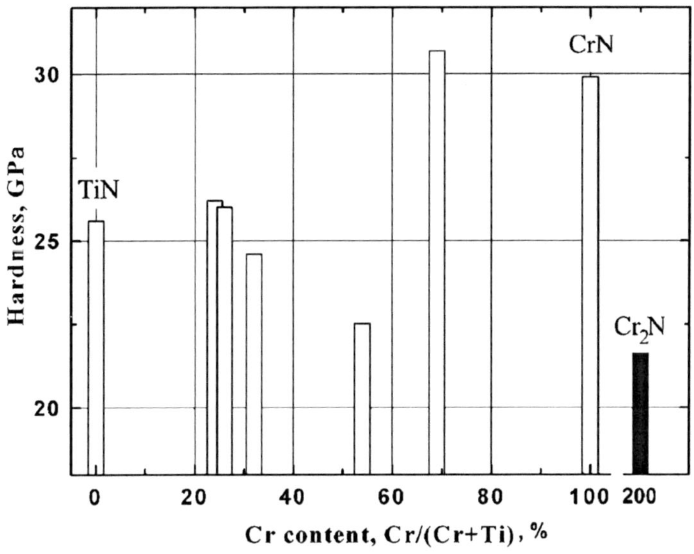

Figure 4. Hardness vs composition for Ti-Cr-N film.
Ion/atom ratio 1/1, 300eV N_2 ions.

2.2. Non-Evaporable Getters

Large particle accelerators and synchrotron radiation sources need to maintain vacuum levels below 10^{-10} Pa. One method of achieving these low pressures is to complete the pumping process by providing a layer of pumping material along the inside surfaces of the beam lines. Ti-Zr and Ti-Zr-V alloys have been shown by Benvenuti et al to provide this pumping action [10,11]. The materials are deposited by sputtering and are activated by heating in vacuum to temperatures between 300-400°C. Any sorbed oxygen, hydrogen or CO is then driven into the bulk leaving a clean metal surface layer for further pumping. Such a material is called a non-evaporable getter (NEG). After regeneration the temperature is reduced to allow further gas sorption and pumping. Clearly the pumping action depends strongly on the microstructure of the alloys since this influences the gas diffusion into the bulk. A major challenge is to produce NEGs which can be reactivated at lower temperatures and which have high pumping capacity. Ion-assisted deposition methods are now being used to influence this microstructure. Fig. 6 shows XPS data for vanadium in a $Ti_{0.39}$ $Zr_{0.33}$ $V_{0.28}$ alloy, prepared by IAD, before and after heating to 350°C for 24 hr. The oxides are seen to disappear leaving the vanadium surface ready for gettering. Similar behaviour was observed for a Ti-Zr alloy.

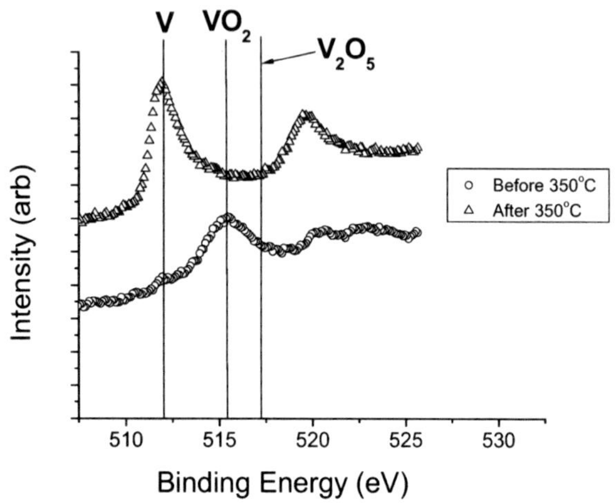

Figure 5. XPS data for vanadium in a $Ti_{0.39}$ $Zr_{0.33}$ $V_{0.28}$ alloy before and after regeneration at 350°C.

2.3. In-Situ Control of Film Stress

It is well known that ion-assisted deposition of coatings modifies the film stress [12]. Stress can be a serious problem both for optical coatings, where high stresses cause delamination and failure, and in engineering coatings, including nanocomposite materials, as this means that the material property can change in use when, for example, annealing may occur during high temperature operation. An IAD system has been built where a double-side polished silicon substrate is clamped at one end. The non-coated side faces a window where it reflects an external laser beam. The flexure of the clamped Si substrate is then monitored by the movement of the reflected laser beam and this can be used to evaluate the stress in the coating.

Initial results have been obtained for SiO_2 thin films and are shown in Fig. 7 [13]. There is a significant reduction in film stress when using oxygen ion bombardment of the growing film. This is at the expense of a reduced growth rate because there is also re-sputtering of the film which, eventually, can be removed completely.

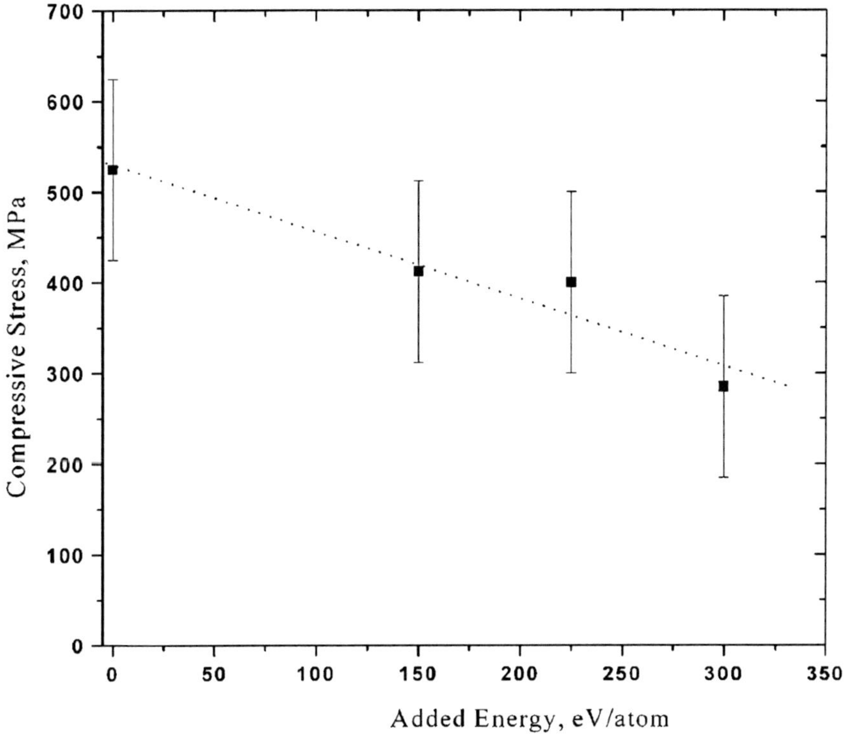

Figure 6. Stress in SiO_2 for IAD using O_2 ions.

2.4. Biocompatibity of Surfaces

A study of biological cell-surface interactions is important so that materials which may be used as implant prostheses can be identified. In an ideal situation one is looking for a surface which encourages body cell seeding so that the implant is accepted by the host. At the same time this surface must resist attachment of micro-organisms. There is wide debate on whether surface chemistry, or, surface topography is the main factor influencing cell seeding and this may depend on the level of surface roughness. There is general agreement that surface hydrophilicity is an important factor. A useful contribution to this debate has been provided by Britland et al [14] who studied surfaces with line groove features which have been further striped with a chemical cue called laminin at right angles to the mechanical grooves. When the mechanical grooves were 500nm deep, or less, the cells reacted mainly to the chemical pattern but, for deeper grooves, the topographic features start to dominate. Verran et al [15] have shown that substratum surface topography with R_a values greater than 900nm is a significant factor affecting the initial stages of surface colonisation by microorganisms. If bacterial cells are retained in a surface feature, then active attachment is not necessary, since the feature provides protection from external shear forces and allows the pioneer micro-organisms to multiply. The resultant micro-colony (biofilm) then provides a focus of infection/cross-contamination.

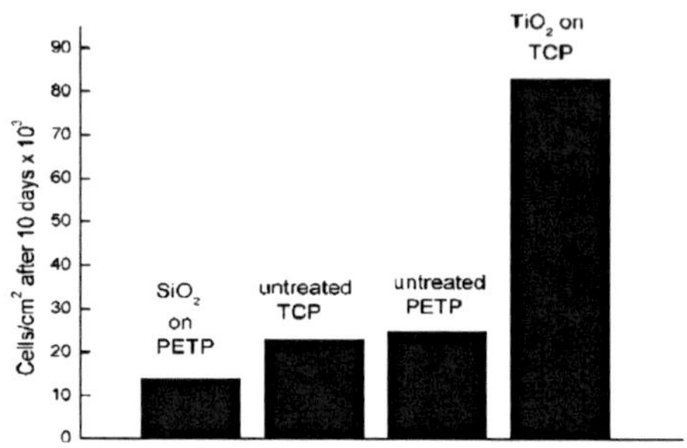

Figure 7. HUVEC Endothelial cell proliferation after 10 days.

One approach to this problem is to study cell-surface interactions on smooth surfaces coated by thin layers of different materials and also to study surfaces with regular features. In Fig. 7 the results of endothelial cell proliferation on TiO_2 and SiO_2 films deposited onto standard tissue culture plastic (TCP) and Polyethylene terephthalate substrates (PETP) are shown [16]. The TiO_2 surface shows increased cell proliferation after 10 days whereas the SiO_2 surface has a much reduced cell count.

In another set of experiments [17] regular arrays of 1μ square Ti islands have been formed on silicon wafers by sputtering through fine grid masks. The Ti will have formed an oxide layer on removal from the vacuum system. This surface was then sprayed with pseudomonas aeruginosa bacteria using an aerosol spray and the bacteria stained with 0.03% acridine orange in 2% glacial acetic acid and viewed using a Nikon Epifluorescence Microscope. The resulting array of deposited micro-organisms shown in Fig. 8 clearly shows a strong preference for the TiO_2 surface. This is further evidence that the chemical nature of the surface (perhaps its hydrophilic nature) is dominant, even on very smooth regions of a surface.

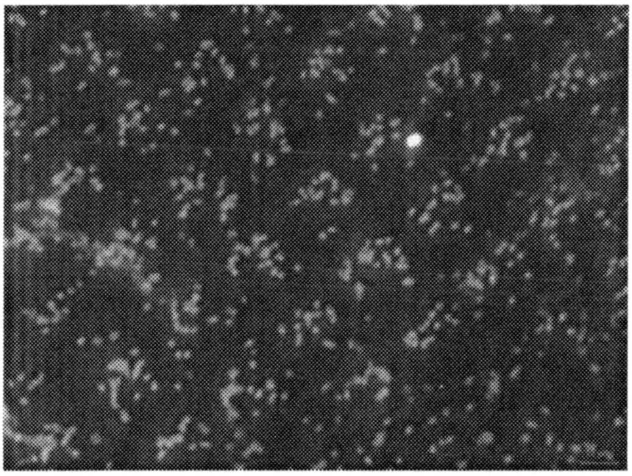

Figure 8: Distribution of pseudomonas aeruginosa cells on an array of Ti-oxide 1μ squares on a Si-oxide substrate.

3. CONCLUSIONS

Ion-assisted deposition of thin film materials using a dual ion beam system, which allows the influence of various deposition parameters to be studied, has been used to study hardness, gettering, stress and biocompatibility properties of several coatings. The hardness was seen to increase with substrate temperature and was related to changes in the nano-structure. Ti-Zr-V coatings were shown to regenerate at temperatures of 350°C. Stress levels of SiO_2 coatings were reduced with increasing ion-assistance from 520 to 280 GPa. Finally, effects of changing surface chemistry on cell-surface interaction have been seen and appear to be more significant than topographical effects for the magnitude of surface feature studied here.

ACKNOWLEDGMENTS

Many colleagues have been involved in the work presented and I would like to express particular thanks to them all, in particular to Dr R Valizadeh and Dr V M Vishnyakov my research fellows. I also acknowledge support of the UK EPSRC, Daresbury Laboratory UK and Oxford Plasma Technology for financial support.

REFERENCES

1. J.S.Colligon, J Vac Sci Technol. A13 3 (1995) 1649.

2. A. Matthews, R Artley and P. Holiday, 2005 Revisited; the UK surface engineering industry to 2010, Published by The National Surface Engineering Centre (NASURF), DERA, Farnborough, UK, 1998, isbn 0 9532872 03.
3. B.Window and N.Savvides, J Vac Sci Technol A4 (1986) 504.
4. I.I.Aksenov, V.A.Belous and V.A.Padalka, Instrum. Exp. Technol, 21 (1978) 322 (original Pribory Tekh.Eksp. 21 (1977) 236.
5. A.Anders, Surf. Coat. Technol. 93 (1997) 158.
6. S.Veprek, J Vac Sci Technol A17 (1998) 2401.
7. J.Musil and J.Vlcek, Surf. Coat. Technol., 142 (2001) 557.
8. J.Patscheider, T.Zehnder and M.Diserens, Surf. Coat. Technol., 146 (2001) 201.
9. J.S.Colligon, R.Valizadeh and I.F.Brunel, Izvestia Akad. Nauk, ser fiz, 62 (1998) 843.
10. C. Benvenuti, J.M.Cazeneuve, P.Chiggiato, F.Cicoira, A.Escudeiro Santana, V.Johanek, V.Ruzinov and J.Fraxedas, Vacuum, 53 (1999) 219.
11. C. Benvenuti, P.Chiggiato, P.C. Pinto, Vacuum, 61 (2001) 175.
12. Y.Pauleau, Vacuum, 61 (2001) 175.
13. N.Sirotkina PhD thesis, Stress in ion-beam assisted silicon dioxide and tantalum pentoxide thin films, University of Salford, UK 2003.
14. S.Britland, C.Perridge, M.Denyer, H.Morgan, A.Curtis and C.Wilkinson, Exp. Biol. Online 1 (1996).
15. J.Verran and R.D.Boyd, Biofouling, 17 (2001) 59 .
16. J.S.Colligon, A.Dekker, A.Curry and C.J.Kirkpatrick, J.Mater. Sci.: Mater. Med. (1996) 119.
17. K.A.Whitehead, MMU, unpublished.

Chapter 31
BIOMEDICAL PROPERTIES OF NANOSTRUCTURED SURFACES FORMED BY ION-PLASMA METHODS

V.M. Elinson, V.V. Sletsov, V.V. Potraysay, A.N. Laymin
"MATI" - Russian StateTechnological University named K.E. Tsiolkovsky, 3, Orshanskaya str., 121552 Moscow, Russia

In the present paper the fields of application for materials with nanostructured surfaces (NSS) in biology, medicine and the food industry are discussed. Investigations of processes for manufacturing NSS on polymers: polyethyleneterephtalate (PET) and polytetrafluorethylene (PTFE) are discribed, a study of their surface characteristics and possibilities for using them, as well as a study of their biomedical characteristics, in particular, antimicrobial activity, are presented.

Key words: nanostructured surfaces, polymers, surface properties, carbon films, superporous films, surface topography, surface charge, antimicrobial activity.

1. INTRODUCTION

Recent investigations have demonstrated the important role of nanostructures in different branches of science and technique (physics, chemistry, machinery, biology, medicine). The main features of materials and substances in the nanoscale range are defined not only by the reduction in structural element dimensions but also by the appearence of quantum dimensional effects, the wave nature of transient processes and the dominating role of interfaces. By changing the dimensions and shape of nanostructures we may provide novel functional characteristics of materials which sharply differ from the properties of bulk materials.

Biology and medicine present the widest areas for application of synthetic nanostructured materials. It is known, for example, that nanoparticles of diamond damage cancer cells [1]. It is also known, that nanoparticles of metals are used as pesticides for protection of seeds and plants from soil pathogens and for stimulation of their growth and development [2].

Materials with nanostructured surfaces (NSS) are a separate class of synthetic nanostructured materials. NSS are very desirable on the surface of implants and matrixes for cell culturing, especially for imbedding in bone;

A. A. Voevodin et al. (eds.),
Nanostructured Thin Films and Nanodispersion Strengthened Coatings, 307–319.

for creation of medicines for external application, in particular transdermal forms providing medicines which are absorbed through the skin; for creation of matrixes for internal delivery [3, 4].

Modern therapy cell technologies are examples of the novel fast developing field of medicine. These technologies are connected with implantation of desirable cell cultures into organs and tissues including stem cells. Address delivery of cell cultures may be achieved on corresponding substrates including two- and three dimensional matrixes.

Matrixes and substrates for cell cultures in vitro must be suitable for qualitative adhesion of cells: their spreading, locomotion and proliferation, and for implantation in vivo to the human organism. So the obligatory properties of matrixes and substrates must be: the absence of toxicity, biocompatibility, dispersed surface and good adhesive properties for cells. Besides that in special cases they must possess additional physical and chemical properties; e.g. biodegradation, or high firmness, or flexibility, or have the porous structure for synthesis of a cell matrix [5, 6].

Consumable materials (substrates, glasses for cell culturing, etc.) for diagnostics, express-analysis and cell culturing must possess very close biomedical properties. Using NSS modified in a reliable way biocatalysts for different biotechnological processes may be realised.

Packaging materials and systems play the important role in biology, medicine and the food industry. Such materials are to possess large complex range of properties, in particular antimicrobial activity. The last quality may be provided by means of nanostructuring and modification of the material surface.

The possibilities for application of materials with NSS in biology, medicine and food industry are presented in Fig. 1.

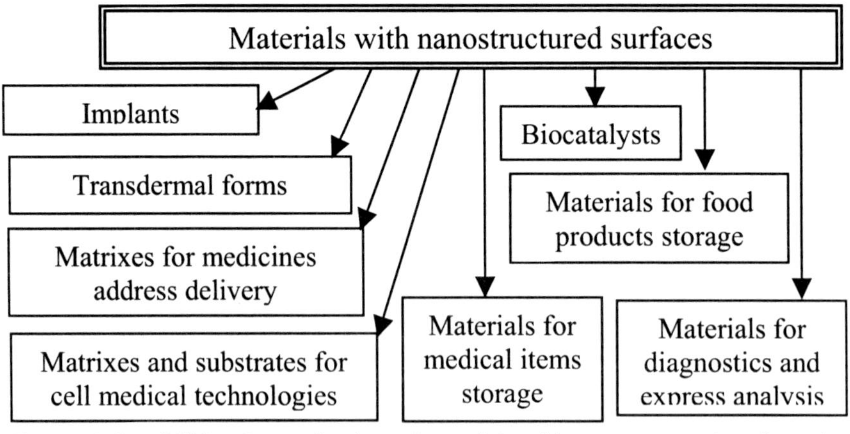

Figure 1. Possibilities for application of materials with nanostructured surfaces in medicine, biotechnology and the food industry.

2. MATERIALS AND EXPERIMENTAL DETAILS

In the present paper results of research on processes for manufacturing NSS on polymers, a study of their surface characteristics and possibilities for using them, a study of biomedical properties of polymers with NSS are presented. The PET and PTFE, which are mostly used in biology and medicine have been chosen as initial materials for study.

The choice of these materials made to explore the wide application of these polymers, and also possibility of a providing a large range of new properties after modification of these surfaces: PET possesses a high energy polar surface, whereus PTFE has a low energy surface, high dielectric properties and a wide working temperature range. NSS on PET and PTFE have been formed by treatment of initial surface by ions of chemically active and noble gases and their mixtures (CF_4, Ar, O_2) [7, 8]. Modification was carried out using two methods: by a deposition of carbon films from directed ion-plasma flows of hydrocarbon vapours with 5 - 120 nm thickness [9, 10] and (b) by magnetron deposition of higly dispersed Al films [11]. In the first case the change in surface properties monitoring is in the form of a change in phase and cluster composition, layer thickness and number of layers with a ratio of real to plan-view surface area in the range 8 - 10: in the second case this ratio may be increased by a factor of 100.

Atomic force microscope “FemtoScan” with maximum scan square 10 x 10 μm has been used for the study of the surface structure. For every sample photos of the surface on different points have been obtained. The main parameters defined were a characteristic horizontal size R, perimeter L and volume V of the surface relief features. In addition the surface roughness R_q, surface area S and fractal dimension D_F were obtained [12].

Gram-positive Staphylococcus aureus ATCC 29213 and Gram-negative Pseudomonas aeruginosa ATCC 27853 have been used for antimicrobial activity study. For this purpose 3 - 4 colonies of microorganisms cultured in 18 - 20 hours have been suspended in 3 ml of saline solution. The suspension had a turbidity standard of 0,5 according to the McFarland scale ($1{,}5 \cdot 10^8$ cfu/ml). The base suspension was then diluted in saline solution serium dilutions until the concentration was 15 cfu/ml.

Antimicrobial NSS was cut into fragments of size 15 x 15 mm^2. One fragment the substrate was put into a tube with liquid nutrition medium for control of the sterility of the coatings. Muller-Hinton bovril was used as a nut-rition media. From every tube with different concentration 25 μl of suspension were extracted using pipetto and were deposited on separate fragments of the polymer substrate with NSS, as well as onto a substrate without NSS for control of cultere growth. Such infected fragments were incubated in wet media with thermostat at 37°C for 2 hours. After incubation all fragments were put into tubes with 4 ml of nutrition media were again incubated in termostate at 37°C for 48 hours Antibacterial activity was estimated by the appearance of visible growth in tubes with nutrition media after 48 hours.

3. RESULTS AND DISCUSSION

3.1. Topography of PET and PTFE Surface after Nanostructuring and Carbon Films Deposition

An AFM study of PET and PTFE after preliminary treatment by ion bombardment and film deposition have shown that film deposition initially leads to smoothing of the surface and then, as the film thickness increases, the height of the roughness begins to increase while the diameter of globules keeps nearly constant. When the film thickness achieves some specific value the dimensions of characteristic relief features also increase (Tables 1 and 3). Preliminary treatment of the surface by means of CF_4 leads to relief growth (Figs 2-4). After prolonged pretreatment a surface with a deposited film has higher value of roughness than without pretreatment. Increase of pretreatment time leads to growth in the roughness obtained after film deposition (Tables 2 and 3, Figs 4 and 7). Surface roughness increases almost by a factor 100 in comparison with that for a film deposited without pretreatment.

Table 1. Surface characteristics of carbon films deposited onto PET (size of frame 2x2 μm)

№	Object description	D_F	R_q, nm	S, nm^2;	$\Delta S = S - S_{xy}$; nm^2	Relief parameters R, nm	L, nm	V, nm^3
1	PET; α-C:H ~10nm	2.000111± 0.000005	1.33	$4.016 \cdot 10^6$	$2.47 \cdot 10^3$	120	677	$3.869 \cdot 10^4$
2	PET; α-C:H ~40 nm	2.00027± 0.00001	2.64	$4.025 \cdot 10^6$	$8.65 \cdot 10^3$	117	888	$4.602 \cdot 10^4$
3	PET; α-C:H ~70 nm	0.000502± 0.000013	3.39	$4.029 \cdot 10^6$	$1.56 \cdot 10^4$	92	686	$1.485 \cdot 10^5$
4	PET; α-C:H ~120 nm	0.00055± 0.00002	8.47	$4.037 \cdot 10^6$	$2.48 \cdot 10^4$	107	899	$1.570 \cdot 10^5$
5	PET (control)	2.000042± 0.000002	2.66	$4.028 \cdot 10^6$	$2.12 \cdot 10^3$	356	–	–

3.2. Study of Surface Charge

Measurement of surface charge has been carried out using the dynamic capacitor method [13, 17, 18]. Experiments have shown that pretreatment of polymers surface, film thickness, energy of particles sufficiently effect the surface charge on the polymers. During the process of carbon film deposition PET and PTFE acquire an excess positive charge of about a few μC/m^2. Preliminary surface treatment of PET and PTFE in mixtures of Ar + O_2 and air produce more surface charge than the pretreatment in CF_4 plasma (Figs 8-10).

Table 2. Surface characteristics of carbon films on PET

№	Object description	Frame size, nm	R_q, nm	R, nm	D_F	S, nm^2	$\Delta S = S - S_{xy}$; nm^2
6	PET, treated CF_4 5 min	X: 2712; Y: 2731	1.327	68	2.000259± ±0.000017	7.422·10^6	1.49·10^4
7	PET, treated CF_4 10 min	X: 2868; Y: 2916	1.169	129	2.000114± ±0.000007	8.369·10^6	5.39·10^3
8	PET, treated CF_4 5 min; α-C:H ~100 Å	X: 3318; Y: 3398	1.133	88	2.000078± ±0.000005	1.128·10^7	7.17·10^3
9	PET, treated CF_4 10 min; α-C:H ~10 nm	X: 2855; Y: 2855	1.519	95	2.000238± ±0.000011	8.159·10^6	8.45·10^3
10	PET, treated CF_4 5 min; α-C:H ~40 nm	X: 2735; Y: 2770	2.283	111	2.000438± ±0.000018	7.591·10^6	1.55·10^4
11	PET, treated CF_4 10 min; α-C:H ~40 nm	X: 3230; Y: 3204	3.137	103	2.000264± ±0.000009	1.037·10^6	1.82·10^4
12	PET, treated CF_4 5 min; α-C:H ~70 nm	X: 2953; Y: 3014	2.858	107	2.000176± ±0.000006	8.917·10^6	1.67·10^4
13	PET treated CF_4 10 min; α-C:H ~70 nm	X: 3064; Y: 2940	5.15	98	2.000723± ±0.000022	9.041·10^6	3.26·10^4

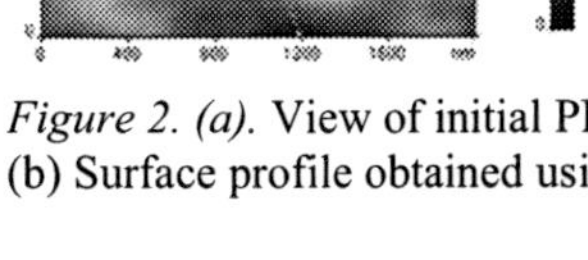
a

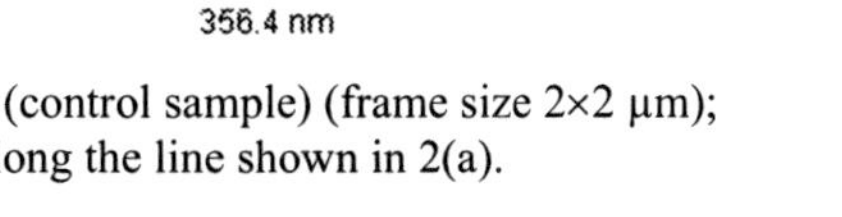

b

Figure 2. (a). View of initial PET surface (control sample) (frame size 2×2 μm);
(b) Surface profile obtained using AFM along the line shown in 2(a).

An investigation of the electrophysical properties of the polymer surface after pretreatment and subsequent carbon film deposition has allowed some features of films formation and growth on polymers with different surface structure to be established. The possibility to influence the electrical properties of polymers by means of nanostructuring their surface has also been shown. Investigation results are described in details in [13, 14].

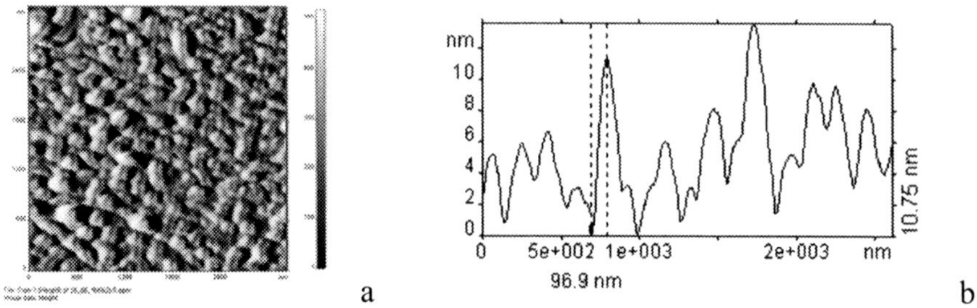

Figure 3. (a) Surface of PET samples treated byCF_4 for 10 min with α-C:H film of 40 nm thickness (frame size ~3×3 μm – see table 3); (b)Surface profile obtained using AFM along the vertical line in 2 (a).

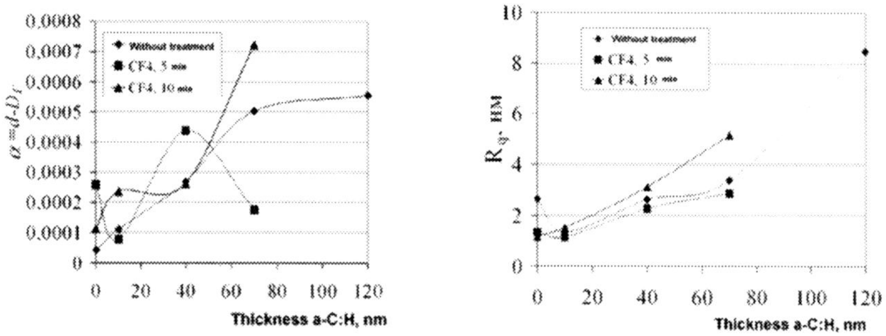

Figure 4. Surface microstructure characteristics of PET in dependence on thickness of deposited carbon film for samples treated by CF_4 during 5 min, 10 min and without treatment by CF_4.

Table 3. Surface characteristics of carbon films on PTFE

№	Object description	Frame size, nm	R_q, nm	R, nm	D_F	S, nm^2	$\Delta S = S - S_{xy}$; nm^2
16	PTFE; α-C:H ~10 nm	*X*: 7614; *Y*: 8344	74.46	925	2.00319± ±0.00010	$6.562 \cdot 10^7$	$2.09 \cdot 10^6$
17	PTFE, α-C:H ~40 nm	*X*: 10080; *Y*: 10080	42.05	1280	2.00185± ±0.00010	$1.040 \cdot 10^8$	$2.52 \cdot 10^6$
18	PTFE, control	*X*: 10080; *Y*: 10080	27.4	768	2.00100± ±0.00003	$1.026 \cdot 10^8$	$1 \cdot 10^6$
19	PTFE, treated CF_4 30 min	*X*: 9824; *Y*: 9705	33.5	1221	2.000477± ±0.000013	$9.588 \cdot 10^7$	$5.36 \cdot 10^5$
20	PTFE, treated CF_4 30 min; α-C:H ~10 nm	*X*: 9232; *Y*: 8088	47.57	768	2.00143± ±0.00003	$7.581 \cdot 10^7$	$1.15 \cdot 10^6$
21	PTFE, treated CF_4 30 min; α-C:H ~40nm	*X*: 9824; *Y*: 9883	80.6	565	2.0059± ±0.0004	$1.033 \cdot 10^8$	$6.17 \cdot 10^6$

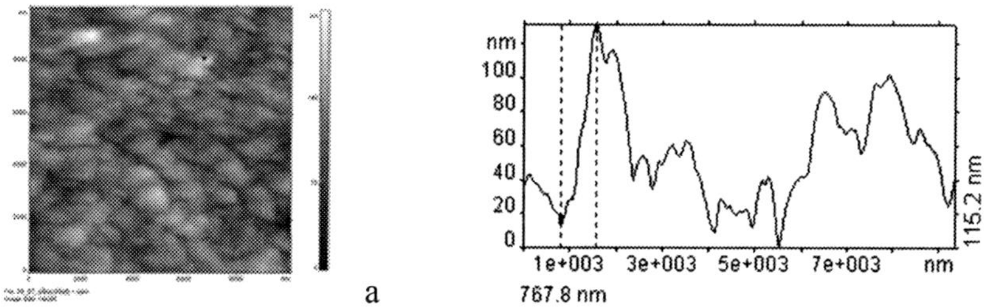

Figure 5. (a) View of initial PTFE surface (scan size ~10×10 μm –see table 3); (b) Surface profile obtained using AFM along the line shown in 5 (a).

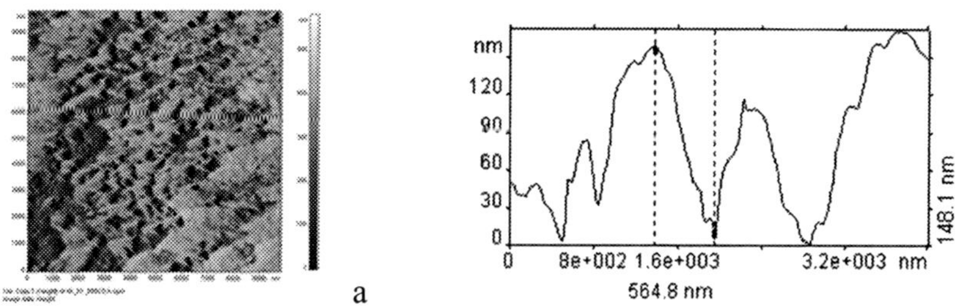

Figure 6. (a) Surface of PTFE samples treated by CF_4 in 30 min with α-C:H film of 10 nm and 40 nm thickness(scan size~9 x 9 μm – see table 3); (b) Surface profile obtained using AFM along the vertical line.

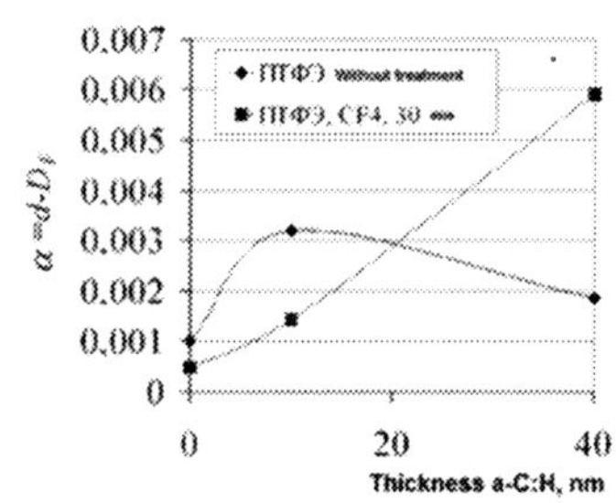

Figure 7. PTFE surface microstructure in dependence on thickness of deposited carbon film for samples treated by CF_4 for 30 min and without pretreatment by CF_4

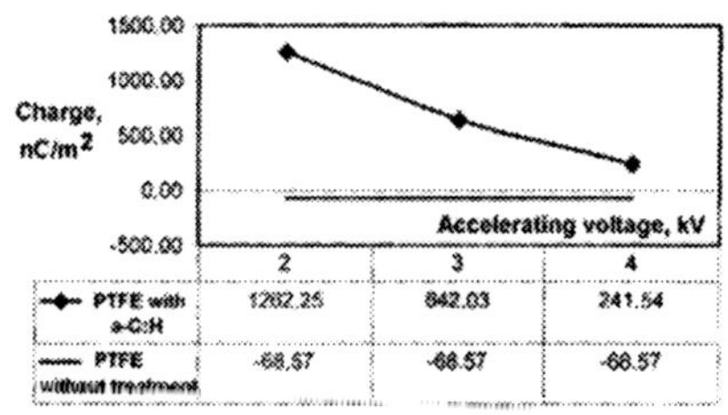

Figure 8. Dependence of PTFE surface charge on particle energy during preliminary treatment

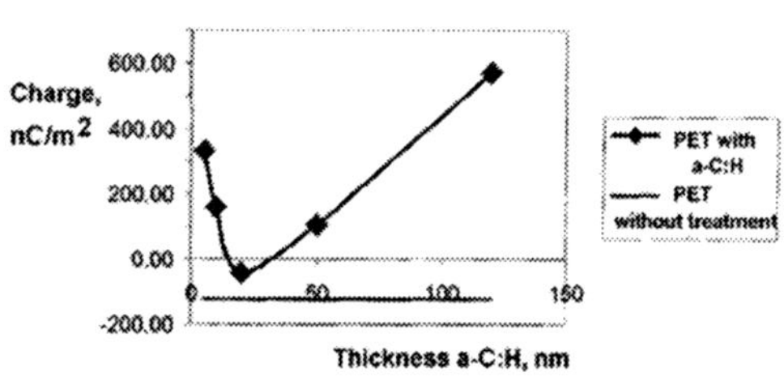

Figure 9. Dependence of PET surface charge on film thickness after preliminary treatment in $Ar+O_2$ mixture.

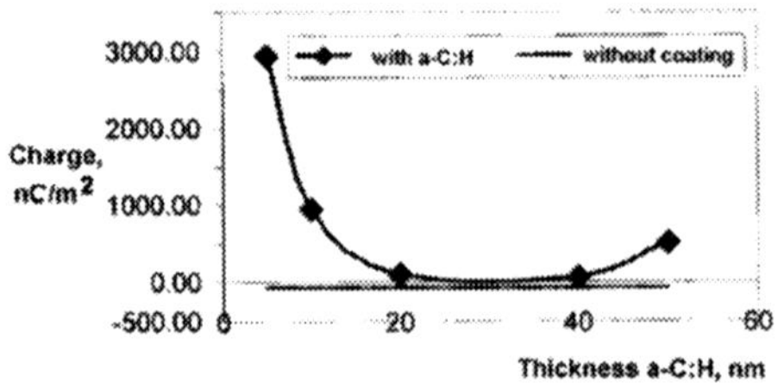

Figure 10. Dependence of PTFE surface charge on film thickness after preliminary treatment in air mixture.

3.3. Composition, Physical and Chemical Properties of Surface

A study of the influence of preliminary treatment of a PET surface by ion beams containing ions of nitrogen and oxygen on the surface composition and its physical and chemical properties has also been carried out. Changes of surface chemical structure were studied using electron spectroscopy for chemical analysis (ESCA). The study of surface characteristics also included a goniometric method for wet contact angles measurement (θ) [15, 16]. Water (bi-distillate) and glycerine (chemically clean) were used as working liquids. From the data obtained values of the total surface energy (γ), polar (γ^P) and dispersion component (γ^d) have been calculated [15, 16]. For these calculations values of the adhesion energy obtained from experimental values of wet contact angles and Dupre-Yung equation for two liquids [16-18]. It was established that carbon C_{1s} and oxygen O_{1s} peaks in ESCA in the spectrum of the initial PET foil have a complex character, which is connected with the presence of different bonding types: C–C (284.6 eV); C = O (288.4 eV and 531.9 eV); C–O–O–C (532.7 eV).

Bombardment of PET foil by a beam of positive ions leads to visible changes of carbon C_{1s}, as well as oxygen O_{1s} peaks in the ESCA spectrum. An especially visible change was seen in the oxygen atom spectrum. In the treated film the C_{1s} peak with maximum 288.4 eV escapes. This peak is connected with carbonyl group (C = O).

At the same time total content of oxygen atoms relative to carbon atoms remains constant (0.4-0.43). It is well known that oxygen atoms of the C = O group during the interaction with active particles (ions, quanta of VUV – irradiation) easily transform into radical form producing group C–O$^\bullet$ [17, 18]. In the air this radical group C–O$^\bullet$ interacts with water molecules of the atmosphere and forms a peroxide group –COOH. In this case the ratio O/C must significantly increase. But according to the calculations based on the ESCA data this ratio remains constant. During the interaction with PET macromolecules part of the ion kinetic energy is spent in the break up of chemical bonds and second part – to increase the energy of oscillations of macromolecule fragments and increase their mobility (local heating)

One can assume that the C–O$^\bullet$ group obtained under the influence of ion bombardment is able in definite moments can move close to the neighbouring carbonyl group as result of its high molecule mobility and interact with it forming transverse C–O–O–C bonds between the chains of macromolecules.

Such interpretation is confirmed by results of a surface wetting study of the initial foil and the foil after ion beam treatment (Table 4). As may be seen from the table the surface of the initial PET foil is characterized by relatively low values of wetting angles and surface energy $\gamma = 39.2$ mJ/m^2.

Table 4. Values of contact wet angles (Θ), surface energy (γ) and its polar (γ^P) and dispersion (γ^d) components for PET foil: initial, treated by beam of nitrogen and oxegen ions and with deposited a-C:H film.

PET foil	Wet angle Θ		γ, mJ/m^2		
	water	glycerin	γ	γ^P	γ^d
Initial	66	55	39.2	163	22.9
Treated by ion beam	89	68	40.0	1.3	38.7
α-C:H film on the surface	77	61	37.3	7.0	30.3

After ion beam bombardment the surface of the PET foil becomes hydrophobic and is characterised by high values of wetting angles (Table 4). At the same time the surface energy value is practically stable because of the increase of its dispersion component. The value of the polar component is sharply decreased to a very low value of a $\gamma^P = 1{,}3$ mJ/m^2, which is typical for non-polar polymers [18]. This experimental result shows that, after bombardment by a positive argon ions with average energy 700 eV the macromolecular structure of the PET foil surface layer becomes non-polar and does not contain polar groups (e.g. –C = O, –OH, –COOH).

The increase of dispersion component of surface energy is connected with the growth of the polymer surface as a result of the etching processes under the treatment of beam of high energy particles (see Figs 2-4 and Figs 5-7, made by AFM). Subsequent deposition of a carbon film leads (as mentioned in 3.1) to smoothing of the sample relief and the dispersion component of surface energy of the foil covered by the carbon film decreases slightly in comparison with a foil simply treated by ion beam (Table 4).

3.4. Study of Antimicrobial Activity

Experimental results are presented in Table 5 (*Notes:* growth – «+»; weak growth – «±»; no growth – «–»). In tubes with PET without antimicrobial coatings we observed growth of microorganisms in all the dilutions, which demonstrated high turbidity of the media. It is necessary to note that, in all the experiments, as well as the control samples for cultures growth, control of sterility of samples was also provided (i.e. cultering of selected control coated samples without microorganisms addition).

In tubes with samples № 2-5 и № 8 an antimicrobial effect was absent both for Pseudomonas aeruginosa and Staphylococcus aureus. In probes with samples № 1, № 6, № 7, № 9 suppression of growth only for Staphylococcus aureous was observed. In this case sample № 10 possessed the highest antimicrobial activity, where suppression of growth was observed even at a concentration 10^5 cfu/ml In relation to Pseudomonas aeruginosa antimicrobial activity was absent (№ 1, № 6, № 7, № 9) (excluding sample № 1 where in tubes with low concentrations ($1{,}5 \cdot 10^2$, $1{,}5 \cdot 10$), slight suppression of growth was observed. The absence of an antimicrobial effect in relation to Pseudomonas aeruginosa may be explained by the fact that this microorganism is more resistant to the influence of different aggressive factors (relative to Staphylococcus aureus). Hence the results demonstrating the antimicrobial activity (samples № 11-16) in relation to Pseudomonas aeruginosa are of great interest. Sample № 15 has suppressed the growth even at concentration of microorganisms of 10^4 cfu/ml. Samples which have suppressed the growth of the both types of microorganisms are of much more interest (№ 11 и № 12) (Table 5).

Analysis of Table 5 has shown that maximum antimicrobial activity was observed for samples with the maximum relief (ratio of real surface to geometrical one is about ~ 100). A comparison between the results from the study of surface charge and careful analysis of Table 5 data show that the interaction between NSS and microorganisms may depend upon two mechanisms. One of them is connected with an electrostatic interaction and a sample of this type is shown on Fig. 11 where real surface was increased by a factor 5-8 and the charge was 1-5 $\mu C/m^2$). The second mechanism is connected with the dispersion parameters of NSS, a sample of this type with a porous AL coating is shown in Fig. 12 where the value of the real surface was increased by factor about 100.

4. CONCLUSIONS

In the present paper the importance of NSS on polymers has been shown and possibilities for application of such polymers with NSS in biology, medicine and the food industry has been demonstrated.

The experimental study carried out on PET and PTFE with NSS formed by ion-plasma methods have shown that the degree of dispersion of the surface and the method of its modification defines its antimicrobial characteristics and effectivity, in particular, antimicrobial activity. The mechanism of interaction between a NSS and microorganisms, influence of surface dispersion (fractal dimension) and methods of surface modification require future complex investigations.

Table 5. Antimicrobial properties of samples

№	Polymer/ deposition conditions	Indicator m/o	Innoculation dose (cfu/ml)				
			10^5	10^4	10^3	10^2	10
1	PTFE: 1. treated CF_4 2. α-C:H (50 nm) 3. Al (sputtered from foil)	Ps. aerug.	+	+	+	±	±
		S. aureus	+	+	±	–	–
2	PTFE, smooth: 1. α-C:H (50 nm) 2. Al (sputtered from foil)	Ps. aerug.	+	+	+	+	+
		S. aureus	+	+	+	+	+
3	PTFE: 1 Treated CF_4 2. α-C:H (50 nm)	Ps. aerug.	+	+	+	+	+
		S. aureus	+	+	+	+	+
4	PTFE smooth: α-C:H (50 nm)	Ps. aerug.	+	+	+	+	+
		S. aureus	+	+	+	+	+
5	PTFE: 1. treated CF_4 2. α-C:H (10 nm)	Ps. aerug.	+	+	+	+	+
		S. aureus	+	+	+	+	+
6	PTFE smooth: 1. α-C:H (10 nm)	Ps. aerug.	+	+	+	+	+
		S. aureus	+	+	±	–	–
7	PTFE: treated CF_4	Ps. aerug.	+	+	+	+	+
		S. aureus	+	–	–	–	–
8	PET(relief) + Al 1. "R" /air. 1′ 2. "R" / α-C:H 10 nm	Ps. aerug.	+	+	+	+	+
		S. aureus	+	+	+	+	+
9	PET(relief) + Al 1. "R" / air1′ 2. "R" / α-C:H 50 nm	Ps. aerug.	+	+	+	+	+
		S. aureus	+	+	–	–	–
10	PET (relief) + Al/Al_2O_3 1. "R" / air 1′ 2. "R" / α-C:H 50 nm	Ps. aerug.	+	+	+	+	+
		S. aureus	–	–	–	–	–
11	PET (relief) + Al/Al_2O_3 1. "R" /air 3′ 2. "R"/α-C:H (50 nm)	Ps. aerug.	+	+	+	–	–
		S. aureus	+	–	–	–	–
12	PET (relief) + Al/Al_2O_3 1. "R" /air. 3′ 2. "R"/α-C:H (100 нм)	Ps. aerug.	+	+	–	–	–
		S. aureus	+	–	–	–	–
13	PET: 1. "R" / air 1,5′ 2. "R"/α-C:H (30+20)nm	Ps. aerug.	+	+	±	±	–
14	PET 1. "R" / CF_4 2. "R" / α-C:H 30 nm	Ps. aerug.	+	+	+	–	–
15	PET: 1. "R" / air1,5′ 2. "R" / α-C:H 40 nm	Ps. aerug.	+	–	–	–	–
16	PET: "R" / α-C:H 40 nm	Ps. aerug.	+	+	+	+	–
17	PET 1. "R" /air 1,5′ 2. "R"/α-C:H (20+20)nm	Ps. aerug.	+	+	+	+	+

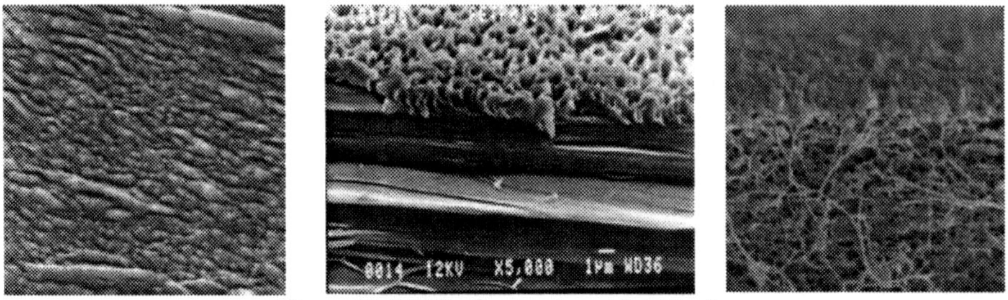

Figure 11. PTFE after treatment (etching).

Figure 12. Microphotograph of superporous AL.

The experimental results obtained point the way to novel industrial approaches to the creation of biologically active systems.

ACKNOWLEDGMENT

These investigations have been supported by INTAS grant № 2001 NANO-2274.

REFERENCES

1. Nanotechnology Research Directions. Vision for Nanotechnology R&D in the Next Decade. Edited bu M.C.Roco, R.S. Williams and P. Alivisatos, Kluwer Academic Publishers, 2000.
2. Proc of 11 Moscow Internatuinal Congress "Biotechnology: State of the Art & Prospects of Development" Moscow, 10-14 Nov. 2003. Part 2, Максима LTD, 2003.
3. M. Shtilman. Polymemers of Medical and Biological Application and Polymer Materials, Science and Technology in Industry (Ecology, Biotechnology, Medicine etc), 3(10)-4(11), 2002 (in Russian).
4. Biocompatibility. Edited by V.I. Sevastyanov, M., 1999 (in Russian).
5. G.E. Afinogenov, E.F. Panarin. Antimicrobial Polymers, SPb, Gippocrat, 1993 (in Russian).
6. N.S. Sergeeva, I.K. Sviridova, V.A. Kirsanova, V.I. Chissov, E.G. Novikova, J. Experim. Clin. Cancer Res. 14 (1995) 81.

7. V.M. Elinson, Microelectronics, 4 (2001) 279 (in Russian).
8. V.M. Elinson, V.V. Sleptsov, S.N. Dmitriev, Technological Equipment and Materials 2 (1998) 42 (in Russian).
9. V.M. Elinson, V.V. Sleptsov, et. al, Diamond Relat. Mater. 8 (1998) 2103.
10. V.V. Sleptsov, V.M. Elinson, N.V. Simakina, A.N. Laymin, I.V. Tsygankov, A.A. Kivaev, A.D. Musina, Diamond Relat. Mater. 5 (1996) 483.
11. V.T. Grinchenko, V.V. Sleptsov, S.A. Fedorov. Forming of structures metall-polymer with highly dispersed surface, Proc. of Russian sci. and techn. Conf "Novel materials and techn.", MATI-RSTU named by K.E. Tsiolkovsky (1988) 261 (in Russian).
12. V.N. Kemenov, S.B. Nesterov, E.N. Kapustin, O.S. Zilova. Study of surface structure for thin film coatings by scanning probe microscopy, Proc. Of Conf. "Vacuum Sci. Technique" (2002) 153 (in Russian).
13. V.M. Elinson, S.B. Nesterov, S.S. Tsibenko, A.I. Drachev, O.S. Zilova Electro-physical properties and surface topography of carbon-polymer nanostructures, Proc. of Conf. "Vacuum Sci. Technique" (2002) 384 (in Russian).
14. S.B. Nesterov, O.S. Zilova, V.M. Elinson, R.I. Kuznetsov, Vacuum Technique Technol., 2004 (to be published, in Russian).
15. A.D. Ziman. Adhesion of liquid and wetting, M. Chemistry (1974) 413
16. B.D. Summ, Yu.V. Gorunov. Physical and chemical foundations of wetting and spreading, M. Chemistry (1976) 231 (in Russian).
17. A.I. Drachev, Fedorova S.S.,. Gilman A.B., Elinson V.M., Chem. High Energies, 2004 (to be published, in Russian).
18. A.B. Gilman, L.A. Rishina, E.A. Visen, L.E. Vengerskaya, Chem. High Energies 31 (1997) 393 (in Russian).

Index

Printed in the United States
45381LVS00002B/11

9 781402 022203